T0226017

# Green Grades

# Green Grades

Can Information Save the Earth?

Graham Bullock

The MIT Press
Cambridge, Massachusetts
London, England

This book was set in ITC Stone Sans Std and ITC Stone Serif Std by Toppan Best-set Premedia Limited.

Library of Congress Cataloging-in-Publication Data

Names: Bullock, Graham, author.
Title: Green grades : can information save the earth? / Graham Bullock.
Description: Cambridge, MA : MIT Press, [2017] | Includes bibliographical
   references and index.
Identifiers: LCCN 2016052870 | ISBN 9780262036429 (hardcover : alk. paper),
ISBN 9780262534901 (paperback)
Subjects: LCSH: Eco-labeling. | Commercial products--Labeling--Environmental
   aspects. | Consumer goods--Labeling--Environmental aspects. | Environmental
   protection--Economic aspects.
Classification: LCC HF5413 .B85 2017 | DDC 363.7/072--dc23 LC record available
   at https://lccn.loc.gov/2016052870

To my son Jordan, whose generation will inherit the environment and the information that we decide to bequeath to them.

# Contents

●

# Contents

# Preface

A local watershed association in central New Jersey designs a River-Friendly Town Certification Program to encourage the adoption of environmental ordinances and policies. An international nonprofit organization in southwest China develops a Green Tourism and Ecotourism Program to assist local government officials and villagers in creating new models of community-based, environmentally friendly tourism. Five U.S. government agencies recruit stakeholder representatives from government, nonprofit organizations, academia, and industry to help choose a set of national environmental indicators. A social venture Internet startup builds a website that provides information to consumers about the social, environmental, and health performance of products and companies.

These four efforts were initiated by different types of organizations, at different scales, in different places, to solve very different problems, but they all employed similar strategies that share a common goal. Each of these initiatives provides information in order to catalyze improved environmental performance. Over the course of my career, I have had direct experience with each of these initiatives, and they have demonstrated to me the increasingly important role that information is playing in the environmental arena. I believe that the emergence of such information-based governance strategies is one of the most important developments in environmental policy and management in recent years. These strategies include eco-labels on products and services, environmental ratings of government agencies, green rankings of private companies, and scientific assessments of environmental quality. They have been employed by a wide range of actors across the economy and around the world.

These initiatives present difficult organizational and strategic questions for policymakers and advocacy groups interested in regulating industry—does information disclosure complement, replace, or distract from more traditional governance approaches? They also present challenges for companies, consumers, and citizens as well—what is the right response to the multitude of green claims that we encounter in the marketplace? On a more theoretical level, information-based strategies raise important questions about the mechanisms of collective action, the roles of state and nonstate actors, and the importance of individual vs. institutional action in encouraging policy innovations.

And yet their emergence and effectiveness is poorly understood and hotly contested. Proponents view these initiatives as useful approaches to tackle problems that are either ill-suited to top-down regulation or not politically urgent enough to drive such regulation. Critics see them as poor substitutes for effective government regulation and a form of "greenwashing" that reduces support for legislative action. I believe the resolution of this debate requires a deep understanding of the many different factors that influence the outcomes of these initiatives. This book is designed to contribute to such understanding, and enable policymakers, corporate executives, civil society advocates, researchers, and individual citizens to help increase the effectiveness of information-based governance. It builds on important foundational research by scholars in a wide range of different fields, including political science, management, psychology, economics, sociology, and engineering. Their contributions have focused on an impressive diversity of specific industries, initiatives, concepts, and issue areas, and have yielded key insights about the nature of these programs.

My research builds on this work by defining the disparate strategies they describe as a new form of governance that has distinctive properties revolving around their value, trustworthiness, quality, and usability. I have analyzed these initiatives using an interdisciplinary mix of qualitative and quantitative methods, including semi-structured interviews and software-based document analysis and coding. By rigorously looking at this phenomenon across a wide range of sectors and issue areas, this book is designed to provide new insights about the dynamics of information-based governance that can inform both broader comparative studies across different contexts as well as more focused studies of particular initiatives.

I hope that my diverse professional and academic background has enabled me to make this book both intellectually rigorous and practically useful. I have been involved with information-based strategies since early in my career, starting with my first job out of college working for the Stony Brook-Millstone Watershed Association. I drafted a plan for a "River-Friendly Town Certification Program," based on research I conducted on environmental ordinances adopted across the state of New Jersey. I then moved to China to work for The Nature Conservancy, and over the course of three years developed a new sustainable tourism program focused on creating green tourism development plans and community-based ecotourism enterprises. Much of this work involved identifying criteria for defining green tourism and ecotourism and then implementing projects that met those criteria.

After returning to graduate school in the United States, I cofounded the social venture startup GoodGuide, and was responsible for identifying and collecting over six hundred data points from a diverse range of product and corporate datasets and then aggregating them into user-friendly ratings for consumers. Through this experience I developed an extensive knowledge of the landscape of environmental certifications and ratings and an awareness of the complex organizational and technical processes behind them. These hands-on experiences at both GoodGuide and The Nature Conservancy give me a practitioner's understanding of the challenges facing environmental policymakers, activists, and entrepreneurs at the international, national, and local levels as they pursue information-based governance strategies.

My academic research complements this professional experience. As an undergraduate ecology and evolutionary biology major at Princeton University, I investigated the environmental and economic effects of tourism development at nine sites in southwest China, and developed a metric to measure the extent of those effects. As a Master in Public Policy (MPP) student at the Harvard Kennedy School, I examined stakeholder engagement in the development of national environmental indicators by the U.S. Environmental Protection Agency (EPA), U.S. Forest Service, U.S. Bureau of Land Management, and the H. John Heinz III Center for Science, Economics and the Environment. And my dissertation research at UC Berkeley's College of Natural Resources used a mixed methods approach to study the popularity and perceived effectiveness of product eco-labels and corporate green ratings. This book is in large part based on that work, as well as my continuing research at Davidson College, where I have served as a professor of political science and environmental studies since 2011. At Davidson, I have also taught several courses directly related to the focus of this book, including Environmental Politics; Citizens, Consumers, and the Environment; and the Politics of Information.

I have therefore been engaged with the environmental information space for nearly twenty years, in a range of both professional and academic contexts. Throughout these experiences, I have observed and participated in many debates about the effectiveness of information-based strategies, and been torn between the arguments presented by the opposing sides. As an academic I have often been skeptical of the optimistic rhetoric I have heard from information entrepreneurs, who often do not seem cognizant of the complexities and limitations of these strategies. But as a practitioner I often found myself frustrated with the knee-jerk negativity of academics and advocates who criticize any and all governance approaches that rely on information. Even those who make reasonable arguments for or against particular programs often seem to be animated by a deeper optimism or pessimism about this form of governance.

I too have felt the push and pull of these opposing perspectives; I remember, for example, explaining my work on the river-friendly town idea with particular gusto and enthusiasm, and with little emphasis on the challenges it faced. I was similarly optimistic about my work with GoodGuide and The Nature Conservancy. And I remember being perhaps overly critical of particular programs in class discussions as both an undergraduate and graduate student. But throughout these experiences, my intuition has been that neither the fully positive nor fully negative perspective are right, as they fail to capture the complexity of both information-based governance specifically and the enterprise of governance more generally. It is from this intuition and these experiences that I have tried to articulate the more nuanced third way of "information realism" that is presented in this book.

My own history of involvement with the four initiatives I describe earlier may be a bit of an outlier, but I believe most readers, upon reflection, will also recognize the ubiquity of such information in their own lives, and acknowledge the complexity of our relationship with that information. Whether it is the air quality index reported in the daily weather report, the fuel economy rating of your car, the instructions on what to recycle at the office, or the panoply of food labels you experience at the grocery store, we increasingly encounter information about the environment (or our impacts on the environment) in both our personal and professional lives. And we understand, at least subconsciously, that this information is not perfectly accurate or absolutely trustworthy, but may nevertheless be useful to us at some level. Rather than deciding whether all of these different forms of information are either unvaryingly good or bad, the real challenge is figuring out which information is most and least useful to specific decisions we have to make.

Building on my background in the field and using a variety of research methods and theoretical concepts, I have written this book to help scholars, practitioners, and the rest of us tackle this challenge and make sense of the panoply of information surrounding us. The book focuses on environmental information about products and companies, but I believe it is applicable not only to other forms of environmental information, but also to information in other sectors and arenas as well. I hope that it not only empowers scholars to do more research in this field of inquiry, but also encourages organizations to design more effective information-based initiatives and consumers to more effectively evaluate and utilize these initiatives.

I have thus tried to write the book for a diverse set of audiences, which is always a challenging task. In order to accomplish this goal, I have organized the book around the accessible but novel concept of an information value chain, which is explained in more detail in chapter 1. Each chapter discusses one component of this value chain and the challenges and opportunities associated with it. Each chapter is also animated by the common theme of information realism, which I suggest is a useful alternative to the information pessimism and information optimism that currently dominate the debate about information-based strategies. Readers are introduced to this debate in chapter 1, and reengage with it in the context of the different components of the information value chain in subsequent chapters.

The chapters themselves are also designed to provide useful information and insights to different audiences while still being broadly accessible to everyone. The introductory chapter introduces the topic and provides some historical background and political context for it, while the final chapter summarizes the book's insights and provides both general and specific recommendations for a range of actors. Each of the main chapters begins with a short vignette about an individual facing a decision about some form of environmental information. These individuals range from corporate executives to government officials to environmental activists, and they help situate and ground each chapter's focus.

These vignettes are then followed by a discussion of key theoretical concepts that are useful in analyzing information-based strategies. By clearly articulating the meaning of these concepts, which are highlighted in italics the first time they are introduced, these theory-based discussions are designed to make these concepts more accessible and useful to all readers. Understanding the meaning of legitimacy and the difference between replicability and reliability, for example, can help practitioners design better information, scholars do more precise research on that information, and users evaluate it more accurately. Within each chapter, I provide a theoretical framework for readers that connects the different concepts discussed. Discussion of these concepts is complemented by related insights from empirical studies by a wide range of political scientists, sociologists, economics, engineers, and scholars of management, marketing, and psychology.

Following these theory-oriented sections, I then present original research from a cross-cutting dataset of 245 cases of product eco-labels and corporate sustainability ratings related to each chapter's primary focus. I created this dataset to provide an overarching cross-sector perspective on these programs, and it provides valuable insights on the patterns of organizational behavior that have developed among existing programs since the first modern examples were created in the 1970s. It also enables the identification of the most promising and problematic practices across a wide range of sectors and initiative types, which are also discussed in each chapter and applied to each chapter's opening vignette.

Before we begin learning about these practices, exploring this dataset, and engaging with the debate over environmental information, a few quick notes about terminology are in order. Throughout the book, I use terms such as "environmental," "green," "sustainable," and "ethical" to describe the programs that are the focus of my analysis. Each of these terms arguably deserves quotes around it whenever it is used to signify its contested and vague nature. Such a convention would quickly become tiresome, however, and so instead I note here that my use of these terms is not an endorsement of their implied meaning (i.e., that a particular product is indeed sustainable) but a placeholder for my longer definition of the phenomenon I am investigating—information-based environmental governance strategies that generate publicly-available evaluations of products or companies (for more on this definition, see chapter 1 and appendix I). These strategies are the "green grades" named in the title of this book (and my dissertation), and make claims of sustainability and environmental benefits that may or may not be valid. Likewise, the terms "evaluation organizations," "labels," "certifications," "ratings," "programs," and "initiatives" all generally refer to this phenomenon.

A couple of other caveats are necessary. While sustainability has generally been recognized to include both social and economic dimensions, the scope of this book is confined to the environmental, or green, dimension of the concept. This decision was not based on my evaluation of the relative importance of this dimension, but the need to narrow my focus of analysis and the fact that my own expertise is in the environmental domain. I do briefly discuss social issues in chapter 2, but they deserve much more extensive treatment than is possible in this book. The same caveat can be made about the book's geographic focus—I chose to focus on initiatives that evaluate products or companies that are generally available in the United States not only because of the disproportional impact that Americans have on the environment, but also because it is the context with which I am most familiar (despite my work in China). But I believe the frameworks and conclusions presented in the book may be useful and relevant to other national contexts as well as other policy domains.

A final note is one of intellectual gratitude and humility. I could not have completed this book without the guidance and advice of a great number of people, from my many academic mentors, professional supervisors, intellectual peers, and organizational colleagues. They include Kate O'Neill, Bill Clark, and Andy Dobson, my academic advisors who expertly guided me through my thesis-writing experiences at Berkeley, Harvard, and Princeton, respectively. My committee members at Berkeley—David Levine, David Vogel, Keith Gilless, John Harte, and David Winickoff—also were immensely helpful to me throughout my dissertation research process.

I received helpful feedback on each chapter of the book from a wide range of both scholars and practitioners. Sarah Anderson, Chris Ansell, Mary Bullock, Susan Caplow, Charlotte Hill, Ben Cashore, Holly Elwood, Elena Fagotto, Archon Fung, Gill Holland, Allison Kinn, Erika Weinthal, and Ken Worthy all read one or more chapters of the book and provided me with useful feedback that was informed by their particular areas of expertise. The book also benefited greatly from comments on individual chapters from a multidisciplinary group of my Davidson College colleagues, including Besir Ceka (political science), Sean McKeever (philosophy), Tabitha Peck (math and computer science), Kevin Smith (biology), and Anelise Shrout (digital studies). I was also fortunate to receive particularly detailed and constructive comments from several anonymous reviewers on the full manuscript. And many others too numerous to list—at UC Berkeley, Duke University, and elsewhere—provided useful advice and assistance on my work over the nearly ten years I have been occupied with various components of this project. I am also grateful for the many engaging and insightful conversations about certifications and ratings I had with my colleagues while I was at GoodGuide (e.g. Dara O'Rourke, Shawn Jeffery, Ryan Aipperspach), Resources for the Future (e.g., Jim Boyd, Leonard Shabman, Roger Sedjo), The Nature Conservancy (e.g., Ed Norton, Rose Niu, Angela Cun), and the Stony Brook-Millstone Watershed Association (e.g., George Hawkins, Noelle MacKay).

My research assistants Hayley Currier, Becky Johnson, and Ashley Page were particularly instrumental in conducting the research presented in the book. They served as tirelessly systematic collectors and coders of the text from hundreds of webpages, contributing tremendously to the compilation of the EEPAC Dataset. Both their work and my work were made possible by the support of the National Science Foundation, the Intel People and Practices Research Group, UC Berkeley, and Davidson College. I am grateful for the comments on several of the chapters from students in my Politics of Information class at Davidson, including Naomi Coffman, Becky Johnson, Chris Johnson, Hannah Lieberman, Kacey Merlini, Sean Vassar, Charlotte Woodhams, and Philip Yu. And I also deeply appreciate the willingness of my interviewees to share their perspectives on eco-labels and sustainability ratings with me. Last but not least, special recognition goes out to my wife Sally Bullock for providing detailed comments on all seven chapters (and sometimes on multiple versions!). Suffice to say, I am deeply grateful for all of the assistance and feedback I have received from so many different and generous people.

This support has enabled me to come to the conclusions and make the recommendations that I do in this book, and I genuinely believe they will be helpful to scholars, practitioners, and citizens. But these insights also require an important caveat that should accompany all credible knowledge claims—they are contingent and open to critique and revision as our understanding improves over time. This is why I have identified practices as "most promising" and "most problematic," as opposed to simply "best" and "worst." The latter implies a finality and certainty that at this point in time is inappropriate and misleading; instead, the former suggests the necessity of continued inquiry and experimentation that is the hallmark of both sound science and democratic governance. The aim of this book is to serve as a catalyst for such inquiry and experimentation, by building on the lessons learned from the past and informing the efforts of the future. Information-based governance, like any form of governance, requires such constant and continuous improvement to truly mature and develop its full potential.

# 1   The Green Debate: Information Optimists, Pessimists, and Realists

## Green Decisions

Mark is standing in the grocery store trying to decide between the organic and conventional milk. He has figured out that the organic option costs more, but is unsure whether it has any additional value for him and his family. He knows that organic food has become increasingly popular in recent years, but what are the real benefits of purchasing the milk with the USDA Organic label, and would doing so reflect his personal values?

Carrie is an environmental activist who just received an email from a newspaper reporter who is doing a story on toilet paper. He wants to know if Carrie trusts either the Forest Stewardship Council (FSC) or Sustainable Forestry Initiative (SFI) eco-labels that can be found on some brands of toilet paper. She knows these two certifications have come under criticism in recent years. How should she evaluate their trustworthiness, and respond to the reporter's inquiry?

Lynn is a college professor attending a planning meeting for a new academic building on campus. Some of her colleagues want the new building to be designed so that it earns a Leadership in Energy and Environmental Design (LEED) green building certification, while others claim that the LEED point system is arbitrary and often nonsensical. As the environmental scientist among them, they turn to Lynn for her opinion. Does she think that LEED is a valid metric of sustainability performance?

Anu was hired recently as the chief marketing officer for a major consumer products firm. A major challenge she faces is to turn around negative perceptions of her company's corporate social responsibility. Fortunately she is convinced that this poor reputation is undeserved, and believes the firm is a leading corporate citizen in many ways. But how can she effectively deliver that message to the public and her key audiences?

Vernon is a senior government official who is responsible for identifying environmentally preferable products for his agency to purchase. The agency needs to order a large batch of new computer monitors, and his boss wants to be able to claim that they actually "make a difference" in protecting the environment. Sounds good in principle, but what exactly does this mean? How should Vernon determine if the monitors under consideration—and the ENERGY STAR, EPEAT, and TCO labels on them—achieve this standard of effectiveness?

## The Green Debate

The decisions that Mark, Carrie, Lynn, Anu, and Vernon face in these hypothetical examples are part of a raging debate in environmental politics. It is both an explicit debate in dueling words and an implicit one through competing actions. While much of the public and even many of those who work on environmental issues are not fully aware of this debate, it has far-reaching consequences in terms of the allocation of scarce resources, the selection of strategic priorities, and the power of competing organizations and individuals. Both sides of this debate are martialing their arguments and resources, and deploying them to win battles both in public and behind closed doors.

This is not, however, a simple rerun of past conflicts between environmentalists and industrialists. Each side has both passionate tree huggers and committed capitalists, and they do not fundamentally disagree about the importance of protecting the environment. Their disagreement is at once both larger and more narrowly focused. It is about both ideology and tactics, about ends and means, and about how society should be governed and how such governance should be achieved.

Put in its simplest terms, this debate is about the role of information in environmental governance. One side believes that information, particularly in the form of ratings, labels, and certifications, is one of the best available options for combatting threats such as climate change, deforestation, and biodiversity loss. These *information optimists* believe that strategies relying on traditional government regulations and international treaties are too slow, too cumbersome, and too bogged down in political gridlock. They include both activists who are disillusioned by this gridlock and executives who view information-based approaches as important opportunities to outperform their competitors. These optimists also believe that information empowers individuals in important ways that other approaches do not. From their perspective, the programs that our five friends just introduced are deliberating about—USDA Organic, FSC, SFI, LEED, ENERGY STAR, EPEAT, and TCO—are all important examples of innovative efforts to use information to promote a more sustainable society.

The other side, meanwhile, asserts that such information-based strategies are a poor substitute for regulation and the rule of law in environmental affairs. In the view of these *information pessimists*, product eco-labels and sustainability ratings distract from more comprehensive approaches that do not depend on the limited capacities of individual consumers and do not give citizens a false sense of progress on pressing environmental issues. Activists, policymakers, corporate executives, and academic researchers should instead be focused on enacting legislation that mandates the creation of real and measurable benefits to society and the environment. Instead of product eco-labels and sustainability ratings, we should be developing and pushing through the Clean Air Acts, Endangered Species Acts, and Montreal Protocols of the twenty-first century. This view is held by environmental advocates who are proud of those legacies as well as business leaders who see them as important ways to level the playing field for companies and penalize the laggards who are not behaving as responsible corporate citizens.

The implications of this debate are far-reaching. On the one hand, if the public is most persuaded by the information optimists, corporations as different as Walmart and Patagonia that have invested millions of dollars in developing and marketing green products will be vindicated, and industry in general can hope to avoid new onerous environmental regulations. More financial support will likely flow to nonprofit organizations such as Rainforest Alliance and Environmental Defense Fund that have developed expertise and reputations around certifications and voluntary partnerships with industry.[1] Consumers will be further empowered to take responsibility for the impacts of the products they buy, and government agencies will be encouraged to develop programs—as opposed to regulations—to help them with this endeavor.

On the other hand, if the information pessimists are perceived as winning the debate, corporations may find consumers developing even greater skepticism about the eco-labels they encounter in the marketplace. Advocacy groups such as the Sierra Club and the Natural Resources Defense Council that have traditionally lobbied for stronger environmental laws will likely have stronger support from the public and policymakers for new legislative action.[2] The resolution of this debate will thus influence the direction of the twenty-first-century environmental movement for years to come. Will it be focused on traditional tried-and-true strategies that rely on government action and mandates or on innovative and cutting-edge approaches that emphasize individual action, corporate initiatives, and the voluntary use of information to protect the planet?

### Green Grades Grow Up

Which side is right? How effective are information-based environmental governance strategies, and what effects have they had, if any? What role do these "green grades" play in the environmental arena, and have they complemented, replaced, or undermined more traditional governance approaches? How should they be conceptualized in order to best understand and analyze their internal dynamics and external impacts? More succinctly, can information "save the Earth?" In this book, I address these questions by using the concept of the *information value chain* as a framework to systematically examine the arguments made by both information optimists and information pessimists. I also present a possible resolution to this debate, or at least a third position to consider. This position is one of *information realism*; information-based strategies are important tools but we must be realistic about their usefulness. Their numbers have grown significantly in recent years and they have played an important role in environmental politics across a wide range of sectors and issues. This role, however, has had fundamental limitations, and these strategies must therefore enter a new phase of maturation and development. In order for these strategies to overcome these limitations, their designers must learn from both the most promising and problematic practices among them. This book examines these practices across five core areas—their organizational associations, content, methods, interfaces, and outcomes.

These five areas are the topics of chapters 2–6, each of which uses concepts such as value, legitimacy, validity, usability, and effectiveness to analyze the practices of existing information-based initiatives. In each of these chapters, I present original research from the Environmental Evaluations of Products and Companies (EEPAC) Dataset, a database that I created of 245 cases of product eco-labels and corporate sustainability ratings, from ENERGY STAR to Greenpeace's rankings of electronics companies. All of these cases generate publicly-available environmental evaluations of products or companies that make products that are generally available in the U.S. marketplace, and include a range of government, nonprofit, and corporate initiatives. They also include both mandatory programs established by legislation that require companies to disclose information (such as the Toxics Release Inventory and ENERGY GUIDE) as well as voluntary initiatives that companies may or may not choose to participate in (such as LEED and USDA Organic).

I also share conclusions from the relevant literature in the fields of political science, economics, psychology, sociology, and management, and report insights from the sixty-eight interviews I conducted with government officials, nonprofit representatives, academic researchers, and consumers about their perceptions of environmental ratings and certifications. In the rest of this introductory chapter, I further examine the contours of the debate over the use of information-based strategies in environmental politics. In the seventh and concluding chapter, I provide specific recommendations to improve the long-term viability and effectiveness of these strategies. These recommendations are all informed by an information realist perspective, and range from developing a more sophisticated sustainability information marketplace to enhancing the linkages between these strategies and other forms of governance, such as traditional regulations and public policy initiatives.

## Understanding Information-Based Governance

In order to better understand the debate about information-based governance strategies, it is important to clearly describe the characteristics they share in common and their relationships to other forms of governance.[3] *Governance* is a broadly used term, but can be defined in this context as the use of power to encourage collective action and create public goods.[4]

*Power*, in its most general sense, is "the capacity to affect results,"[5] or as nineteenth-century German sociologist Max Weber famously described it, the probability that an actor will be "in a position to carry out his own will despite resistance."[6] Governance is therefore distinct from its opposite, *anarchy* (i.e., the use of power in the absence of governance and collective action), and the broader and more inclusive concept of *politics* (i.e., the use of power for any means).

Traditional forms of governance typically exercise power through either physical force or economic incentives.[7] As table 1.1 shows, these two dimensions can illuminate an important range of governance strategies that actors can pursue. Governance strategies that only use the stick of physical force (or threat thereof) are examples of traditional command-and-control *regulation-based governance*. Strategies that only use the carrot of economic incentives (e.g., tax credits, market opportunities, low-interest loans) are examples of *voluntary market-based governance*. Strategies that use both carrots and sticks, such as "sin" taxes on smoking or pollution cap and trade programs, are examples of *mandatory market-based governance*. Strategies that use neither economic incentives nor physical force are examples of *information-based governance*.[8] Some approaches may combine elements of more than one of these strategies and exist along a spectrum between these different types of governance; for example, mandates for government agencies to purchase products with particular eco-labels not only depend on information but also have the force of law, but only for public institutions.

Not all uses of information, however, should be classified as information-based governance, as they are not all necessarily oriented toward the creation of public goods and collective action. Information can be deployed, for example, to enhance the management of private firms and the creation

**Table 1.1**

Types of governance

|  | No use of force | Use of force |
| --- | --- | --- |
| **No use of incentives** | Information-based governance: e.g., eco-labels | Regulation-based governance: e.g., food safety regulations |
| **Use of incentives** | Voluntary market-based governance: e.g., sustainable agriculture subsidies | Mandatory market-based governance: e.g., soda/sugar taxes |

of private or club goods—that is, goods that are excludable and do not suffer from the same collective action problems associated with nonexcludable public or common goods.[9] Although often proprietary and not visible to the public, such information is frequently used internally by private actors to govern their own businesses, through employee evaluations, ratings of supply chain performance, and balanced scorecards.[10] These are all examples of *information-based management*—the provision of information to create private or club goods (e.g., products, productivity, profits) for private firms. To the extent that these internal processes are motivated by a desire to create public and common goods (e.g., environmental protection), they are instead examples of *information-based responsibility* efforts. Such efforts include the use of information to encourage employees to reduce their carbon footprints or exercise more, and follow in the tradition of corporate social responsibility initiatives that are focused on decreasing the negative impacts and increasing the positive impacts of particular firms.

Private firms can also deploy information more widely and publicly to encourage and enable public actors (citizens, policymakers, civil society organizations) to assist them in creating private or club goods. Such a strategy can be considered a form of *information-based politics*, in the sense of politics being about "who gets what, when, and how" and the "actions concerned with the acquisition or exercise of power, status, or authority."[11] Such efforts can include the publication of reports demonstrating the need for industrial policy, subsidies, or other support for the private sector. While such support may be justified in terms of the value of certain companies or industries to society, its overarching motivation and function is to stimulate the creation of private or club goods.

*Information-based governance*, in contrast, deploys information to encourage and enable the creation of public and common goods—nonexcludable goods that are often underproduced by private markets.[12] In this sense, governance is any attempt by either state or nonstate entities to mobilize individuals or institutions toward goals that transcend their own immediate private interests. It is also not limited to government activities, but can be entirely nongovernmental in nature. The goals of governance initiatives might include providing for the common defense, helping the poor, or protecting the environment. Information itself is a classic example of a public good—once it is produced it is both nonexcludable (i.e., it is difficult to keep people from accessing it) and nonsubtractable (i.e., use of it does not reduce its value to others). Governance does not include actions that have no orientation toward specific public or common goods, such as everyday personal interactions, management activities, and market transactions.

As table 1.2 summarizes, information—and green grades specifically—can thus serve as important instruments of corporate responsibility, internal management, transactional politics, and public governance. On the one hand, in information-based governance and information-based responsibility initiatives, information plays a primary role in the creation of public and common goods. On the other hand, in information-based politics and information-based management, information plays a primary role in the production of private or club goods—either directly through private firms or with the assistance of public entities. While these distinctions may be difficult to make in some instances, it is usually possible to identify the general orientation of an information-based strategy by the language it uses and the extent to which its information is made available to the public. Such distinctions are helpful in analyzing the motivations behind these programs, their primary purposes, and how they define effectiveness, a subject to which we return in chapter 6.

Information-based strategies can be deployed by a range of different actors, including advocacy organizations, corporations, government agencies, academic researchers, individual citizens, and coalitions of all or some of theseactors. In studying these strategies, it is important to consider the activities of the full range of these social stakeholders. Past studies have often taken a relatively narrow and disciplinary approach to analyzing

**Table 1.2**
Types of information-based strategies

|  | Club goods and private goods focus | Public and common goods focus |
|---|---|---|
| **Internal organizational orientation** | Information-based management (e.g. balanced scorecards) | information-based responsibility (e.g., information designed to encourage employees to reduce their carbon footprints) |
| **External public orientation** | information-based politics (e.g., data-driven reports calling for industry subsidies) | information-based governance (e.g., product eco-labels, public ratings of corporate environmental performance) |

governance dynamics, with separate emphases on government, business, social movement, and consumer perspectives.[13] While these bodies of work have contributed to our understanding of environmental certifications and ratings, such an approach hides the commonalities across these divisions and obscures the relationships between these different groups and their use of information-based strategies. This book instead takes a broader perspective, and investigates this phenomenon as it has developed across a broad range of government agencies, private companies, and civil society organizations.

## Something Old: Histories of Information

The use of information-based strategies for management, political, and governance purposes is certainly not a new phenomenon. The first newspapers, dictionaries, encyclopedias, libraries, and museums were all projects to deploy information for specific ends and have had profound effects on society over at least the past hundreds of years.[14] Indeed, the information studies literature discusses the revolutionary novelty of these new forms of information in their own times and challenges the notion that the current age is a distinctly "information age."[15] Each age has witnessed sophisticated and contested efforts by private, public, and civil society actors to use information to advance their interests.[16]

Trade associations and guilds, for example, have used information-based strategies like examinations and certifications to regulate the quality and size of their own industry for centuries. The bar in the legal profession and the boards in the medical profession are two well-known modern examples of information-based management strategies, but historically smaller guild-based professions often utilized these approaches well. Guilds in medieval Europe, for example, overcame the problem of widespread counterfeiting and adverse selection by inspecting their members' products and prohibiting the sale of low-quality products.[17] They also sold merchandise with conspicuous characteristics—unmistakable colors, weaves, or other attributes—that served as signals of quality.[18] The words used to describe these products—Damascus steel, Parmesan cheese, London pewter—helped establish the reputation of these guilds and differentiate them from counterfeiters.[19] Guilds also used specific marks—such as abstract symbols and allegorical figures—to communicate information about their products to consumers, as did individual craft masters.[20]

Public authorities have also used information for their purposes historically as well. A widespread example is the provision of statistical information
about various social, demographic, and economic trends in society—Great
Britain's census reports and the United States Statistical Abstract are two
prominent examples.[21] William the Conqueror commanded the first population survey of England in 1085, and its results—documented in the
Domesday Book—were likely used to calculate taxes and feudal duties owed
to the Crown.[22] The modern British census was initiated in 1801 to address
concerns about the population increases and food shortages predicted by
Thomas Malthus, and it quickly became a powerful instrument of the rising state bureaucracy.[23] The data provided by the census not only served
as a tool of surveillance and control, particularly of migrant people and
the urban poor, but also as a mechanism of empowerment and representation for the public.[24] The census also contributed to a growing sense of
national identity and identity with specific social subgroups, and it enabled
reformers within and outside the government to advance their agendas to
improve the working and living conditions of the British people.[25]

Information-based strategies have also been utilized by advocacy organizations to advance their causes for quite some time as well. One of the
earliest examples is the nineteenth-century product marks that were used
by labor unions to differentiate union-made goods and boycott nonla-
beled products.[26] The first union label was adopted by the Cigar Makers'
International Union of America in 1874 to distinguish products made by
white unionized workers from those made by Chinese immigrants.[27] By
the 1890s, union labels were being used by unionized printers, bakers, barbers, brewers, and other "first class" workmen.[28] While it can be argued that
these marks increased profits for shops and therefore created private goods,
the broader benefits that these labor marks promised—higher wages for
(generally white) workers who would then have a higher standard of living
and be able to buy more goods themselves—is a distinctly nonexcludable
public good (and similar to today's Fair Trade label).

Each of these historical examples of information-based strategies were
highly debated in their respective eras, and consumers, government officials, businesspeople, and social activists in those eras faced decisions similar to those facing Mark, Carrie, Lynn, Anu, and Vernon in the examples
given earlier. As a manufacturer, how should I use information to distinguish my product from those of my competitors and counterfeiters? As a

consumer, should I buy a product certified as union-made? As a policy-maker, what census information—if any—should I collect from the public, and how should I make use of that information? As an activist, should I support and use these information-based strategies as mechanisms to advance social progress, or should I criticize and oppose them as dangerous instruments of government control and private interests?

## Something New: Outbreaks of Information

While information-based strategies have been around for centuries, they have become more common and prominent in recent years. Consumer products, colleges, hospitals, movies, stocks, and much more are being rated, ranked, certified, boycotted, and labeled by nonprofit organizations, for-profit companies, media outlets, government agencies, and consumers all over the world. Examples include U.S. News and World Report's college rankings, Morningstar stock ratings, Consumer Reports product ratings, Nielsen's television ratings, Freedom House's democracy ranking, Charity Navigator's nonprofit ratings, and FICO's credit scores.[29] The subjects of these efforts are not only products and companies, but also entire countries and individual politicians, doctors, professors, and citizens.

Building on earlier signaling efforts by craft masters and guilds, corporations have increasingly used information to differentiate themselves within their industries through trademarks, branding, marketing, awards, corporate reporting, and certifications. On average, U.S. firms spend 11 percent of their budget on marketing (an amount that increased 3 percent between 2011 and 2014), and the average American consumer sees 247 commercial images per day.[30] Nearly 640,000 trademarks were registered with the U.S. Patent and Trademark Office between 1870 and 1962; that number increased to over five million between 1982 and 2012. There are more than 300 professional certifications available in the United States for professions as diverse as archivists, fire code inspectors, and food executives.[31]

Advocacy organizations have also expanded their use of information-based strategies since the first union labels, and become prominent supporters of a wide range of product certifications, boycotts, and rating systems that advance their causes. Consumers Union rates products on their quality and safety, Transparency International rates countries on their levels of corruption and transparency, Freedom House rates countries on the quality of their democratic institutions, and Charity Navigator rates nonprofits on their financial health, accountability, and transparency.[32] Some of the more well-known boycotts in the past century include the anti-Nazi boycott of German goods before World War II, the Montgomery, Alabama, bus boycott, the United Farm Worker boycott of grapes and lettuce, and the boycott of Nestle for its promotion of breast milk substitutes in less developed countries.[33] Media organizations and academic researchers have also become involved in similar information-based governance strategies. For example, U.S. News and World Report has published rankings of colleges and universities since 1983, while the National Research Council produced rankings of graduate research institutions in 2010.[34] All of these information-based governance efforts share a common interest in creating public and common goods for society.

Government agencies are also collecting or mandating the disclosure of information across a wide range of sectors and issues. The U.S. federal government has required public companies to disclose their financial performance and risks, packaged food companies to provide nutrition labels on their products, banks to report on their mortgage lending practices, and employers to inform their workers about workplace hazards.[35] States such as New York and Pennsylvania have required hospitals to publicly report data on cardiac surgeries, and cities such as Los Angeles have mandated that restaurants post their hygiene grades at their entrances.[36] The Consumer Product Safety Commission manages a website (www.SaferProducts.gov) that the public can use to report harm caused by particular consumer products; these reports are publicly available on the same website, as is information about recalls issued by manufacturers and the Commission.[37] Government agencies are also awarding companies and certifying products for strong performance on issues the government has prioritized, through programs such as ENERGY STAR and the Green Power Partnership.[38]

Indeed, nowhere is this increased use of information-based strategies more apparent than in the environmental field. Between 1975, when Congress created the EnergyGuide Program to label products in eleven household categories, and today, nearly four hundred environmental certifications of products, or *eco-labels*, have been introduced around the world.[39] These include the U.S. Government's ENERGY STAR label, the Forest Stewardship Council's (FSC) wood product label, and the U.S. Green Building Council's LEED building certification.[40] As the Global Ecolabeling

Network explains, these eco-labels identify products or services that have been determined to be environmentally preferable, compared to similar products or services within their categories.[41]

During this same period, environmental ratings of companies have proliferated as well. Nonprofit organizations such as Greenpeace, World Wildlife Fund, and the Union of Concerned Scientists have issued corporate environmental ratings across a wide range of sectors, as have socially responsible investment firms such as KLD and Innovest (both now part of MSCI) and media outlets such as Newsweek and Fortune magazine.[42] Meanwhile, Yale University and Columbia University have developed a rating system that ranks the environmental performance of 192 countries, and a myriad of organizations have developed websites to calculate the carbon footprints of consumers.[43] Similarly, sustainability rankings of cities have been developed by German manufacturing company Siemens, the media company SustainLane, and Kent Portney, a political scientist at Texas A&M University.[44] The environmental performance of products, organizations, cities, countries, and individuals are thus all being evaluated by these initiatives.

This development represents a radical shift in emphasis and strategy for these organizations. While environmental organizations have traditionally focused on government regulation as their primary strategy, many now are dedicating their resources to initiating these information-based strategies, whether they are boycotts, eco-labels, or green ratings.[45] In response to such efforts, corporations have engaged in similar approaches, such as the forestry industry's Sustainable Forestry Initiative and the chemical industry's Responsible Care Program.[46] Government agencies, recognizing the popularity and merits of these nonregulatory mechanisms, have also initiated programs to reward strong performers and provide more information to the public.[47] The U.S. Environmental Protection Agency (EPA), for example, has more than sixty such voluntary programs, including WasteWise, WaterSense, and Safer Choice.[48]

## The Promise of Information

This proliferation of information-based environmental strategies is likely due to several factors. Technology has of course played an important role—the rise of personal computers and the Internet have made the processing and sharing of information easier for all of the actors mentioned. While technology has supplied the means, politics and ideology have provided the motivation for organizations and individuals to embrace these approaches. Shrinking resources and political challenges at the international and domestic scales of government have frustrated traditional governance strategies and created a strong pragmatic "push" toward more creative efforts to encourage collective action. Meanwhile, economic, populist, and media logics have generated an unmistakable ideological "pull" toward policy solutions that are perceived as improving the efficiency of the marketplace, supporting grassroots social movements, and enabling individuals to make their own decisions based on "objective" information. These forces have provided a host of both strategic and philosophical arguments to justify the move toward information-based governance strategies.

**The Push of Politics**
After the excitement of the historic negotiation of the Montreal Protocol in 1987 to confront the challenge of the depleting ozone layer, efforts to coordinate and harmonize environmental protection across national boundaries appeared to have a hopeful future.[49] Instead, however, these efforts suffered several major setbacks only a few years later at the Rio Earth Summit. While leaders from around the world at the Rio Summit made many nonbinding declarations of intentions such as Agenda 21, commitments to binding international environmental treaties with specific targets and timetables were limited.[50] Particularly disillusioning for environmental advocates was the lack of a binding agreement on protecting the world's forests, which was vetoed by developing countries led by Malaysia.[51] Also disappointing was the United States' blocking of a binding accord on combatting climate change at the Summit.[52] As Ohio State sociologist Tim Bartley summarizes, "Environmental groups viewed Rio as a nearly complete failure on forestry issues," and some became "especially disenchanted with intergovernmental areas at this point. ... [They] seem to have interpreted Rio as one more piece of evidence that private rather than intergovernmental, initiatives were the place to focus their energies."[53]

Progress was not much better on the domestic front. The heady days of the early 1970s had witnessed the passage of the Clean Air Act, Clean Water Act, and Endangered Species Act under Richard Nixon, but the implementation of these seminal laws had numerous technical challenges, mixed results, and legal and political resistance from many state officials, businesspeople, landowners, and conservatives.[54] Due to claims from companies that the legislated deadlines were too onerous and unrealistic, the EPA granted a host of extensions to companies to comply with the Clean Air Act in the 1970s, and was the subject of more than 150 lawsuits brought by companies regarding its implementation of the Clean Water Act.[55] The agency's budget was slashed by a third by the Reagan administration, and efforts to revise the Clean Air Act were blocked by Democrats in Congress with strong ties to the automobile and coal industries.[56] While levels of most major pollutants had decreased significantly between 1970 and 1987, nearly 102 million Americans still lived in counties that exceeded at least one air quality standard in 1987.[57] Likewise, that same year—fifteen years after the Clean Water Act was passed—30 percent of river miles assessed were found to be at least partially impaired and not fully supportive of beneficial uses.[58]

These challenges and mixed results motivated some actors to redouble their efforts to strengthen these regulatory approaches. Their work led to amendments to the Clean Water and Clean Air Acts being passed by Congress (in 1987 and 1990 respectively) that made significant improvements to these laws.[59] Others became increasingly disillusioned with the standard command-and-control model and began advocating for more collaborative, integrative, and innovative strategies.[60] These included the Toxics Release Inventory, which was passed by Congress in 1986 and tracks and publishes chemical releases to the environment by companies, which are required by law to provide this information to the EPA.[61] Other new approaches included a federal program embedded in the 1990 Food, Agriculture, Conservation, and Trade Act (otherwise known as the "farm bill") that would develop and enforce a national definition of organic food and a market-based cap-and-trade system incorporated in the 1990 Clean Air Act Amendments that enabled companies to buy and sell pollution credits.[62]

Partisanship and polarization over environmental policymaking continued into the 1990s and 2000s, providing further justification for regulation skeptics to continue their quest for more effective governance strategies. President Bill Clinton gave up on the BTU tax he initially proposed as part of his deficit reduction package in 1993 after facing resistance to it from within his own party.[63] His tenure became particularly punctuated with conflicts over environmental regulations once Republicans took control of Congress in 1994.[64] As a strategic response to these partisan clashes, his administration strongly emphasized voluntary programs and public-private partnerships with industry.[65] This emphasis on collaboration as well as further conflict over environmental regulations and international agreements continued during George W. Bush's time in office.[66] As exemplified by its development of the Clean Power Plan and support for the Paris Climate Agreement, the Obama administration has placed more emphasis on regulatory approaches, but gridlock in Congress and lawsuits in the courts continue to drive many actors toward exploring alternative governance strategies.[67]

## The Pull of Information

As the possibilities for passing far-reaching domestic environmental regulations or international treaties dimmed in recent decades, optimism about the power of information has skyrocketed. Such optimism is undergirded by a deeper ideology—in the sense of ideology being a "systematic scheme of ideas, usually relating to politics, economics, or society, and forming the basis of action or policy"—that has multiple roots and manifestations.[68] This "ideology of information" has grown in popularity as the appealing concepts of the "information society" and the "information age" have become increasingly commonplace since the 1970s.[69] The foundation of this ideology is a belief in the ability of information to transmit knowledge, catalyze change, and improve social conditions. For example, Google Executive Chairman Eric Schmidt and Google Ideas Director Jared Cohen claim in their book, *The New Digital Age*, that "everyone benefits from digital data": governments, the media, nongovernmental organizations, companies such as Amazon, and even an illiterate Maasai cattle herder in the Serengeti.[70]

This information optimism spans the political spectrum. For example, George Gilder, Ronald Reagan's most quoted living author and a proponent of supply side economics, posits that information—not incentives—drives economic growth in his book *Knowledge and Power: The Information Theory of Capitalism and How It Is Revolutionizing Our World*.[71] Meanwhile, Cass Sunstein, former administrator of the White House Office of Information and Regulatory Affairs in the Obama administration, discusses the power and potential of new information technologies—such as wikis, prediction markets, and blogs—to facilitate the development of beneficial products and activities in his book *Infotopia: How Many Minds Produce Knowledge*.[72] Such bipartisan support has enabled information-based governance strategies to proliferate while many traditional governance initiatives have been mired in political gridlock.

Each of these authors would agree with journalist Simon Rogers, who asserts that "information is power" in the title of his article about the British government's release of a "tsunami of data" and government-held datasets. This belief in the power of information is rooted in Sir Francis Bacon's older and more well-known statement, "Knowledge is power" (*"Scientia potentia est"*).[73] Nearly four hundred years later, French philosopher Michel Foucault reiterated this relationship, but reframed it as being more complex and bidirectional—"in knowing we control and in controlling we know."[74] As its means of communication, information is central to this power of knowledge.[75] Information represents an individual's ability to translate tacit, internalized knowledge into explicit, articulated words and symbols that can influence the knowledge and actions of other individuals.[76] As the capacity of individuals and organizations to create such information has increased with the advent of new technologies, recognition of the power of information to influence others has increased as well.

Awareness of this power is also rooted in economic theory, which considers "perfect information" to be one of the four prerequisites of well-functioning markets.[77] Friedrich Hayek considered price to be the most relevant information for markets to function efficiently,[78] but more recent work has revealed that other forms of information—signals of quality or willingness to pay, for example—are also relevant to buyers and sellers.[79] Such information is often underproduced because the public good nature of information limits incentives for its dissemination.[80] Thus information asymmetries can develop where buyers know more than sellers (or vice versa), which can cause individuals with more information to increase risks for those with less information by behaving differently (i.e., moral hazards) or by acting on hidden information only available to them (i.e., adverse selection).[81] Different schools of economic thought, from public choice theory and public finance theory to new institutional economics, cultural economics, and behavioral economics, offer a range of arguments for why the provision of relevant information (by either public or private actors) can eliminate these issues, enable markets to work more efficiently, and increase everyone's overall utility—without infringing on their rights.[82]

A third contribution to the rising interest in information-based strategies comes from populist social movements that view information as a liberating and powerful weapon against injustice and repression. These movements have mobilized around concerns about civil rights, public health, environmental degradation, government corruption, and other social issues, and participants in these movements believe that information on these issues should be provided to citizens, consumers, and society at large.[83] This is the basis of the grassroots "right to know" campaigns that have created laws requiring public disclosure of government-held information in seventy countries around the world.[84] Harvard Kennedy School scholar Mary Graham has explored how these campaigns are a form of "technopopulism" that—like the initiatives and referendums of the original populism movement—embrace direct democracy over deliberation.[85] They build on Supreme Court Justice Louis Brandeis's insight that just as "sunlight is said to be the best of disinfectants" and "electric light the most efficient policeman," "publicity is justly commended as a remedy for social and industrial diseases."[86] As Graham's coauthor and Harvard colleague Archon Fung has pointed out, such publicity has the potential to create an "infotopia" that enables citizens to exercise their democratic rights of self-government, make better choices that advance their "own welfare and flourishing," and, "collectively, to control the organizations that affect their lives."[87]

Another basis of this belief in information is the relatively modern concept of objectivity.[88] As information became more widely available with the invention of the printing press, contestation over its veracity and value increased, and by the late nineteenth century, had contributed to the development of a "norm of objectivity."[89] In 1922, the American Society of Newspaper Editors adopted a Code of Ethics that demanded the "sincerity, truthfulness, accuracy ... and impartiality" of journalists.[90] Such a norm of objectivity has become a fundamental tenet of the journalism industry, and now pervades the rhetoric that the industry uses to describe itself. Scholars such as Donna Haraway, a feminist studies scholar at the University of California, Santa Cruz, have pointed out that absolute objectivity is not possible, given the human, "situated" biases that inevitably are incorporated into any form of knowledge.[91] Yet the public still expects and assumes that the information it consumes from the media can be, should be, and is objective. This expectation and assumption further undergirds the new faith in information as a promising basis for effective governance.

Each of these logics of objectivity, technopopulism, and market efficiency contribute to an ideology of information that is an important explanation for why such a wide range of actors have chosen information-based strategies over other governance strategies. For different reasons, this ideology resonates with liberals, conservatives, and libertarians alike, as it respects individual liberties, reduces the need for government regulations, and empowers civil society. Because it requires neither the carrots or sticks that are the basis of the three other forms of governance discussed previously (command-and-control-based regulation, mandatory market-based governance, and voluntary market-based governance), it is perceived as being less costly for society and less demanding of both individuals and companies. It is also seen as more personally empowering, more objective, less "political," and for all of these reasons, more politically feasible. Proponents of information-based strategies thus find them attractive for both ideological and pragmatic reasons, and contend that they can transcend the polarization and partisanship that has dominated environmental politics for decades.

## The Pitfalls of Information

As support for this ideology of information has grown and initiatives based on its doctrines have proliferated, the concerns of skeptics have grown as well. These information pessimists have strongly criticized the new emphasis on the use of information to encourage collective action on both strategic and philosophical grounds. Chief among their concerns is that eco-labels and sustainability ratings are part of a broader corporate takeover of the environmental and social justice movements. University of British Columbia researchers Peter Dauvergne and Jane Lister, for example, assert that the "eco-business" model has allowed "big-brand retailers and manufacturers" to "use 'sustainability' to protect their private interests."[92] They argue that this model cannot on its own solve the ecological challenges facing humanity, as it is focused on "sustainability for big business, not sustainability and the planet" and in many cases is "actually increasing risks and adding to an ever-mounting global crisis."[93] In their view, certifications and ratings are primarily a tool for corporations to better control their supply chains and increase their productivity and profitability.[94] As Queen Mary University scholar Frances Bowen explains, these critical perspectives consider information-based programs to be examples of symbolic corporate environmentalism that maintain existing power structures and delay meaningful environmental action.[95]

Other critics have raised a host of additional concerns about efforts that rely on consumers "shopping for good."[96] As Scott Nova, executive director of the Worker Rights Consortium, explains, such consumers are often comparing results from small-scale and limited corporate initiatives that obscure the deeper environmental and social problems with modern supply chains.[97] Consumers are also confronted with a host of deceptive and fraudulent claims that amount to disinformation and *greenwashing"*—intentional attempts by companies to mislead the public about their environmental performance.[98] The structure and format of these claims can further mislead consumers by grading on a curve that emphasizes the relative rather than absolute and often poor performance of companies.[99] More fundamentally, critics are concerned that eco-labels promote a dangerous myth of the power of consumer influence and "private politics," and weaken the motivation for new environmental regulations and international agreements that make deeper change and progress possible. This is the heart of the skeptic's argument; they believe that despite the time and resources they require, government action is necessary to tackle society's most pressing sustainability challenges. Labels and ratings, however, are often only relevant to the rich and well-to-do, and offer marginal environmental and social benefits, if any at all.[100]

Örebro University sociologist Magnus Boström and Lund University sociologist Mikael Klintman add several other reasons for skepticism about eco-standards and product labels. In their research, they found that some stakeholders believe labeling adds an unwanted additional layer of rules for "already overregulated" industries and gives too much power and control to certain advocacy groups, large companies, and developed countries.[101] Stakeholders also complained that labeling initiatives disregard the experience of public authorities, provide only a "shallow transparency" about products, put too much responsibility on the consumer, and are not an effective or "radical enough" instrument of change.[102] Boström and Klintman also cite skeptical arguments relating to the validity of green labels—that they are "pseudo-science" marketing tools that do not "take environmental and social consequences sufficiently into account."[103] For these reasons, information pessimists would advise our five friends introduced earlier in this chapter to treat the green claims facing them with extreme caution, and instead pursue other, more effective avenues of political action.

The Obama administration, reversing a longstanding policy orientation promoted by both Democratic and Republican presidents, has accepted many of these critiques of information-based governance. In 2009, Lisa Jackson, President Obama's first administrator of the EPA, announced the termination of the National Environmental Performance Track Program, a voluntary program that had recognized companies for their "continuous environmental improvement" and rewarded them with public recognition and fewer enforcement inspections.[104] A 2007 report on the program by EPA's Inspector General found that "only 2 of 30 sampled Performance Track members met all of their environmental improvement commitments."[105] In her letter announcing the program's termination, Jackson stated that such "stewardship initiatives should not take the place of our regulatory framework but should augment it by bringing to bear advanced technology and innovation to achieve progress on emerging problems that our regulations do not yet address."[106] Two years later, EPA also announced the termination of the Climate Leaders program, another voluntary initiative designed to recognize corporations for their efforts to inventory and reduce their greenhouse gas emissions.[107]

The reasons for this increased skepticism about information-based governance strategies are diverse and complex, but can be divided into five main categories. The first is a fundamental distrust in the information being provided by these initiatives. Why should the public trust the organizations behind these initiatives or the processes by which their standards are developed? The second is a perception of the irrelevance of this information. Even if the public trusts these organizations, why are the ratings and labels they are providing important and worth paying attention to? The third set of reasons for skepticism relate to uncertainty about the quality of the information provided. How does the public know if the information is accurate and valid? Even if this uncertainty can be overcome, the public can also develop a sense of information overload from the complexity and diversity of the claims they encounter. How can the public effectively identify and make use of the most trustworthy, relevant, and accurate information available? Finally, and most importantly, information pessimists are concerned that information-based governance is ineffective and impotent—both in absolute terms and relative to other forms of governance. How can the public know whether these efforts are leading to positive and meaningful environmental outcomes?

## Theoretical Framework: The Information Value Chain

These five areas of concerns about information-based governance strategies raise important questions about their efficacy as solutions to society's environmental challenges. Proponents of these strategies generally acknowledge these concerns, but point out that other forms of governance have their own sets of problems.[108] A fundamental concern about regulation-based governance, for example, is its political feasibility in the current domestic climate. Alternative forms of governance, as Lisa Jackson argued in her letter terminating the Performance Track program, can demonstrate the feasibility of new technologies that such regulations would require and build support for their future passage.[109] In the meantime, information optimists have argued that information-based strategies can and have created significant shifts in the market toward more environmentally friendly products.[110]

This is true, although in most cases these strategies are fundamentally limited by their voluntary nature. Even in cases where disclosure of environmental information (e.g., the Toxics Release Inventory) or agency procurement of green products (e.g., Executive Order 13693) is required by the government, the use of disclosed information by the public and other stakeholders is still voluntary. Given the public goods nature of the benefits that the use of this information creates, individuals can benefit from others buying green products without themselves doing so. This is a classic free rider problem, and suggests that a large number of people will not "buy green" on their own.[111] Some environmental information is a blended good (i.e., impure public good) that has both public and private benefits (e.g., an energy efficiency label that indicates both reduced energy costs and environmental impacts), and thus may attract more interest among the public.[112] But even this information faces the five categories of concerns listed previously, each of which may short-circuit the public's interest in paying attention to environmental certifications and ratings. Designers of these programs must address these concerns if they hope to maximize the potential of information-based governance, both as a mechanism to facilitate more far-reaching regulations and as a means of creating direct environmental benefits themselves.

This book provides a theoretical framework that can help designers accomplish these dual goals. It can also assist researchers, consumers, and other actors to better compare and evaluate these strategies. The framework builds on the idea that green ratings and eco-labels are tangible objects that are produced by specific organizations and individuals and have particular characteristics. In order to more clearly analyze and visualize these initiatives, I have conceptualized the information that they deliver as products themselves. These *information products* have four basic components—the organizations that are affiliated with them, the issue areas they cover, the data they are based on, and the interfaces through which they are delivered.

These components are developed in an *information value chain* through four distinct processes—institutional engagement, issue identification, data analysis, and interface design. In this sense, they are similar to supply or value chains that begin with raw materials and produce a final product for end consumers to utilize,[113] but differ from traditional supply chains in the sense that they pull together more intangible resources (e.g., ideas, organizations, data, delivery mechanisms) and create a more intangible asset (i.e., information).

Corresponding to each of the value chain's main components is a set of four basic attributes that are common to all information-based governance strategies—organizational trustworthiness, content relevance, methodological credibility, and cognitive usability.[114] The processes, components, and attributes of this information value chain are shown graphically in figure 1.1. Each of these four components and their attributes influence how different audiences perceive and make use of the information created by these value chains.

## Outline of the Book: A Path toward Information Realism

Using a broad range of social science concepts and empirical examples, the next four chapters describe these attributes in more detail. While recognizing other types of information-based initiatives are important in their own right, given the controversies that surround them I have focused on

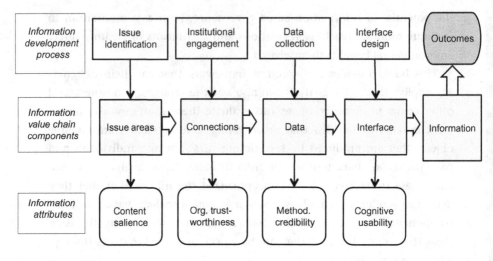

**Figure 1.1**
The information-based governance development process. *Note:* While these process-es are displayed linearly, they can all have important feedback effects on each other.

information-based governance strategies in this book. The empirical data presented comes from my EEPAC Dataset, mentioned earlier in this chap-ter, which contains more than 2,500 webpages and PDF files that were downloaded from the websites of 245 examples of information-based gov-ernance in 2009 and 2010. These pages were analyzed using a rigorous cod-ing process that generated nearly ten thousand coded text segments across 223 different codes. These codes represent characteristics that are relevant to each of the five core areas mentioned earlier. This original dataset pro-vides a unique perspective on the landscape of information-based environ-mental governance strategies and the prevalence of different characteristics and practices. More information about the methods used to construct and analyze this dataset is provided in appendix I.

Building on this data, chapter 2 focuses on the relevance of the informa-tion provided by these initiatives, with particular attention to the issues they cover—from climate change to toxic waste—and the values they activate. Chapter 3 discusses the trustworthiness, legitimacy, and account-ability of organizations that must be recruited to help develop information-based initiatives, by providing expertise, funding, data, and other types of assistance. Chapter 4 analyzes the validity and reliability of the methods used to develop the information provided by these initiatives. Chapter 5 explores the prominence, intelligibility, and feasibility of the interfaces by which the public makes use of the information—the final component of the information value chain.

Chapter 6 then focuses on the effects and effectiveness of the information itself, once it emerges from this value chain. Because of the basic voluntary nature of information-based governance, audience responsiveness to these programs is the primary mechanism through which they act. If audiences respond positively to the information, they may then pursue complementary governance strategies that may make use of new or existing markets, regulations, technologies, information, or moral arguments. Consumers and institutions may change their purchasing behavior, manufacturers may introduce new technologies, government agencies may enact new regulations, and advocacy organizations may begin new campaigns, all in response to the information provided by these information-based strategies. The environmental performance related to the original issue—whether it was climate impacts or some other issue—may then be improved and a public or common good created. Chapter 6 examines these different pathways of effectiveness, summarizes stakeholder perceptions of their viability and usefulness, and investigates examples that demonstrate both their strengths and weaknesses.

Chapter 7 synthesizes the results from the previous chapters in a discussion of the idea of information realism and its cousin, *green realism*. It acknowledges that both sides of the debate about the role of information-based governance strategies have made important points about the promise and pitfalls of these approaches. It also highlights the shortcomings of these two opposing perspectives. While it is true that information-based strategies have made important contributions to protecting the environment, these contributions will be short-lived and overshadowed by their unintended negative effects if they do not further grow into more mature governance initiatives. Likewise, information-based strategies do indeed have important limitations, but they can serve as a critical foundation for other governance approaches that may not come to fruition without the catalytic effect that information-based strategies can provide.

In order to fulfill the expectations of the information optimists and overcome the concerns of the information pessimists, information-based initiatives must become more trustworthy, salient, valid, and usable sources of information. They must directly confront their limitations and tangibly demonstrate how they complement more far-reaching governance strategies. While it will not resolve all of the philosophical and strategic differences in the debate discussed earlier, this book presents a realistic appraisal of the role of information in environmental governance and its relationship to other forms of governance. Based on an interdisciplinary review of the relevant literature and an analysis of over 240 initiatives and over 60 stakeholder interviews, the book distinguishes between the different purposes that these information initiatives serve, and provides guidance for those facing decisions like the five individuals described at the beginning of this chapter.

The book's final chapter outlines a range of recommendations that designers of these programs can implement and users of them can follow that will facilitate the further maturation of information-based governance strategies. Based on the analyses and insights of the previous chapters, these recommendations not only can enhance the effectiveness of information-based environmental governance strategies, but also are relevant to the use of information to encourage collective action on a host of other issues facing society. All members of society—policymakers, corporate executives, civil society advocates, researchers, and individual consumers and citizens—have an important role to play in ensuring that the information we use to help us make decisions also helps us solve our most pressing social and environmental challenges.

## 2 Valuing Green: The Content of the Information

### My Food, My Values

So let's return to our friend Mark, who is standing in the dairy section and deciding between the organic and conventional milk. He has figured out that the price difference is about two dollars more (1.5 cents more per fluid ounce)—an organic gallon of milk costs $5.97 at his grocery store while a conventional gallon costs $3.99.[1] Is the organic option worth it? What is he getting for that extra two dollars? He has a few minutes to spare, so he pulls out his smartphone to do some real-time research.

The top Google results for "should I buy organic?" and "why organic?" list a host of reasons for buying organic. LifeHacker tells Mark that the contents of products with the USDA Organic label must be "at least 95 percent certified organic, meaning free of synthetic additives like pesticides, chemical fertilizers, and dyes, and must not be processed using industrial solvents, irradiation, or genetic engineering."[2] Consumer Reports states that buying organic helps "support farming methods for plants and animals that are healthier for the earth's soil and water supply in the long run."[3] It also cites a study showing that organic milk has 60 percent more "heart-healthy omega-3 fatty acids" than nonorganic milk and does not contain growth hormones.[4] Organic.org lists ten reasons for buying organic, including (1) reducing the toxic load in our air, water, soil, and bodies; (2) reducing farm runoff of fertilizer and the creation of "dead zones"; (3) protecting "future generations" and infants in utero; (4) building healthy soil; (5) tasting better flavor; (6) assisting family farmers; (7) avoiding "hasty and poor science" in your food (such as genetically modified organisms); (8) eating with a sense of place; (9) promoting biodiversity; and (10) celebrating the "culture of agriculture."[5]

Other sites, however, raise several questions about these pro-organic arguments. A CNN article cites a Stanford University study casting doubt on the nutritional benefits of organic food, while a WebMD webpage claims that pesticide residues in nonorganic milk are "far below safety cut offs."[6] It also notes that "some organic foods come from multinational companies and have been trucked across the country" and their environmental footprint may be greater than locally grown, nonorganic foods.[7] Many of the sites Mark visits emphasize that buying organic is a "personal choice," and one that "depends entirely on you, your budget, and what you expect to get from these foods."[8] So how will Mark—and other consumers—respond to these competing reasons and arguments? How will he consider the other sustainability-related choices that he may be confronted with elsewhere in the store, such as fair trade tea, biodynamic wine, locally grown fruit, Rainforest Alliance-certified chocolate, bird-friendly coffee, and Food Alliance-certified meat?

Many factors, from the trustworthiness of these certification organizations to the validity of their claims, will influence Mark's decisions about these products. Arguably the first question that a consumer or anyone else will ask about sustainability-related information, however, is whether they care about the particular focus of that information. If they are not concerned about the underlying issues in question, then the quality of the methods, interfaces, and other components of the information supply chain is doomed to irrelevance. Our story therefore begins with an investigation of the factors that determine how different audiences respond to environmental certifications and ratings, from the nature of the information provided to the personal preferences and background of the individuals receiving that information.

We begin with an exploration of several key concepts that will help us understand how people evaluate the importance of different forms of information content. Building on research from a wide range of fields, this chapter examines different forms of value and different types of values, from use and existence value to egoistic and biospheric values. It discusses a schism that has emerged between scholars and practitioners over which values are most closely linked to pro-environmental behaviors and should be the focus of information-based strategies. The chapter concludes that while the jury is still out on this debate, important normative and strategic reasons exist for information initiatives to encompass as comprehensive a set of values and issues as possible. It then presents data on the extent to which the 245 cases in the Environmental Evaluations of Products and Companies (EEPAC) Dataset accomplish this goal of comprehensiveness, across product categories, geographic scales, types of goods, and parts of the value chain. It finds that while some programs are relatively comprehensive, the vast majority are very narrow in both their focus and scope. The chapter presents both information pessimist and optimist perspectives on these results, and identifies promising and problematic practices for increasing the desirability and communicating the importance of the content of information-based strategies. It then returns to our friend Mark, and applies the insights of the chapter to his quandary in the dairy section.

## Understanding the Nature of Value

Fortunately, social scientists and humanists have been rigorously studying this question of what really matters to people for many years, and have developed several concepts that can help us understand how different types of sustainability claims are valued. Insights on framing and mandatory disclosure policies from political science, value theory and brand salience from marketing and consumer studies, and values theory, construal theory, and the theory of planned behavior from cognitive and social psychology are all relevant to this area of research.[9] Unfortunately, this research is dispersed across different disciplines and research communities that often do not communicate with each other. Thus the concepts they use frequently overlap and conflict with one another. They use the same words—such as "value"—to signify different meanings, and different words to signify the same meaning. In order to effectively build on this past work, this section explores these different concepts and draws important connections between them. From this discussion, an interdisciplinary conceptual framework emerges that brings conceptual clarity and precision to our understanding of the perceived value of competing sustainability claims.

## The Varieties of Value

Let us begin with the concept of value. While most commonly used in the context of market transactions ("Did I get a good value for this product?"), the term is derived from the broader Latin term *valere* ("to be strong, be well; be of value, be worth") and is used extensively in the fields of both economics and philosophy.[10] Indeed, classical economics is based on the labor theory of value, which in the words of eighteenth-century English political economist David Ricardo, posits that "the value of a commodity ... depends on the relative quantity of labour which is necessary for its production."[11] Neoclassical economics, on the other hand, is based on the subjective theory of value, which asserts that value is determined by the relative supply and demand for an object.[12] According to this theory, something is valuable not because someone spent a lot of time making it, but because people really want to buy it.

These two different types of economic value connect to several distinctions made by philosophers about value.[13] The subjective theory of value relates to the ideas of extrinsic and instrumental value, while the labor theory of value can be linked to the concepts of intrinsic and final value. As Harvard philosophy professor Christine Korsgaard explains, *instrumental value* is the value that something has for the sake of something else, while *final value* is the value that something has for its own sake.[14] On the other hand, *extrinsic value* is the relational value that something "gets from some other source" (such as being desired), while *intrinsic value* is the nonrelational value that something has "in itself" (such as being beautiful).[15]

While admittedly somewhat abstract, these distinctions are highly relevant to our understanding of the different types of value embedded in sustainability claims. Environmental activists, economists, and philosophers have long differentiated between the environment's different forms of value, from its usefulness to humans today (its *use value*), to those same humans in the future (its *option value*), or to entirely new and future generations of humans (its *bequest value*). These all represent extrinsic and instrumental forms of value—they are derived from humans and are for the sake of humans, not the environment itself. Commentators also speak of the environment's *existence value*—the value of nature independent of humans. This is a form of intrinsic value and final value that comes from nature and is for nature's sake itself.[16]

A particular endangered species, for example, may be perceived as having existence value because it is intrinsically beautiful or because it has a right to exist for its own sake, independent of its beauty. It may also be perceived as having use value because tourists enjoy observing it in the wild, option value because it may have medicinal uses in the future, or bequest value because it may be valuable to our grandchildren.[17] Some people may perceive an endangered species as having all of these forms of value, while others may focus on either its intrinsic or extrinsic value, or its instrumental or final value. These perceptions of value can inform people's actions either as citizens when they decide who to vote for or what organizations to support, or as consumers when they decide what products to buy (or not to buy).

### The Role of Values

With this understanding of value, we can move to the concept of *values*, which has received a significant amount of attention in sociology, political science, and psychology. Values help people determine what has value to them, whether that value is final vs. instrumental or intrinsic vs. extrinsic. While definitions of the term "value" vary across these disciplines, Hebrew University psychology professor Shalom Schwartz and University of Münster psychology professor Wolfgang Bilksy identify five characteristics of values that are commonly found across the relevant literature, which are that they represent "(a) concepts or beliefs, (b) about desirable end states or behaviors, (c) that transcend specific situations, (d) guide selection or evaluation of behavior and events, and (e) are ordered by relative importance."[18]

Schwartz has developed a values framework that is now widely used in the social sciences and has been validated across a large number of countries and cultural contexts.[19] This framework consists of ten values that are categorized along two dimensions—self-enhancement versus self-transcendence; and conservation vs. openness to change (shown in figure 2.1).[20] The self-enhancement category includes values emphasizing power over others and personal achievement, while self-transcendence includes caring about people you regularly interact with (benevolence) and caring about all people, strangers and friends alike (universalism). The conservation category encompasses security, conformity, and tradition, while the openness to change category includes stimulation and self-direction. The tenth value, hedonism, falls between self-enhancement and openness to change. Following Schwartz's initial work, National Research Council scholar Paul Stern and George Mason University professors Thomas

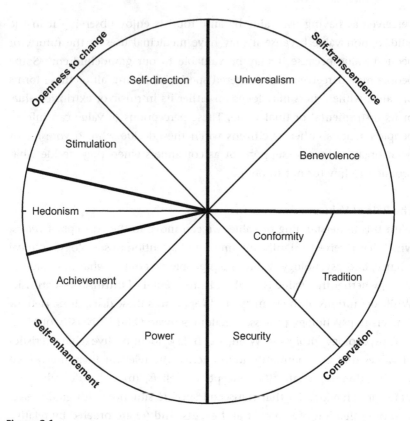

**Figure 2.1**
Schwartz's theoretical model of relations among ten motivational types of value.
Adopted from Schwartz, "An Overview of the Schwartz Theory of Basic Values."

Dietz and Gregory Guagnamo subdivided the existing category of self-
transcendent values in Schwartz's framework into *altruistic* values focused
on social justice, peace, and equality among humans and *biospheric* values
associated with preserving nature, fitting into nature, and "respecting the
earth."[21] They also classified Schwartz's self-enhancement values category
as covering *egoistic values*.[22]

Given its broad acceptance among scholars and direct relevance to our
discussion, I will be returning to the work of Schwartz, Stern, and their colleagues throughout this chapter. Their work on values is important in light
of research that has demonstrated that values are central to understanding
and predicting human behavior. This work provides empirical support for
Stern's value-belief-norm theory, which posits that values are an important

determinant of not only people's beliefs and personal norms, but also their behavior, and particularly pro-environmental behavior.[23] This theory helps explain why people holding certain values and conceptions of value often choose to purchase products certified as environmentally sustainable.

Thus far in this chapter, we have covered a lot of these values and conceptions. Figure 2.2 provides a simple map of these different ideas to help us keep track of them all, and see the connections between them. It is divided into three sectors corresponding to Stern's distinction between egoistic, biospheric, and altruistic values. The white boxes correspond to Schwartz's four main values categories, while the ovals

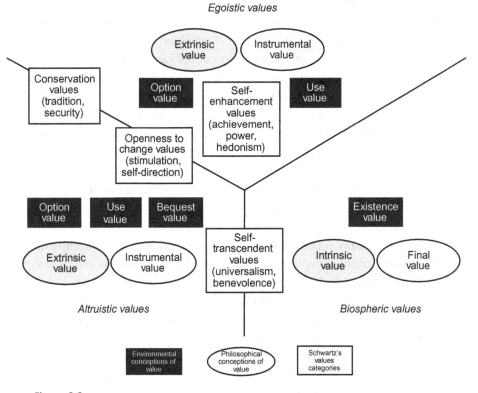

**Figure 2.2**
A conceptual framework for understanding value. *Note:* The white boxes correspond to Schwartz's four main values categories, while the ovals correspond to the philosophical conceptions of intrinsic/extrinsic and instrumental/final value. The dark boxes represent the four environmental conceptions of value—use, option, bequest, and existence value.

correspond to the philosophical conceptions of intrinsic/extrinsic and instrumental/final value. The dark boxes represent the four environmental conceptions of value—use, option, bequest, and existence value. As figure 2.2 shows, on the one hand, existence, intrinsic, and final value are generally only linked to biospheric values, bequest value is only associated with altruistic values, and self-enhancement values are only connected to egoistic values. On the other hand, conservation and openness to change values and use and option value can have both egoistic and altruistic orientations, while self-transcendent values can have either altruistic or biospheric orientations.

We can use this map to analyze why people might value certain things and make certain decisions. Let's consider something in nature that Mark and other consumers might consider to have some value to them. Fisheries are a good example that might be affected by Mark's decision about organic food products. Conventional agriculture can lead to high levels of fertilizer runoff, which can lead to the expansion of hypoxic dead zones in downstream water bodies that have oxygen levels that are "too low to support many aquatic organisms including commercially desirable species."[24] One of the largest such zones exists where the Mississippi River flows into the Gulf of Mexico, and has been associated with a decline in economically important fisheries. National Oceanic and Atmospheric Administration researchers Thomas O'Connor and David Whitall, for example, found a significant correlation between the size of hypoxic zones in the Gulf of Mexico and the catch of brown shrimp in Louisiana and Texas.[25] Thus buying conventionally produced food may be contributing to the decline of Gulf shrimp populations, and if Mark values these populations and fisheries he might consider buying organic products when he can.

As figure 2.2 shows, these shrimp populations may have value to Mark for many reasons. He may perceive them to have use value (he likes shrimp cocktail), option value (he may want to be able to eat shrimp cocktail 20 years from now), or bequest value (he may want his grandchildren to have the option to eat shrimp cocktail). These types of value associated with shrimp may be important to Mark because of his strong egoistic or altruistic values (shrimp for himself or shrimp for society). More specifically, they may resonate with particular values from Schwartz's framework, such as hedonism, achievement, tradition, or benevolence. But the activation of these values is borne out of a sense of the fisheries' extrinsic and instrumental value. The shrimp serve an end beyond themselves—providing tasty nourishment or maintaining a family tradition, for example—and their value depends on something external to themselves. These forms of value therefore reflect the subjective theory of value and depend on the supply of and demand for the shrimp.

Alternatively, Mark may value these shrimp populations for their existence value and believe they have a right to exist independent of their value to humans. He may perceive them as having both intrinsic and final value because they represent an important form of life, created through a certain amount of (nonhuman) energy and labor. This type of value is therefore connected to the labor theory of economics discussed earlier, and resonates with the biospheric and self-transcendent values that Stern and Schwartz describe.

### The Values Debate

This background on values provides the context for understanding a debate that has emerged about the relationship between values and proenvironmental behavior. Survey research has generally found that on the one hand biospheric values are most commonly associated with actions oriented toward protecting the environment, followed by altruistic values. Self-enhancement and egoistic values, on the other hand, have generally been found to be negatively correlated with personal attitudes, norms, and behaviors that protect the environment.[26]

Despite this research, an increasing number of both scholars and practitioners are calling on marketers to use claims based on self-interest to advertise their green products.[27] For example, Jacquelin Ottman, a prominent green marketing consultant, claims that most consumers "want to know how even the greenest of products benefits them personally."[28] She gives several examples of green marketing campaigns that emphasize how the featured products protect consumers' health and save consumers money. Nancy Schneider, another sustainability consultant, asserts that markets can reach mainstream consumers by positioning green products "as better, modern, or optimized" because "their purchasing behavior is based more on personal benefits."[29] Thus there appears to be a divide between marketing practices that are increasingly emphasizing the private benefits of environmentally friendly products and academic research that suggests most people buy those products because of their public benefits.

This divide is likely due to several factors. The first factor is that the two groups are focused on two different types of people. On the one hand, the academic research is generally trying to explain what motivates the greenest of consumers, who may indeed be focused on the public benefits of green products. On the other hand, the practitioners are trying to figure out how to motivate everyone else to become greener consumers, who may be more focused on the private benefits of the goods they purchase. The second factor is that both sides in this debate are focused on values as static psychological traits whose order and importance are fixed and difficult to change. This assumption ignores the possibility that effective communication efforts can either change a person's values or make existing ones more salient. Schwartz's norm activation theory suggests that both norms and values can be activated with focused stimuli that can cause people to make decisions that they previously would not have.[30] University of Tromsø psychology professor Bas Verplanken and University of Nijmegen psychology professor Rob Holland, for example, show that environmental values can be experimentally activated to increase consumer intentions to purchase products with more favorable environmental characteristics.[31] University of Freiburg psychology researcher Ulf Hahnel and his colleagues also find that activating environmental values with nature photos causes consumers to both evaluate the costs of electric vehicles more positively and to be less sensitive to price increases in those vehicles.[32]

### A Broader Values Framework

These studies provide important insights into the complex price effects of a value activation intervention and the important relationship between value centrality and these interventions. They do not, however, distinguish between or test the effects of activating different types of values. This highlights the third factor contributing to the divide between academic researchers and professionals working on pro-environmental consumption—they conceive of the benefits associated with green products in just two dimensions (public or private) when in fact those benefits have multiple dimensions that can appeal to consumers (and their values) in diverse ways.[33] Schwartz's values theory is a helpful framework for thinking about these different dimensions and how people respond to them. Eco-labels may not only activate benevolence, universalism, and security values by referencing their public and private benefits, but may also connect with a broader and more complex set of values associated with hedonism, openness to change (stimulation and self-direction), and self-enhancement (power and achievement).

Such associations have increasingly been investigated in empirical research studies. For example, with regard to hedonism, in one study, German consumers cited their perception that organic food tasted better than conventional foods as one of the four main reasons why they purchase organic food.[34] Another study found that American wine drinkers who expressed a strong interest in living a hedonistic life were more likely to purchase organic wines than those who did not express such an interest.[35] Openness to change values, which include self-direction and stimulation, have also been found to be associated with pro-environmental behaviors. For example, 23 percent of a sample of Sicilian consumers cited curiosity as a motivation for buying organic products.[36] Participants for whom curiosity is a particularly important motivator had consumed organic products for only a short time (less than six months) and tended to be younger and female. A survey of 10,000 U.S. female consumers found that purchasers of fair trade products place more importance on openness to change values, such as freedom of choice, independence, self-respect, creativity, and curiosity.[37] Boston College sociologist David Karp found that openness to change values are most strongly associated with pro-environmental behaviors—ranging from recycling to volunteering to buying organic produce—when individuals also hold strong self-transcendence values (benevolence and universalism).

Documented associations between green consumption and self-enhancement values—achievement and power—are less common in the literature, and some studies have found the relationship to be negative (i.e., individuals for whom these values are particularly important are less likely to buy green products). These survey-based studies, however, frame these values in generic ways that are not necessarily relevant to the context of environmental behavior. Using a mixed-methods research design using focus groups, in-depth interviews, and questionnaires, University of Glasgow professor of marketing Deirdre Shaw and her colleagues found that consumers with a general interest in ethical issues do indeed view their ethical purchases as a form of achievement, and particularly as a means of being influential and having a positive impact on both other people and the environment. They also discovered that these consumers are motivated by both a negative and a positive sense of power—to resist the power of capitalism and multinational companies and to embrace their own personal power as an ethical consumer.[38]

Each of the three values in the conservation category—tradition, conformity, and security—have also been associated with green purchases. Consumers have cited traditional food production, for example, as an important factor in their food choices. They want their locally produced food to be additive-free, traditional, and organically grown, and "to taste as they believed it did in the past."[39] Along these lines, University of Catania economist Gaetano Chinnici and his colleagues identified a small but stable segment of organic consumers as "nostalgic" because they associate "the consumption of organic produce with the genuineness and tastes of the past."[40] And research has shown that many consumers are more likely to buy organic foods if they perceive that people important to them think that they should do so, demonstrating the effect of a strong conformity value.[41]

A wide range of studies have also found that many consumers buy organic products because they perceive them to be healthier and safer than conventional products, which is directly linked to the security value.[42] This appears to be particularly true for occasional organic purchasers and consumers who have recently had children or received distressing news about conventional food products.[43] In other product categories, the security value can be activated when the purchase of green products results in cost savings (as opposed to health benefits), such as in the case of energy-efficient appliances. The importance of the security value found in these studies resonates with the emphasis that sustainability practitioners have recently placed on highlighting the personal benefits of green products for consumers.

The research showing a positive relationship between self-transcendent values and pro-environmental behavior, as mentioned earlier, is extensive and extends across a wide range of contexts.[44] An important question that this literature raises is the value of the distinction between benevolence (concern for those one frequently comes into contact with) and universalism (concern for all humans and nature).[45] While Schwartz's original work provides ample justification for making this distinction, much of the empirical work utilizing his framework has found that they are highly correlated.[46] But as Shaw's focus groups explained, ethical consumers are

motivated by both a desire to have honest, helpful and "benevolent" interactions with companies and their employees as well as broader "universalis-
tic" concerns about equality, social justice, and environmental protection.[47]
Similar to the research on biocentric and altruistic values, results are mixed
as to whether these are distinctions that are universally significant.[48] Given
that they have been shown to be useful distinctions in some contexts
related to green consumption, it makes sense to keep track of them for the
time being.

### Aspiring to Value Comprehensiveness and Clarity

This point is related to several important caveats about the research cited
previously. The first is that for the most part the methods used in these
studies are only identifying correlational and not causal relationships. So
while an association between universalism and organic purchases may
exist, we do not know if those purchases are being caused by the universalism value. They also often reflect the stated preferences of consumers rather
than their preferences as revealed by their actual behavior. Also, these studies are conducted in particular places in particular contexts (e.g., apple consumers in Greece), and their conclusions may not be true elsewhere.[49] And
to the extent they do not find a particular association between a value and
a behavior, this result may be due to inappropriate measures of either the
value or the behavior, and the association may indeed exist. Nevertheless,
despite these limitations, in the absence of better data, this research suggests important working hypotheses about the importance of a diversity
of values in people's decision making about environmental evaluations of
products and companies.

The emphasis on values in this literature and in this chapter also builds
on particular theories of behavior (e.g., the value-belief-norm theory, norm
activation theory), but these theories and other theories of behavior are not
mutually exclusive.[50] For example, values may be activated either through
deeply entrenched (or "frozen") habits through the peripheral route (which
involves less conscious processing) or via the central route (which involves
more conscious processing) of the elaboration likelihood model, a particularly popular theory in psychology. They may become manifest as either
emotional/intuitive or rational/cognitive expressions of approval or concern about a product or company. These expressions may be mediated by
norms about a behavior or the extent of control individuals perceive they

have over a particular behavior, which are key components of another psychological model, the theory of planned behavior. These other models primarily focus on the cognitive processes by which behavioral decisions are made, and how these cognitive processes respond to different forms of information is a major focus of chapter 5. This chapter is instead focused on the desirability of the substantive content that flows through those processes—why do people find some eco-labels more desirable and important than others?

Following the logic and results of these previous studies, values do indeed appear to be a critical component of that content and a key factor in determining whether people find different sustainability labels and ratings to be desirable. While debates rage about whether public or private interests are most associated with pro-environmental behaviors, it is clear that they are not mutually exclusive. Research has clearly shown that not only are both self-transcendent and security values linked to purchases of organic food, but so are values associated with openness to change, tradition, conformity, power, achievement, and hedonism. Past scholarship has also shown that individuals prioritize these values differently; for example, some people care more about tradition while others care more about achievement.

While this research suggests that strategies that positively reference and reinforce all four types of values (self-transcendent, self-enhancement, openness to change, and conservation) are most likely to be attractive to the largest and most diverse audiences, this hypothesis has not yet been confirmed empirically. It is possible that narrowly focused initiatives, due to the simplicity of their message, may become more popular than more broad-based ones. Nevertheless, from both a normative and strategic perspective, the breadth of values that are important to the public demands that environmental certifications and ratings embody as many of the preceding values as possible (i.e., value comprehensiveness) and to clearly communicate those values that they do indeed embody (i.e., value clarity). Normatively, it is highly problematic for a certification to focus on a single issue to the exclusion of others. The marketing consultancy firm TerraChoice identifies this practice as the "sin of the hidden trade-off"— presenting, or greenwashing, a product as environmentally friendly based on an "unreasonably narrow set of attributes without attention to other important environmental issues."[51]

Such a practice is misleading to the public, and is akin to fraudulent marketing. It undermines the credibility of individual initiatives as well as information-based environmental governance strategies collectively. From a strategic perspective, therefore, it also makes sense for programs to be as comprehensive as possible. Until there is robust evidence demonstrating that such comprehensiveness somehow undermines the salience and effectiveness of information-based governance strategies, these programs should be designed to include as many relevant issues and dimensions of performance as they can. Likewise, they should also attempt to activate as many values as they can as well. It is these values that ultimately will determine whether different audiences see value in the claims of these initiatives, whether that value is extrinsic or intrinsic, final or instrumental, important to them in the present (having a high use value) or the future (having a high option value), or important to future generations (having a high bequest value)—or to no one in particular at all (but nevertheless having a high existence value).

## The Content of Information-Based Environmental Governance Strategies

Given this discussion, how well do existing environmental evaluations of products and companies perform at communicating the substance and value of their content? How well do they convey their importance to people like Mark, as they decide whether these evaluations are worth paying attention to? In order to address these questions, we can consult my EEPAC Dataset of 245 different eco-labels and sustainability ratings. This database contains nearly ten thousand coded text segments from the websites of these programs, and is described in more detail in chapter 1 and appendix I. The codes and text segments most relevant to this chapter fall into two general categories—the scope (i.e., what range of entities are being evaluated) and the focus (i.e., what criteria are being used to evaluate these entities) of these evaluations. The initiatives covered in the database cover a wide range of product categories and economic sectors that are relevant to U.S. marketplace.

### Public Benefits: Pollution, Energy, and Climate Predominate

Let's start with the focus of these 245 cases and the criteria they use in their ratings and certifications. Following the literature's overarching focus on the private and public benefits of green products, these criteria can initially be classified by these two broad categories. While we will see later how this is a limiting categorization, it is a useful place to start our investigation. Public benefits include the social and environmental contributions measured by the initiative, while private benefits include health, quality, and economic benefits that accrue primarily to the individual consumer. The specific criteria of these two benefit categories will be explained in more detail later in this section. Approximately 47 percent of the cases claim to use criteria that can be categorized as measuring both public and private benefits, while 40 percent measure only public benefits and 2 percent measure only private benefits. Just over 11 percent (twenty-six cases) do not claim to measure either type of benefit.

It is important to note that these statistics are measures of what these initiatives claim about themselves on their websites, not what they actually are doing. It is therefore possible that some of these twenty-six initiatives do indeed measure a particular type of benefit but do not describe it on their website. Despite this possibility, however, these statistics suggest that if trained researchers cannot identify any criteria that measure public or private benefits on an initiative's website, it is likely that most consumers will not either. Any values they hold that might resonate with those benefits therefore will in turn not likely be activated when they encounter information from that initiative.

The publicly oriented criteria of these evaluations can be further divided into social and environmental categories. Approximately 30 percent of the cases include criteria that relate to social issues (such as diversity, employee health and safety, or human rights issues), while 86 percent of the cases include criteria relating to environmental challenges.[52] These environmental criteria can be further classified by the types of environmental issues they claim to cover. Figure 2.3 shows that pollution and emissions are the most commonly mentioned issues, followed by energy and climate change, materials use, biodiversity and wildlife, and water use. Only 6 percent of the cases include criteria related to all five of these areas, while just over 40 percent cover between two and four of these areas. Approximately one quarter cover only one of these areas, and another 30 percent cover none of them. The correlations between these criteria vary substantially—pollution/ emissions and energy/climate are the most strongly correlated (0.44) while pollution/emissions and biodiversity/wildlife are the least correlated (0.07).

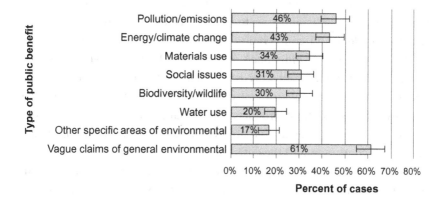

**Figure 2.3**
Criteria coverage (by public issues mentioned). *Note:* Figure shows the percentage of cases that mention criteria that were classified in the listed categories. Error bars indicate 95 percent confidence intervals for each sample proportion.

Approximately 17 percent of the cases also include criteria about a specific aspect of general environmental performance not directly related to any of these issue areas (e.g., use of an environmental management system). Another 13 percent make vague and general claims of green or environmentally friendly performance without referencing any specific environmental challenge, benefit, or policy.

These summary statistics provide several important insights about eco-labels and sustainability ratings (at least as they presented themselves in 2009–2010). The first is that one out of eight of them are committing what TerraChoice's "Sins of Greenwashing" report called "the sin of vagueness"—making a claim that is "so poorly defined or broad that its real meaning is likely to be misunderstood by the consumer."[53] For example, the Big Green Purse states that it lists "eco-friendly products" on its website, while National Geographic's list of the 50 Top Ecolodges claims to have identified the "most earth-friendly retreats." Neither of these cases explains how it make these assessments. Not only is this problematic from a validity and reliability perspective (the subject of chapter 4), but it also represents a missed opportunity to tap into the specific values and concerns that consumers have.

The fact that pollution/emissions, climate/energy, and materials use are the most commonly mentioned environmental criteria may reflect the fact that they are the most likely to be perceived as "blended goods"—goods that have both public and private characteristics that may activate both self-transcendent and conservation values. Pollution can have an effect on both the health of individuals and the broader public, while wasteful energy use can impact not only the climate but also individual pocket-books. Likewise, materials use not only refers to recycled, biodegradable, and sustainably produced products that enhance environmental conservation, but also to products that are free of hazardous chemicals that can harm both consumers and the environment.

Many of these cases are helping consumers make the connection between these private and public benefits. The correlations between mentions of pollution/emissions, energy/climate, and materials and health, economic, and quality benefits (to be discussed in more detail in the next section) are primarily positive, with materials and health having the strongest relationship (correlation coefficient = 0.37). Initiatives such as Green Seal, Cradle to Cradle, and Rainforest Alliance, for example, are at least implicitly linking health concerns with materials use on their sites. Programs such as ENERGY STAR, LEED, and Home Depot's Eco-Options are doing the same for the energy-economics linkage. These cases are working to activate a range of values that can increase the desirability of their certifications.

The less commonly mentioned environmental criteria—biodiversity/wildlife and water use—may be less likely to be perceived as blended goods. Biodiversity is generally conceived of as only motivating to people with strong biocentric values, which might explain why biodiversity and wildlife criteria were negatively associated with economic and quality claims in the EEPAC Dataset. Similarly, water use is seen as important to people with either altruistic or biocentric values. Likewise, social issues are often perceived as only engaging those with altruistic concerns. These cases reinforce this orientation, as mentions of biodiversity, water use, and social issues are less strongly correlated—and in some cases negatively correlated—with mentions of private benefits. They also often do not overlap with the more commonly mentioned environmental criteria already discussed (energy/climate, materials use, and pollution/emissions). Only 16 percent of the cases include criteria related to both biodiversity and energy/climate, for example. This lack of overlap tells us a lot about the beliefs and motivations of the designers of these initiatives, and shows how the emphasis on private benefits by sustainability practitioners mentioned earlier is reflected in the focus of these initiatives. This not only demonstrates the failure of most of these programs to address such pressing issues as biodiversity loss and unsustainable water use, but for several reasons also represents a missed opportunity to use these issues to connect with consumers and their values.

The first reason is that while biocentrism and altruism are not necessarily the central values for many consumers, they nevertheless can be activated to have an effect on their actions and behavior.[54] The second is that for many consumers these values are increasingly important and central to their identity. Gallup polls consistently show that more than one-third of Americans consistently report that they personally worry a great deal about the loss of tropical forests and the extinction of plant and animal species, and this percentage increased from 2015 to 2016.[55] The third, and perhaps most overlooked reason, is that biodiversity, water, and most social issues do provide important private benefits to individuals. More than 50 percent of human medicines were originally derived from natural sources, and as the recent drought in California has shown, inefficient water use can result in increased food prices for consumers.[56]

While most of the cases in the dataset do not make these connections, some at least have attempted to do so. For example, the website of the PEFC Sustainable Forest Management certification explains that certified forests are managed using practices that ensure that the rights of workers and indigenous peoples are protected, "biodiversity of forest ecosystems is maintained or enhanced," and "the range of ecosystem services that forests provide is sustained," including their role in the water and carbon cycles and the provision of food, fiber, biomass, and wood. The EPA's WaterSense website clearly explains the need for water conservation and makes the connection between the environmental and economic benefits of their label explicit, stating that "by using water-efficient products and practices, consumers can help save natural resources, reduce their water consumption, and save money." As an example of a case that emphasizes the intrinsic value of responsible consumption, the grocery store chain Whole Foods Market states that the company believes it has "a responsibility toward all entities involved in our business: our customers, shareholders, Team Members, suppliers, the environment and, not least of all, our community." The company has therefore created its Whole Trade Guarantee certification program as "an extension of our values that lets you rest assured that you are buying the best for you, for your community, and your world."

### Private Benefits: Health First, Then Savings and Quality

I also assessed the extent to which these cases claim to evaluate areas of environmental performance associated with private benefits. Such benefits accrue to individual consumers in three general categories—economic savings, health, and product quality. Given the variety of these claims, they were coded as strong, limited, or implied, with implied being used for references to specific programs that implicitly relate to a specific benefit (e.g., ENERGY STAR and cost savings, USDA Organic and health benefits). Strong codes were used when specific and extensive claims were made, while limited codes were used for more vague and general claims.

As figure 2.4 shows, health benefits (i.e., reduced toxicity, greater product safety, organic) were the most commonly mentioned private benefits. They were mentioned by one third of the initiatives, with 11 percent being strong claims, 18 percent being limited claims, and 4 percent being implied claims. An example of a strong health claim is the Blue Ocean Institute's statement that its seafood guide "incorporates human health recommendations" about fish "that contain levels of mercury or PCBs that may pose a

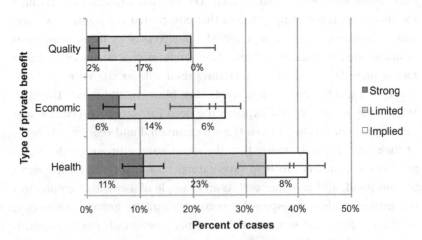

**Figure 2.4**
Criteria coverage (by private benefits mentioned). *Note:* Figure shows the percentage of cases that mention criteria that were classified in the listed categories. Error bars indicate 95 percent confidence intervals for each sample proportion.

health risk," while an example of a limited health claim is the HIP Score-card's statement that their "analysis assessed the share of a company's prod-ucts and services that contributed a net positive benefit to customers' and employees' health and wealth."

The second most commonly mentioned private benefit was economic benefits (i.e., cost savings, lower prices). They were mentioned by 21 per-cent of the cases, with 6 percent being strong claims, 12 percent being lim-ited claims, and 3 percent being implied claims. An example of a strong claim is EPEAT's statement that its electronics certification has saved the healthcare company Kaiser Permanente $4,784,598, while an example of a limited claim is Eco-Crown's statement that they provide "concrete green measures to reduce costs."

Product quality benefits (i.e., improved quality, better overall perfor-mance) were mentioned by the smallest number of cases—19 percent of the programs, with 2 percent being strong claims, 17 percent being limited claims, and none being implied claims. An example of a strong product quality claim is Whole Food's Premium Body Care statement that "all items that meet our Premium Body Care standard are made with ingredients that must be necessary for the product to function well and look appealing while providing real results," while an example of a limited product quality claim is Earth Check's statement that its certification process helps compa-nies improve the "guest experience."

Overall, approximately 49 percent of the cases mentioned either eco-nomic, health, or quality benefits connected with the use of their infor-mation. Only 16 percent mention two types of such benefits and only 5 percent mention all three types. This final group includes initiatives such as Design for the Environment, Veriflora, and Food Alliance. Given that these benefits relate to a range of different values, such a strategy is likely to attract a wider audience. More specifically, people who value a more hedo-nistic lifestyle are likely to be attracted to discussions of product quality, while people for whom security is a central value are likely to be interested in claims about health benefits and cost savings.

These results highlight another important point about these appeals to personal values. Not only should these appeals reference a diverse range of values, but they also need to do so clearly and explicitly. Among the cases that claim to provide private benefits, over 85 percent only make implied or limited claims of such benefits. This statistic suggests that they are not making the strongest case possible regarding the value of their product and company assessments to consumers and other stakeholders.

More generally speaking, while this data reflects sustainability practitioners' emphasis on private benefits, it does not capture the full range of values that might be activated by these programs. Environmental certifications and ratings can also activate openness to change values, such as stimulation and self-direction. While specific criteria and benefits do not map as easily or as directly to these values, a lexical search of terms related to "innovation" can serve as a proxy measure of the extent to which these cases are activating openness to change values. Such a search revealed that only 14 percent of the cases use the term "innovation" or "innovative" more than five times on their sites (56 percent used it once). The Calvert Social Index, for example, describes various aspects of their approach as "innovative" seven times, stating that they have "been a leader in developing innovative ways to meet the financial needs of our shareholders and contribute to the well-being of society at large."

Information-based strategies can also attempt to activate people's self-enhancement values, such as achievement and power. As discussed, being influential, making a difference, and resisting the power of corporations are also important expressions of personal self-enhancement values for many consumers.[57] A similar lexical search for "make a difference" found that 41 initiatives (17 percent) use this phrase on their website. Rainforest Alliance's certification website, for example, states that "purchases of sustainable forest products make a difference," while Staples' EcoEasy program's motto is "making it easy to make a difference." These initiatives are using a mix of both rational and emotional appeals to activate the concerns and values of their audiences.

**Horizontal Scope: Many Sectors Avoid the Spotlight**
We can now turn to the scope of the environmental evaluations made by the 245 cases in the dataset. Their scope can be understood along three dimensions. The first is what I will call the "horizontal" dimension—the breadth of product categories and economic sectors that a particular certification covers. This dimension is important for two reasons—first, all other things being equal, consumers are most likely to be attracted to and remember a certification or rating that can be found on a broad range of products and companies. This tendency is most closely related to a sense

of both cognitive convenience and the self-enhancing value of achievement. For example, is it more impressive to be rated as the most sustainable
company among the ten most sustainable food companies (as Unilever has
been by Oxfam), or as the most sustainable food company across 3,000
companies representing a multitude of economic sectors (as Unilever has
been by Forbes)?[58]

In order to develop a sense of how broad information-based environmental governance strategies are with regard to this horizontal category/
sector dimension, I classified the cases in my dataset by their North American Industry Classification System (NAICS) industry sectors, which are
widely used in the public and private sectors to categorize businesses.[59]
Figure 2.5 shows the number and percentage of cases in the sample by
their NAICS sector. These NAICS sectors encompass the thirty-eight product categories (listed in table 2.1) that are covered by the cases and were
identified during the coding process. Nearly 45 percent of the cases cover
only one product category, while just over 55 percent cover more than
one product category. Approximately 30 percent cover a broad range of
sectors (ten or more) or are not limited to a select group of sectors.[60] Manufacturing is the most common NAICS sector covered by this sample of
cases, followed by Agriculture, Forestry, Fishing, and Hunting. Within the
NAICS sectors, the most commonly covered product categories are Food
(covered by 19 percent of all cases), followed by Household Products (16
percent), Apparel (15 percent), Electronics (15 percent), and Personal Care
(15 percent). The NAICS sectors not covered are Administrative and Support, Waste Management and Remediation Services, Arts, Entertainment,
and Recreation, Management of Companies and Enterprises, Mining,
Quarrying, Oil and Gas Extraction, Public Administration, Retail Trade,
and Wholesale Trade.[61]

Such uneven coverage of economic sectors and product categories suggests that information-based governance strategies are failing to shine a
spotlight on a huge number of companies and products. Such an uneven
focus might be justified by variances in consumer concern about sustainability by sector. Research has indeed shown that such concern about different issues can vary substantially depending on the product category. One
study of consumers across six European countries found, for example, that
over 20 percent of survey participants associated poor working conditions
and use of child labor with chocolate and sweets, a level much higher than

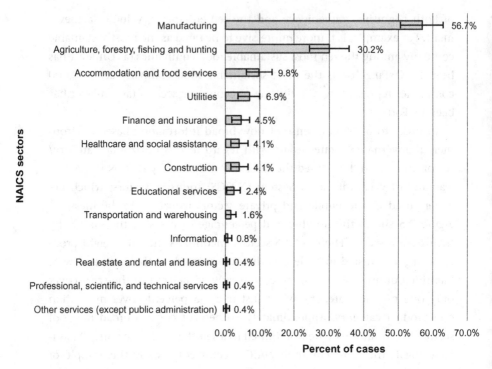

**Figure 2.5**
NAICS economic sectors coverage. *Note:* Figure shows the percentage of cases in the EEPAC Dataset that evaluate products or companies from different economic sectors. Error bars indicate 95 percent confidence intervals for each sample proportion.

the other products surveyed.[62] Meanwhile, nearly 40 percent of these participants also listed the use of pesticides as one of the top three concerns they had when choosing a breakfast cereal. Similarly, Leeds University business professor Timothy Devinney and his colleagues have shown that consumers who are concerned about social issues associated with athletic shoes are not necessarily concerned about similar issues associated with AAA batteries.[63]

These studies bring up the important issue of *materiality*, which has been a hot topic among sustainability practitioners in recent years.[64] In general terms, materiality reflects the extent to which particular issues are "material" or relevant to an entity, although practitioners differ on how such materiality should be defined and who it should be defined for (financial investors or a broader set of stakeholders).[65] To some extent, these questions

**Table 2.1**

Crosswalk between NAICS sectors and product categories covered by the case sample

| NAICS sector | Coded product categories |
| --- | --- |
| Accommodation and Food Services | Restaurant, Travel |
| Agriculture, Forestry, Fishing and Hunting | Forestry, Fishing, Hunting, Carbon Offsets, Flowers, Food, Seafood, Wood Products |
| Construction | Housing |
| Educational Services | Education |
| Finance and Insurance | Banks |
| Healthcare and Social Assistance | Healthcare |
| Information | Media |
| Manufacturing | Apparel, Appliances, Automobiles, Building Products, Carpet, Chemicals, Electronics, Flooring, Furniture, Garden, Household Products, Laundry, Luxury Goods, Materials, Office Products, Paint, Personal Care, Pets, Pharmaceuticals, Toys |
| Multiple | Multiple |
| Other Services (Except Public Administration) | Dry cleaning |
| Professional, Scientific, and Technical Services | Labs |
| Real Estate and Rental and Leasing | Real Estate |
| Transportation and Warehousing | Airlines |
| Utilities | Energy |

*Note:* Table shows how the product categories that are evaluated by these cases are classified by economic sector.

are methodological ones that I will return to in chapter 4, but on a deeper level they relate back to the concepts of values we have been discussing. The materiality of a particular issue in a particular context to a particular person is dependent on that person's perceived connections between that issue and that context and their own personal values. If those connections are perceived as strong, the issue is more likely to elicit a strong cognitive or emotional reaction, or both.

The fact that consumers are more concerned about pesticides in cereal therefore likely reflects a perception that the ingredients in cereal are more likely to have pesticide residues than ice cream or soft drinks. Likewise, consumers who are concerned about the social issues associated with athletic shoes may not care or know as much about the social issues associated with AAA batteries. Other characteristics of these products—their cost, the frequency and form of their use—might also explain these differences in levels of concern. Such differences should therefore not necessarily be interpreted as reflecting value inconsistency. Given the limited knowledge of most consumers about these issues, they should also not be used as justification for the major discrepancies in sector and category coverage shown earlier. As I will discuss in chapter 4, the onus is on certification initiatives to use methods such as Life Cycle Assessment and Hotspots Analysis to credibly demonstrate that particular issues are less material to particular product categories. But until they do so, the expectation should be that they will and should all be covered by these initiatives.

**Vertical Scope: Products First, Then Companies and Facilities**
The second "vertical" dimension of the scope of these programs relates to what entity they are evaluating. Imagine a company as a pyramid, with the base signifying all of the products the company produces, the middle signifying the facilities that produce those products, and the top representing the policies and practices of the company as a whole. Different criteria are more or less relevant to each of these levels, and different audiences may be more interested in particular levels of the pyramid. Generally speaking, personal economic and health concerns that connect to security values are more prominent at the product level, while environmental and social concerns that connect to self-transcendent values are more salient at the facility and company levels. Following the suggested practices of clarity and comprehensiveness discussed previously, initiatives that clearly cover all three levels in their evaluations avoid the sin of the hidden trade-off across these different entity levels. They may also activate the values and attract the interest of the broadest audience.

How common is such breadth among these programs? To answer this question, I coded the cases in the EEPAC Dataset according to the focus of their evaluation—are they evaluating products, companies, or facilities? As figure 2.6 shows, the most common evaluation focus among these programs is products (63 percent of the cases), followed by companies (43 percent) and facilities (17 percent). One quarter of the programs cover more than one level of analysis—12 percent assess both products and companies, 7 percent assess both companies and facilities, and 6 percent assess both

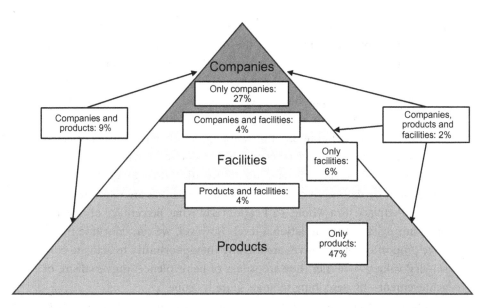

**Figure 2.6**
Evaluation focus (products, companies, or facilities). *Note:* Percentages represent the number of cases in the dataset that fall within each category of evaluation focus.

products and facilities. Only six programs (2 percent) cover companies, products, and facilities. Three of these initiatives (Certified Organic, the Global Organic Textile Standard, level BIFMA Sustainable Furniture Standard) evaluate all three levels within one certification while the other three (Cleaner and Greener, Green Shield, Scorecard) have separate certifications for different levels.

**Geographic Scope: Most Cases Miss an Opportunity**
The final dimension of these evaluations is their geographic scope. Do they evaluate and compare products and companies around the world, within one continent or country, or a subnational state or region? This dimension may be interpreted differently by different audiences. Perhaps possessing a strong benevolence value and altruistic interest in connecting with and supporting people in their immediate sphere of influence, some people may find greater value in a local or regional certification, such as the Bay Area Green Business Program. Others, perhaps having more central achievement and universalism values, may have a higher regard for programs that have a broader reach, such as the B Corporation certification program. Just as a

strong rating across many sectors is more impressive than a rating within only one sector, an evaluation as the most sustainable company across many countries is more impressive than as the most sustainable company in a single state or city.

To assess their geographic scope, I classified the cases in the EEPAC Dataset by whether their websites show a clear global, North American, or U.S. orientation. As figure 2.7 illustrates, approximately 26 percent claim to have a global scope, 14 percent a U.S. scope, and 8 percent a North American scope, while 52 percent do not mention their geographic scope. The dataset by definition only includes initiatives that are relevant to the U.S. marketplace as a whole, so I cannot say what percentage of existing programs exist at the subnational level. However, we can conclude that a large proportion of initiatives are missing the opportunity to activate consumers' values—whether they are values of benevolence, universalism, or achievement—by describing their geographic scope.

A further insight from these data is that these cases appear to be evenly balanced in terms of a global vs. U.S./North American scope. While those in the latter category may be missing opportunities to more strongly activate universalism and achievement values among consumers, they may also be benefiting from a sense of U.S. nationalism (i.e., we're "Made in

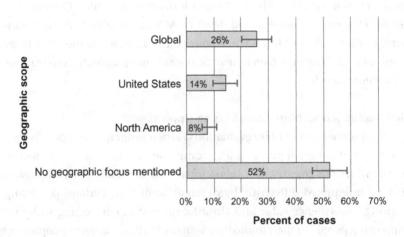

**Figure 2.7**
Geographic scope (global, North American, or U.S.). *Note:* Figure shows the percentage of cases that mention their geographic scope. Error bars indicate 95 percent confidence intervals for each sample proportion.

America") or a North American "warm glow" effect. Nevertheless, they are also guilty of misleading their audiences because their comparison set is not as comprehensive as it could be. If a magazine in the 1970s had rated a dozen American car models as "Most Fuel Efficient" without comparing them to similar Japanese models, that would have been highly misleading. Thus, in order for their comparisons to be meaningful, information-based initiatives have a responsibility to be as geographically comprehensive as they can be.

### The Content Landscape

We can combine at least some of this information about these 245 cases to develop a sense of the overall content landscape of information-based environmental governance strategies.

Figure 2.8 provides a visualization of this landscape, showing how many of the three types of private benefits are covered by the cases on the horizontal axis and how many of the five types of environmental benefits are covered on the diagonal axis. The vertical axis and height of the bars indicate how many cases fall into each particular category. The highest column in the back left-hand corner represents the fifty-one cases (20 percent) that do not claim to cover any of the private or environmental criteria coded for in the dataset. Overall, the relative lack of cases in the front right-hand corner of the figure reveals how few cases cover a broad range of both private and environmental criteria. However, one program, Pristine Planet, covers all five environmental criteria and all three private benefits (Veriflora). One other case covers all three private benefits and four environmental benefits (Greenspecs), while four other cases cover two private benefits and all five environmental benefits (Rainforest Alliance, B Corporation, Green Hotel Ratings, the Green and Natural Store, and LEED).

While too complicated to show in a single figure, we can also identify cases within the broader content landscape that cover not only multiple types of benefits but also multiple entities and product categories. The Cleaner and Greener and USDA Organic certifications, for example, cover multiple product categories, multiple types of benefits, and all three entity levels. The Global Organic Textile Standard, Scorecard, and the BIFMA level Standard (for furniture) includes four of the environmental criteria and one private benefit (health) in evaluations of product, facility, and

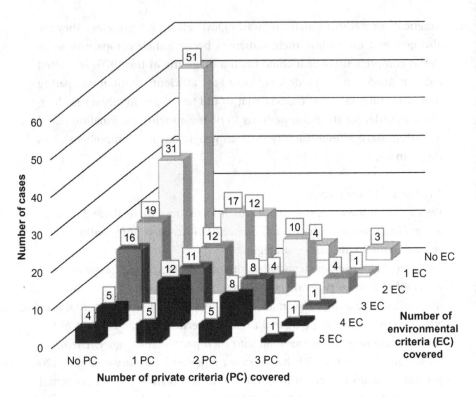

**Figure 2.8**
The content landscape of information-based environmental governance. *Note:* The fifty-one cases in the upper left-hand corner of the figure do not cover any environmental or private criteria. The one case in the lower right-hand corner cover all three types of private criteria and all five types of environmental criteria. The five environmental (public) criteria are pollution/emissions, energy/climate change, materials use, biodiversity/wildlife, and water use. The three private criteria are quality, price/cost savings, and health benefits.

corporate performance (both focus on one product category). ENERGY STAR covers all three private benefits and two environmental issues (pollution and climate change) in evaluations of both product and company evaluations.

Such a landscape analysis also can reveal cases that have a relatively narrow focus and scope. Eleven cases cover only one product category and do not identify any of the three classified private benefits or any of the five specific environmental criteria. Two of these programs, Green Format (for building products) and the Natural Food Network, appear to have been terminated, while six others—Oikos (also for building products), the Drycleaning and Laundry Institute's Certified Environmental Drycleaner program, Travel & Leisure's Global Vision Awards (for tourism-related organizations), Management & Excellence's ranking of the world's most sustainable and ethical oil companies, National Geographic's list of the 50 Top Ecolodges, and the Aspen Institute's Beyond Grey Pinstripes ranking of business school programs have not changed appreciably since their websites were originally coded. The eleventh case, however, Sustainable Travel International's Eco-Directory, does appear to have evolved and its criteria have become more specific and comprehensive. There are forty other cases in the dataset (16 percent of the total) that cover multiple product categories but also do not describe any private or environmental criteria in their evaluations.

### The Information Realist Perspective

Reviewing the literature discussed previously and looking at this snapshot of the landscape of information-based environmental governance strategies, the information optimists mentioned in chapter 1 will likely see much to celebrate. Research has consistently shown that a range of values, from those that are grounded more in concerns about society or the environment to those more centered on the needs and interests of the individual, are either already associated with green consumption or can be activated to increase the probability that people will purchase green products. The vast majority of the cases in the EEPAC Dataset are using criteria that can contribute to such value activation, and nearly half are referencing both public and private benefits to do so. They are covering a range of economic sectors, types of entities, and geographic scales, and the field has progressed significantly from the early days when only a few programs focused on single issues (e.g., ENERGY STAR), single industries (Responsible Care), and single scales (primarily the national scale).

The information pessimists, on the other hand, likely view many of these results as reinforcing their concerns about this form of governance. Here the relevance of the different forms of value and values discussed earlier in the chapter becomes manifest. The orientation of sustainability professionals toward private benefits and egoistic values, for example, is highly problematic, as it distracts from efforts to inform and mobilize the public about the pressing environmental challenges facing society. Perhaps at the margins, a focus on a range of values, from stimulation to tradition, may be effective at influencing a few consumers, but it risks losing the larger battle to activate the self-transcendent values that are more directly connected to the threats of pollution, climate change, and biodiversity loss.

These pessimists would also likely argue that this focus reinforces a paradigm of consumerism that inappropriately elevates the values of the market—cost, efficiency, and the neoclassical (and some would add "neo-liberal") emphasis on the shallow egoistic, instrumental, and extrinsic forms of value discussed earlier. Indeed, this is the primary message of the Common Cause Foundation, which argues that charities should collectively focus on activating intrinsic values (e.g., a world of beauty) and avoid using language that references extrinsic values (e.g., wealth).[66] These information pessimists might also add that the emphasis of the reasoned action approach (an often-cited psychological theory) on perceived behavioral control should not be overlooked. Consumers who perceive that they have limited influence over product supply chains have a point, and we need to focus on strategies that do indeed exert such influence—like stronger laws and regulations.

As for the sample of cases, the fact that 11 percent do not mention any public or private criteria and that 10 percent are guilty of the greenwashing sin of vagueness should be highly alarming to information pessimists. Also concerning should be the results showing that 70 percent of the cases do not cover any social issues, and that less than half include criteria related to climate change/energy, pollution, biodiversity, water use, and materials use. Furthermore, the 30 percent of cases that cover only one environmental issue are guilty of the second form of greenwashing mentioned earlier—the "sin of the hidden trade-off."[67] By only focusing on one issue, programs are implying that the issues they are not covering are not important—a company certified as carbon-neutral, for example, may still be emitting enormous amounts of toxic pollutants into the atmosphere.

Pessimists would likely also see the emphasis on product performance instead of company performance as myopic and misleading. A company may sell a few green products, for example, but still have very poor environmental and social performance as a company as a whole. The fact that over half of the sample does not discuss their geographic focus reveals another problem—without knowing what their comparative set is, how can their audiences evaluate the scope and relevance of these programs? For the 22 percent that have only a U.S. or North American focus, some pessimists might point out that the standards of these programs may be much weaker than their European or global counterparts, further undermining their already-limited effectiveness. And last but not least, this sample of 245 cases is missing entire economic sectors, from waste management to mining to entertainment, which suggests that some industries appear to be largely escaping the gaze of these initiatives. This problem is probably even greater in developing countries.

These concerns are legitimate, and proponents of information-based strategies should take heed of them. Nevertheless, the information optimists' faith in this form of governance is also not without basis—some of the initiatives do have a relatively wide focus and scope, and cover a wide range of issues and product categories. To some extent, the identified gaps in coverage represent real methodological challenges and trade-offs for these governance efforts, and a narrower focus may be justified until those challenges are overcome. For example, the data and skills necessary to certify individual products can differ significantly from those needed to effectively assess a facility or entire company. This is an issue we will return to in chapter 4, but it is not an argument against aspiring to comprehensiveness whenever possible.

Information optimists might also argue that a narrow focus is necessary to effectively market to specific consumer segments and concisely explain the utility of the information being provided. This argument is sometimes made for single-issue programs such as ENERGY STAR, which focuses only on energy efficiency. The problem with this reasoning is that information-based governance strategies are not marketing strategies designed to sell products, they are governance strategies designed to create public goods. If an ENERGY STAR-certified product contains high levels of toxic chemicals, it is undermining those public goods, regardless of how much energy it is saving. And again, it remains to be shown that consumers respond less strongly to multi-attribute certifications. Indeed, such broad-based initiatives may appeal to a broader audience and have a greater impact than those that are more narrowly focused, all other things being equal.

The information realist perspective therefore is that while comprehensiveness in the short term may be a challenging goal, in the long run a holistic approach is necessary and critical to avoid the shortsightedness and tunnel vision associated with focusing on a narrow set of criteria and values. Such a limited approach ignores the problem of the hidden trade-off and the long-term implications of excluding important dimensions of performance. Information realism recognizes that while existing strategies currently have significant limitations in terms of their relevance to a broad-based set of stakeholders and their coverage of important public and private benefits, some incorporate a much wider range of concerns and present them much more clearly than others.

For example, a lexical search of the term "public policy" in the EEPAC Dataset revealed that six cases (the Carbon Disclosure Project, Ceres's ratings of the banking sector's climate change performance, Climate Counts, As You Sow's Beverage Container Recycling Scorecard, the Electronics Take-back Coalition's TV Companies Report Card, and Citizen's Market) include criteria measuring the extent to which companies positively engaged on public policy issues relevant to the focus of the evaluation. Climate Counts, for example, includes a criterion asking, "Does the company support public policy that could require mandatory climate change action by business?" While these cases only represent 2 percent of the EEPAC sample, they also represent important examples of how information-based governance can contribute to the crafting of the laws and regulations called for by the information pessimists, a theme to which we will return in chapters 6 and 7.

As a group, these initiatives have also broadened both their focus and scope significantly over the last three decades, which offers hope that they can and will continue to do so in the future. If this trend persists, coverage of product categories such as mining and criteria such as public policy positions should expand and deepen over time. In the meantime, however, consumers, activists, policymakers, and business leaders need to recognize this content diversity and reward those programs that clearly articulate a comprehensive approach both to the range of entities they are evaluating and the criteria they are using to evaluate those entities. Such clarity and comprehensiveness are the most promising practices with regard to the content relevance of information-based governance strategies, and should be strongly supported.

## Promising and Problematic Practices

More specifically, these initiatives should use criteria and language that activate as many of the ten values included in Schwartz's human values framework as possible. They should also try to appeal to all of the types of value that are highlighted in the interdisciplinary theoretical framework shown in figure 2.2, from the intrinsic to the extrinsic and the egoistic to the biospheric. Sustainability ratings and environmental certifications should not only make the appropriate connections to private benefits such as improved health, lower costs, and higher quality, but also to public benefits related to existence value, option value, bequest value, and use value. They should cover the full range of social and environmental challenges whenever possible, and they should use arguments that rely both on reason and emotion. Their goal should be to cover product, facility, and company performance, and cover as many product categories as they can. Initiative designers should think hard about the geographic scope of their evaluations, and carefully consider the costs and benefits of a more local vs. global approach. And they should communicate all of these characteristics of their content clearly to the public.

While the EEPAC Dataset does not directly measure all of these practices, it highlights several important ones and enables us to identify a few programs that are doing relatively well in implementing these practices and a large number that are not. As our discussion of the content landscape identified, eleven cases provide particularly narrow and unclear coverage of different product categories, entity levels, and substantive areas of focus. Another thirty-one cases do not describe any private or environmental criteria in their evaluations. Other cases—Pristine Planet, Veriflora, Greenspecs, Rainforest Alliance, B Corporation, Green Hotel Ratings, the Green and Natural Store, and LEED—have a relatively broad focus and scope in their evaluations, and have the potential to connect to a broader audience as a result.

## The Value of Eco-Labels on Food

Speaking of promising practices, what about Mark? Can this analysis help him as he considers whether to purchase a gallon of organic milk? What value would he get for that two extra dollars? The pro-organic websites he visits presents him with a range of claims that connect directly to several of the types of value and values we have discussed. If the security value is central to him, hearing that organic milk has more omega-3s and fewer pesticide residues that can harm his health may resonate. If he has strong altruism, biocentrism, and universalism values, the benefits for air and water quality, biodiversity, and soils may make the organic option more desirable. While the skeptical websites are right to assert that organic does not necessarily mean local or small, it does harken back to historical forms of agriculture that existed before the introduction of chemical fertilizers and pesticides. In this sense, the claims about organic food contributing to family farms, a sense of place, and a culture of agriculture may effectively activate Mark's values relating to tradition and benevolence. Likewise, Mark might interpret the positive effect of organic products on future genera-tions as having bequest value. On the other hand, none of the sites he vis-its, including the USDA Organic site, clearly articulate the use, option, or existence value of buying organic, even though the case could be made for each.

So all that for two dollars? Or approximately $100 a year if he buys milk once a week? Perhaps he'll decide it is worth it, but what about the other labels he might encounter later in the store? Their value claims are similar to those of USDA Organic, but differ in subtle but important ways. Fair Trade, for example, has a much stronger orientation toward social issues, and its environmental standards are not as specific or as broad as those of USDA Organic. The certifications of Food Alliance and Rainforest Alliance distinguish themselves from the USDA Organic standard by their added focus on the treatment of workers and the conservation of wildlife habitats—and by allowing some use of synthetic chemicals. Food Alliance also includes animal welfare-oriented criteria, which are the sole focus of programs such as Certified Humane, Certified Vegan, and the American Vegetarian Association. The Smithsonian's Bird Friendly Coffee certifica-tion emphasizes the improved quality and taste of its certified products (all of which are 100 percent certified organic and shade grown) as well as its biodiversity, water, climate, and community benefits. The Biodynamic certification is similar to the organic certification but focuses on the whole farm as opposed to individual crops and has several additional criteria, including a biodiversity set-aside requirement. Consumer Reports only considers the Biodynamic and Certified Naturally Grown (a program in which inspections are generally conducted by other farmers) labels to prohibit the same range of toxic pesticides that are prohibited by the USDA Organic program.[68]

By investigating the content of these and other eco-labels, Mark and other consumers can identify which claims resonate with their values and decide whether they are getting a good value for the price premium associated with the certification. One way to think about this decision is whether this added cost creates more or less value than an equivalent contribution to an environmental advocacy organization. For example, is spending $100 on organic milk for a year providing more public and private benefits for Mark, society, and the environment than a $100 contribution to Greenpeace, or The Nature Conservancy? While buying organic and being philanthropic are not mutually exclusive and citizens can choose to pursue both strategies, this is a good test for the certification organizations and companies that are developing and using these labels. They need to make the case that the certification they are promoting does indeed deliver such value to people like Mark.

We might ask who exactly are "people like Mark," and assume that only white, wealthy and well-educated consumers would even consider buying organic food. Nielsen Homescan data from 2013, however, show that African Americans and Hispanic consumers buy similar proportions of organic food to Caucasians, and Asians purchase greater amounts than all three other ethnic groups.[69] While information on the relationship between education levels and organic purchases is not available for 2013, the same dataset shows that consumers making less than $20,000 and consumers making more than $100,000 are the most likely income groups to buy organic items, while those making between $20,000 and $50,000 are the least likely to do so.[70] Thus consumers' ability to pay is not directly correlated with organic purchases; the most economically disadvantaged consumers are purchasing organic food in the same proportion as the wealthiest members of society. This data contradicts an earlier analysis using 2004 data that concludes nonwhite, low income, and high-school educated consumers are generally less likely to purchase pre-packaged organic vegetables.[71] And even that study shows that some consumers in these demographic groups do purchase organic products.

The broadening of the organic market is reflected in surveys of consumer preferences as well. A 2015 study by the Organic Consumers Association found that the percentages of organic consumers who are African American (14 percent), Hispanic (16 percent), and white (73 percent) are very similar to their percentages in the American population as a whole.[72] The number of African-American and Hispanic consumers stating that they regularly choose organic products have more than doubled since 2009 and 2011, respectively.[73] Overall, 84 percent of consumers report buying organic food, and 45 percent report actively seeking it out and buying it once a month or more.[74] Interest in organic food does not vary much by political affiliation—40 percent of Republicans actively seek it out while 48 percent of Democrats do.[75]

Such surveys of stated preferences may not accurately reflect actual consumer behavior, but they do reflect the aspirations of a broad cross-section of the American public. While organic sales still only represent a small fraction of all the food sold in the U.nited States (5 percent in 2015), they are growing across the board (11 percent increase in 2015 organic sales vs. 3 percent increase in sales of the overall food market) and particularly fast for some items (40 percent of all spinach sold in the United States is now organic).[76] This is all happening without much direct marketing or outreach to consumers at all, who in turn may not fully understand the benefits of the label. A BFG survey found that only 20 percent of consumers could correctly define what "organic" really means.[77]

Indeed, a key conclusion from this chapter's analysis is that the organizations that are implementing information-based environmental governance strategies can do a much better job of clearly communicating the range of benefits associated with their particular certification or rating. From the images and taglines embedded in their labels to the text and video on their websites, they should be clear and comprehensive about how their value connects with specific human values. Manufacturers of products with these labels can provide more information on their packaging about what these labels mean, while retailers selling these products can provide similar information through in-store guides in the aisles and on the shelves. Such efforts can not only further expand the already fast-growing organic market but increase interest in other green products as well. By appealing to a wide range of values and interests, on the one hand, broadly salient labels and ratings can attract support from both liberals and conservatives who are not always well represented in the environmental movement. Vague and narrowly oriented labels, on the other hand, risk remaining obscure, unknown, and unused by the vast majority of society.

While all of these efforts to enhance the clarity and comprehensiveness of the content of information-based governance strategies are important, the focus and scope of these initiatives is only one part of our story. In the next chapter, we will explore an equally important component of the information value chain—the trustworthiness of the organizations behind these initiatives. Even if the content of an eco-label is comprehensive and presented clearly, it will likely not have much of an impact if it is perceived as coming from an untrustworthy source.

# 3 Trusting Green: The Organizations behind the Information

## Trusting Toilet Paper

Carrie, as you'll remember from chapter 1, is an environmental activist pondering how she will respond to a newspaper reporter's inquiry about environmental certifications of toilet paper. She remembers a campaign from a few years ago led by the environmental organization ForestEthics against the Sustainable Forestry Initiative (SFI) contending that SFI's claims of "independence" were misleading and deceptive. SFI was created in 1994 by the trade association of the forestry industry, the American Forest and Paper Association, and spun off as a nonprofit organization in 2001 with representatives from several environmental organizations on its board of directors. But ForestEthics documented how these representatives either quickly departed their positions or had strong economic ties to the forestry industry.[1] Twenty other environmental organizations also asserted in a letter to SFI that its claims of "being fully independent' are "false, deceptive, or misleading" because it refused to reveal the sources of its funding.[2]

Carrie also recalls concerns that were raised about the primary alternative to SFI, the Forest Stewardship Council (FSC). Founded in 1993 by a "group of businesses, environmentalists and community leaders," the FSC is governed by the FSC General Assembly consisting of three chambers, each representing environmental, social, and economic interests.[3] Despite this diverse representation, several environmental organizations have criticized FSC for a host of governance issues. In 2008, Greenpeace issued a report outlining its problems with some of FSC's practices, Friends of the Earth UK ended their support for the organization, and one of the founders of FSC, Simon Counsel, said that the FSC had become the "Enron of Forestry."[4] Counsel helped create FSC-Watch, which lists conflicts of interest as one of the "Ten Worst Things" about FSC: "certifying bodies (assessors) are paid by the companies wanting to get certified."[5] FSC-Watch explains that "it is in the assessors' interest not to get a reputation for being too 'difficult,' otherwise they will not be hired in future."[6] As for direct financial support, FSC lists the eleven organizations that donated over $20,000 in either 2012 or 2013 on its website (six of which are companies that sell paper products), but does not provide an overall breakdown of where their funding comes from.[7]

These concerns about both FSC and SFI give Carrie, and many others, reason to pause before they endorse either organization, or choose to buy particular products because they have FSC or SFI seals of approval on them. Even if she is convinced that purchasing sustainably produced paper products is an important way for her to express her values, Carrie may be uncertain whether these particular labels are credible and whether the organizations behind them are trustworthy. While many of the issues raised about these two initiatives center on the validity of their specific methods, this chapter focuses on the trustworthiness, accountability, and legitimacy of the organizations behind these initiatives. Methodological validity is a key component of these information value chains and is the focus of chapter 4, but most people do not have the time, expertise, or motivation to systematically analyze and compare their validity. Research discussed in the sections that follow has shown that individuals often rely instead on cognitive shortcuts that focus on the organizations behind these programs to determine whether they will utilize them. Just as the content of these initiatives must be desirable to their audiences, the organizations behind them must also be perceived as trustworthy sources of information.

However, unlike content desirability, I argue in this chapter that organizations do not necessarily need to directly appeal to a broad range of audiences in order to be effective. Instead, they need to send clear signals of credibility that demonstrate their accountability to particular stakeholders, whether they are advocacy organizations, scientific experts, businesses, or particular segments of the public. In order to make this point, the chapter first defines several relevant concepts and weaves them together into a theoretical framework that maps out the pathways by which trust is communicated between these organizations and their audiences. The chapter then presents originally coded data from my Environmental Evaluations of Products and Companies (EEPAC) Dataset on the extent to which existing information-based environmental governance strategies are utilizing these communication pathways.[8] It then concludes with a discussion of the most promising and problematic trustworthiness communication practices, and also provides a further analysis of Carrie's forest certification quandary.

## Understanding the Nature of Trust

Carrie's central question is, on the surface, a simple one—does she trust either SFI or FSC? Beneath that surface, however, this straightforward dilemma actually is quite complicated and raises further questions. What does it mean to trust these organizations? How does she know that she can trust them? In order to answer these questions, it is helpful to understand the nature of trust and several other related concepts. While it is easy to view it as relatively commonplace and unremarkable, trust is one of the most important features of human society. A wide range of social scientists have argued that it is a critical form of social capital that has enabled the development of modern social, economic, and political institutions. Francis Fukuyama and other scholars, for example, have argued that countries characterized by a high degree of social trust have been able to create large-scale corporations and capitalist economies more effectively than those with low levels of such trust.[9] Following this logic, economics research has consistently found a strong relationship between levels of trust and both income per capita and economic growth.[10] Warren Buffet succinctly summarizes this importance of trust: "Trust is like the air we breathe. When it's present, nobody really notices. But when it's absent, everybody notices."[11]

## From Trust and Trustworthiness to Mistrust and Distrust

Given its importance, it is not surprising that trust has been studied by scholars from a wide range of disciplines. Economists, sociologists, political scientists, and psychologists have investigated the dynamics of trust in social, political, and economic contexts and as both a psychological state and a behavioral choice.[12] While these scholars often have different understandings of the meaning of trust, it is possible to identify several areas of agreement about the phenomenon. *Trust* is generally viewed as a relational concept in which one trusting party becomes vulnerable to harm by having a positive expectation about the behavior of a second trusted party.[13] Such expectations can be generalized to large groups (or even all human beings) or particularized to individual family members, friends, or leaders.[14] They may be the result of rational decision-making processes or more unconscious moral intuitions and perceived social norms.[15] They may also be based on either an implicit or explicit belief that the trusted party will take into account (or "encapsulate") the interests of the trusting party and will not intentionally injure that party, if at all possible.[16]

Regardless of how they are formed, these expectations create the potential for betrayal—the trusted party may not live up to these expectations and will harm the trusting party.[17] This point brings us to the concept of *trustworthiness,* which is a measure of someone's or something's likelihood of fulfilling the expectations others have of them. Some scholars distinguish between trustworthiness and competence as two separate dimensions of *credibility*—the former encompassing characteristics such as kindness, friendliness, and honesty and the latter encompassing attributes such as expertise, ability, and qualification.[18] These two dimensions relate to distinctions made by Stephen Marsh (University of Ontario) and Mark Dibben (University of Tasmania) between mistrust and distrust. *Mistrust* is a measure of misplaced trust—a trusting party made a mistake in placing trust in a trusted party because that party did not fulfill the trusting party's expectations, either because of a lack of trustworthiness or competence. *Distrust,* in contrast, is a measure of how much a trusting party believes the (dis) trusted party will actively work against the trusting party's interests.[19] Thus Carrie may mistrust SFI and FSC because she feels they have not competently evaluated forestry operations in the past and are providing incorrect information, or *misinformation.* Alternatively, she may distrust them because she believes they are beholden to the forestry industry and are undermining efforts to protect forests and biodiversity. She is therefore convinced they are intentionally providing false information to deceive her, or *disinformation.*[20]

A perception that the public's levels of both mistrust and distrust in a wide range of institutions are rising has driven research on trust and trustworthiness over the past several decades.[21] The percentage of U.S. citizens who state they can trust the government to do what is right declined from over 70 percent in the mid-1960s to below 50 percent after the mid-1970s.[22] Such a decline in perceptions of governmental trustworthiness has been attributed to the Vietnam War, Watergate, and the media's frequent report-ing on corruption, scandals, and unsolved social problems.[23] Such low lev-els of trust extend to other institutions as well—the 2015 Edelman Trust Barometer revealed that "trust in government, business, media and NGOs in the general population is below 50 percent in two-thirds of countries, including the U.S., U.K., Germany and Japan."[24] Experimental research has also shown that even though individuals generally have a norm of being trustworthy (and expect that they will be punished if they are not), they do not expect others to be trusting of the people around them.[25]

### The Roles of Accountability, Credibility, and Legitimacy

These declining levels of trust extend to the environmental arena and make information-based governance initiatives across all policy areas more chal-lenging. As a result, scholars and practitioners alike have shown increased interest in the related concepts of accountability, credibility, and legiti-macy. As generalized trust in institutions has declined, demands for stricter tracking of their accountability have increased. As London School of Eco-nomics legal scholar Julia Black explains, *accountability* is a type of rela-tionship "between different actors in which one gives account and another has the power or authority to impose consequences."[26] These accounts can enable actors to overcome the lack of trust between them,[27] and can include descriptions of methodological processes (the subject of chapter 4), reports on performance outcomes (the subject of chapter 6), or signals of organi-zational credibility (the focus of this chapter). These credibility signals can emphasize an actor's own characteristics—particularly independence and lack of conflicts of interest—that directly communicate the actor's trust-worthiness or expertise. Or these signals can focus on an actor's organiza-tional associations that indirectly lend the actor an aura of credibility. Thus, even though Carrie may currently have low levels of trust in FSC or SFI, she can look for any of these signals from them that might increase her sense of their credibility.

*Credibility* has been defined as "the quality or power of inspiring belief,"[28] or as believability or authoritativeness.[29] This definition is instructive because it suggests that accepting a claim's credibility is to "take it on faith," even absent more tangible and direct evidence of actual outcomes. As many scholars have pointed out, credibility is also a relational concept, and must be understood in relation to the perceptions of relevant stakeholders.[30]

Thus some characteristics may be more credible to some stakeholder groups than others. This raises the important possibility that agents are strategically sending particular signals in order to attract the support of specific stakeholders. Given this potential dynamic, it is important to understand how perceptions of legitimacy drive stakeholders' responsiveness to these signals of credibility.

Building on Max Weber's original conception, Brown University sociology professor Mark Suchman defines *legitimacy* as the belief that "the actions of an entity are desirable, proper, or appropriate,"[31] while Cornell University government professor Norman Uphoff describes how legitimacy is granted to individuals or organizations "in keeping with the beliefs people have about what is right and proper."[32] Legitimacy theory suggests that organizations depend on legitimacy for their survival and will use strategies such as information disclosure to ensure its continued supply, while stakeholder theory further suggests that organizations will disclose information that is salient to stakeholders they perceive as particularly important sources of legitimacy.[33] Such disclosures can earn an organization several different types of legitimacy. Information that contributes to the self-interest of stakeholders can enhance an organization's pragmatic legitimacy, while information that enhances the welfare of society can enhance its moral legitimacy. Likewise, information that encourages stakeholders to view an organization as a natural and inevitable part of their lives can enhance its cognitive legitimacy.[34]

Stakeholders may evaluate these forms of legitimacy in terms of either an organization's actions and "outputs" or its essence and "inputs."[35] *Output legitimacy*, or "rule effectiveness," is the extent to which initiatives "effectively solve the issues that they target," and requires comprehensive coverage of the relevant actors, strong rule efficacy, and effective enforcement.[36] Such outputs, which are the subject of chapter 6, can be difficult to systematically quantify,[37] and so audiences may instead focus on the *input legitimacy* of green claims and whether the process by which the claims were generated is perceived as justified.[38] This form of legitimacy derives from a concern in democratic theory that "political choices should be derived, directly or indirectly, from the *authentic preferences* of citizens."[39] From this perspective, process matters as much or more than outcomes. Sébastien Mena and Guido Palazzo, business school professors at the City University of London and University of Lausanne, respectively, suggest that input

legitimacy requires stakeholder inclusion, procedural fairness of delibera-
tions, promotion of a consensual orientation, and transparency of an organization's structures and processes.[40] In essence, both who is involved and
how they are involved are relevant to determining the input legitimacy of
an organization or initiative.

The different signals of credibility discussed previously can help organizations earn these various types of legitimacy. For example, organizations can gain pragmatic and moral legitimacy if stakeholders view them as
trustworthy and competent enough to deliver information that is relevant
to either themselves or society at large. Likewise, they can gain cognitive
legitimacy if stakeholders sense their traits of trustworthiness and competence are culturally appropriate and perceived as "predictable, meaningful,
and inviting."[41] These traits generally help agents earn input legitimacy;
trustworthiness and expertise are all traits that stakeholders may value as
important inputs in the process of developing sustainability information.
The one exception is transparency about outcomes, which, as we explore
further in chapter 6, can also earn agents output legitimacy.

These grants of legitimacy are often coupled with a transfer of resources
to the organization. These resources may be either tangible or intangible,
and can include grants of authority, influence, information, economic
resources, or social prestige (the sections that follow provide a more detailed
description of these different types of resources). If the organization fails
to continue to send relevant signals of credibility, does not demonstrate
its accountability to the stakeholder providing these grants, or otherwise
undermines the stakeholder's perceptions of its legitimacy, trustworthiness, or competence, then that stakeholder may discontinue these grants
of resources. Stakeholders, however, may disagree over which of these signals of credibility are most important, and may be willing to grant legitimacy for some traits more than others. In this case, which stakeholders
and which signals do evaluation organizations and firms prioritize in their
pursuit of legitimacy? Do they focus more on signals of trustworthiness or
competence, for example?

The rest of this chapter addresses these questions in the context of information-based environmental governance strategies. Using the EEPAC Dataset of 245 cases of environmental certifications and ratings discussed in
earlier chapters, it explores not only what signals of credibility these initiatives are sending via their websites, but also what grants of legitimacy they

are advertising as additional reasons to trust them. These grants of legitimacy are manifested by a transfer of resources from a range of public, private, and civil society organizations, and also represent a signal of credibility and "trustworthiness by association" in their own right. Figure 3.1 provides a graphical depiction of the flow of credibility signals and legitimacy grants between stakeholders (the trusting parties) and the organizations seeking their trust (the trusted parties—firms and evaluation organizations making sustainability claims). The signals of validity and effectiveness that it depicts are discussed in chapters 4 and 6, respectively.

### Green Trust Deficits and Opportunities

The trustworthiness communication pathways shown in figure 3.1 are particularly important for eco-labels and sustainability ratings. The information

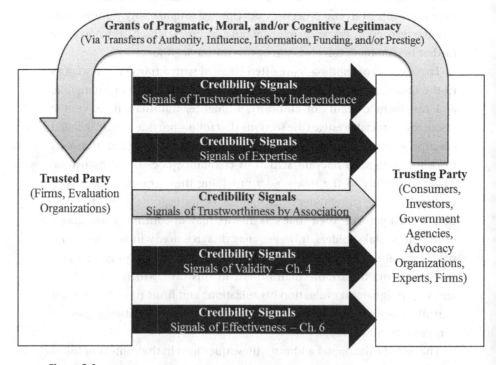

**Figure 3.1**
Pathways of trustworthiness communication: signals of credibility and grants of legitimacy. *Note:* Adapted from Bullock, "Signaling the Credibility of Private Actors as Public Agents," 181.

that these initiatives provide is usually a "credence good," which requires trust in its quality even after its use because it is difficult to know how accurate it is.[42] This partially explains why 56 percent of Americans do not trust companies' green claims.[43] These distrusting consumers may suspect that such claims are not authentic examples of improved environmental performance, but rather are the result of efforts to deflect environmental criticisms, superficially jump on the green bandwagon, earn revenue from products marketed as green, or get credit for minimal regulatory compliance (rather than going above and beyond legal requirements).[44] Whether these efforts are the result of market, organizational, or individual psychological drivers,[45] consumer concerns about them can culminate in a belief that increased prices associated with eco-labels are not due to legitimate differences in production costs. Thus more than 50 percent of American shoppers, for example, believe organic food is too expensive and organic certification is "an excuse to charge more."[46]

These beliefs likely stem from both a specific distrust in the organizations behind these sustainability claims or from more generalized social mistrust. On the one hand, Aarhus University professor of economic psychology John Thøgersen and his colleagues found that Danish consumers, for example, are less likely to consider the Marine Stewardship Council seafood eco-label in their purchases if they have relatively low levels of trust in World Wildlife Fund (WWF), one of the organizations behind the certification.[47] On the other hand, in a survey of citizens across eighteen European countries, University of Konstanz professor of corporate social responsibility Sebastian Koos found that participants who generally consider other people to be trustworthy are more willing to purchase products with eco-labels. Koos concludes that people living in countries with relatively low levels of such generalized trust may be particularly suspicious of green claims. In 2008, the United States ranked as the tenth least-trusting industrialized country, suggesting that environmental certification programs face relatively high levels of distrust among Americans.[48]

Nevertheless, some types of distrust and mistrust can paradoxically lead to greater trust in these information-based governance initiatives. Ken Peattie, a professor of marketing and strategy at Cardiff Business School, asserts that green labels can address the loss of trust among consumers due to media coverage of greenwashing controversies by providing consumers with reliable information about product ingredients, production methods, in-use resource efficiency, and their lifespans.[49] Research by University of Nantes scholar Dorothée Brécard and her colleagues shows that people who do not trust that governments are adequately protecting fisheries are more likely to buy eco-labeled seafood products. Their mistrust and/or distrust of fishery regulations is apparently greater than their mistrust and/or distrust of seafood eco-labels.[50]

Thus trust can vary by sector and organization, and not all sources of information are uniformly perceived as untrustworthy. Over two-thirds of Americans report that information provided by word-of-mouth discussions, the news media, food retailers, and food companies helps them learn how food companies promote human and environmental well-being and the safety of food sources.[51] Sustainability experts generally trust nongovernmental organizations (NGOs) more than governments to evaluate a company's sustainability performance, which parallels the higher levels of trust that NGOs enjoy over both national governments and global companies among the general public, both in the United States and abroad (an Edelman survey, for example, shows that NGOs are the most trusted institution in twenty-three of twenty-six countries).[52] Mario Teisl, an economist at the University of Maine, confirmed that consumers have a strong positive bias toward eco-labels provided by NGOs by experimentally comparing their evaluations of labels provided by four different types of organizations. The label attributed to the Sierra Club garnered the highest ratings of environmental friendliness and satisfaction, among labels attributed to the Forest Stewardship Council, the EPA, and a fictional Maine Wood Products Association.[53]

However, other studies have shown that government involvement can significantly enhance consumer acceptance and the outcome effectiveness of environmental certifications. In a review of five energy labels, for example, Abhijit Banerjee (MIT) and Barry Solomon (Michigan Technological University) conclude that "government support proved to be crucial in determining a program's credibility, financial stability, and long-term viability."[54] And Kim Mannemar Sønderskov and Carsten Daugbjerg, political scientists at Aarhus University and Australian National University, respectively, find that consumer confidence that products marketed as organic are indeed organic is generally stronger in countries (such as Denmark) where the government plays a strong role in the organic certification process.[55]

These findings suggest that different audiences are making different evaluations of the trustworthiness of the organizations behind environmental evaluations of companies and products. The reputations of these organizations undoubtedly play a critical role in how stakeholders perceive different eco-labels and ratings, but those perceptions may also be influenced by other signals of credibility broadcast by these initiatives. This is particularly true of online initiatives. As Michigan State accounting and information systems professor Harrison McKnight and his colleagues find, website trustworthiness and website quality (which we will to return in chapter 5) are both important determinants of consumer trust.[56] Other scholars suggest more specific criteria for evaluating label legitimacy, including stakeholder inclusivity (along the entire supply chain), independence, expertise, discursive quality, democratic control, and transparency (including auditability).[57]

Several surveys suggest that independence, transparency, and expertise are particularly important criteria for both consumers and sustainability professionals. For example, I conducted an online survey that asked 428 consumers to identify their most preferred characteristics of eco-labels.[58] From a set of thirty-two attributes that included affiliations with specific types of organizations (media, corporate, nonprofit, government, and academic) and specific content areas, independence and transparency were the two most preferred characteristics of eco-labels. The inclusion of energy/climate change criteria and expertise were the third and fourth most preferred characteristics. Similarly, a survey of more than a thousand sustainability professionals found that the three most important factors for this audience, in order of importance, were objectivity/credibility of the data sources, disclosure of methodology, and experience and size of the research team.[59] These three top factors map well to the dimensions of transparency (disclosure), independence (objectivity), and expertise (research team experience) identified in my consumer study.[60]

## Signals of Credibility

While other characteristics may also influence stakeholder perceptions of these programs, these results suggest that transparency, independence, and expertise are among the most likely to effectively serve as specific signals of credibility for these initiatives for a broad range of audiences, from sustainability experts and professionals to the public at large. These three characteristics may overlap and complement one another (e.g., experts can be independent and transparent), but nevertheless represent distinct and independent characteristics that initiatives can choose to signal to their audiences. While I discuss transparency in chapter 4 in the context of methodological validity and replicability, the sections that follow describe the nature of expertise and independence as signals of credibility. They then present empirical data showing the extent to which the 245 cases in the EEPAC Dataset are sending these signals to their respective audiences.

### Independence: Signaling Distance

Independence of the assessment organization and its lack of conflicts of interest is perhaps the most commonly mentioned proxy for trustworthiness in the literature on sustainability claims.[61] INSEAD professor of ethics and social responsibility Craig Smith and his colleagues argue that a claim's credibility is particularly undermined "where consumers perceive firm-serving motivations rather than motivations to serve the public good."[62] Anita Jose, a management professor at Hood College, and Shang-Mei Lee, a finance professor at St. Edward's University, find that "companies are using third party external audits to establish the credibility of their commitment to environmental management practices."[63] The underlying logic is that companies should not be the principals for independent assessments. In other words, the more objective and distant the source of an assessment is from the source of the product, the better. The assumption is that signals of credibility either sent directly by firms or by evaluation organizations associated with those firms are inherently unreliable.

Independence maps well to scholars' definitions of trustworthiness as a form of safety.[64] People may be more likely to trust and feel "safe" using information coming from third parties that have fewer conflicts of interest. If they are professional certification organizations, academic institutions, or government agencies, they may be perceived as more "fair" and "calm." If the third parties are nonprofit organizations, they may also be perceived as more "altruistic" and "kind."[65] Policies promoting independent data verification or generation by third parties may also express a normative belief in the value of civil society organizations as advocates of the public's interests. Likewise, they also imply that the critical locus of power and accountability should be with these organizations because of their public orientations, watchdog status, and focus on social welfare. In this sense, agents emphasizing their independence may be recruiting grants of moral input legitimacy from principals who value the role these intermediary organizations play in society. Such an approach is justified by surveys that consistently find that nongovernmental organizations are society's most trusted institutions, both by the public in general terms and by sustainability professionals as evaluators of corporate sustainability performance.[66]

Despite its importance, most studies of independence have ignored the multiple dimensions of the concept. The first of these dimensions is the *type of independence*—has the data been generated by independent organizations or only verified by such organizations? Independent generation implies full control of the data from collection to analysis to delivery, while verification indicates external monitoring of a self-assessment process that has higher potential for fraud. For one of its investigations, Greenpeace, for example, sent computers to an independent lab to be analyzed for toxic chemicals, rather than rely on company reports.[67] A second important dimension is the *source of the independence*—is the data generation or verification performed by the evaluation organization or firm itself, or is it conducted by an organization that has been accredited or contracted by a third organization? Increasingly, "third party" certification systems are assigning the strategic roles of standard-setting and administration and the operational roles of monitoring and assessment to separate organizations. A third dimension is the *level of independence*—is all of the data independently generated or verified, or only some of it?

Three sample text segments provide examples of each of these different characteristics and demonstrate how they were coded. The website of the EPA's WaterSense program states, "All products bearing the WaterSense label must be tested and certified by an approved third party laboratory to ensure they meet EPA water efficiency and performance criteria."[68] This is an example of a text segment that was coded as *all data* ("all products ... must be tested"), *independent generation* (external labs, not EPA or the firms themselves, are conducting the tests), and *contracted/accredited organization* ("an approved third party laboratory"). As a second example, the website of B-Corp states, "When a company becomes Certified they must submit documentation for approximately 20 percent of their answers to the B

Survey ... 10 percent of B Corporations are audited every year ... [by B Lab auditors]." This text segment was coded as *some data* (only 10 percent are audited), *independent verification* (data is submitted by the company), and *evaluation organization* (B Lab auditors). In cases where the source or type of independence was unclear, such as the phrase "third-party, independent validation and verification" found on Rainforest Alliance's website, the text segment was coded as *evaluation organization* and *independent verification* by default.

Almost 40 percent of the cases in my EEPAC Dataset verified or generated at least some of their data. Slightly over 14 percent of the cases generated their own data independently of the organizations being evaluated, and slightly over 33 percent had mechanisms in place to verify the accuracy of the data they received from the organizations they were evaluating. Almost 30 percent of the cases verified or generated all of their data, and nearly 10 percent verified or generated some of their data. Approximately 28 percent of the cases have other organizations generate or verify their data, while just under 18 percent generate or verify their information themselves. Figure 3.2 presents a more granular view of these data. The proportion of cases that use independently verified or generated data was not significantly different for cases implemented by firms than for cases implemented by evaluation organizations.[69]

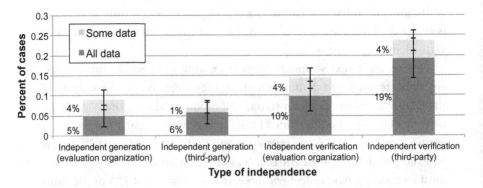

**Figure 3.2**
Types and levels of data independence (by percentage of cases). *Note:* Error bars indicate 95 percent confidence intervals for both the "some data" and "all data" sample proportions. Because of overlap between categories, the statistics presented in the figure do not always add up to the percentages mentioned in the text. *Note:* Adapted from Bullock, "Signaling the Credibility of Private Actors as Public Agents," 200.

An additional dimension of independence is the *type of peer review*, if any, that is used in the evaluation process. Both the methods used in the evaluation and the data collected can be peer reviewed, and the review can be conducted by individuals with varying levels of expertise who work inside or outside the firm or evaluation organization. An example of *data peer review* comes from the Rainforest Alliance, which states that a team of trained specialists writes an assessment report of a farm or forest that has applied for certification, and this report is then "evaluated by an independent, voluntary committee of outside experts (i.e. peer reviewed)." An example of *method peer review* comes from Protected Harvest, which states that its "standards are peer-reviewed by the scientific community and then must be approved by the distinguished environmentalists on the Protected Harvest board." Approximately 5 percent of programs mention peer review processes for their methods, and 4 percent mention peer review processes for their data.[70] Less than 2 percent of the cases specified the expertise of the individuals conducting the peer review process. For example, one text segment states that "BASF's eco-efficiency was carefully examined and evaluated by David R. Shonnard, PhD, an independent expert in green engineering," and goes on to describe his academic credentials.

### Expertise: Signaling Knowledge
Expertise has also been cited as an important aspect of legitimacy,[71] which is not only an evaluation of particular decisions but also the suitability of those who make those decisions.[72] Thus the people implementing these initiatives may have varying levels of knowledge that make them more or less qualified to determine the sustainability of a particular product or company. Assurance statements for corporate social responsibility reports therefore often provide "commentary from high profile experts deemed trustworthy by the public."[73] In some cases, regulatory agencies may even delegate policy-making authority to private agents because of their preexisting specialized expertise in particularly complex and technical issue areas.[74] There is a rich literature on the subject of expertise, and it discusses the phenomenon both generally as well as in the specific context of environmental politics.[75] One important distinction that this literature reveals is the difference between expertise from academic training ("book learning") and expertise from professional experience ("learning by doing").

Expertise is one of the core dimensions of academic typologies of credibility, and is a primary reason why the public might accept an organization as legitimate in the absence of more direct evidence of output legitimacy.[76] An emphasis on the expertise behind an assessment process may represent a commitment to scientific knowledge as the best way to ensure the validity of an evaluation (and dealing with sustainability challenges more generally). From this perspective, it is the scientists and experts who should be trusted to solve society's environmental problems and evaluate claims of greenness. Following this logic, organizations that hire experts with relevant expertise are more likely to produce valid environmental assessments.

An emphasis on expertise may also represent a normative commitment to the rigorous pursuit of truth as a fundamentally important social value. It may also signify an attempt to activate a sense of cognitive input legitimacy; like technical evaluations in other domains, assessments of sustainability should naturally be conducted by experts with relevant technical knowledge, and to think otherwise is "unthinkable."[77] Such a dynamic would explain why over a thousand sustainability experts rated the experience and size of the research team as one of the three most important factors in determining the credibility of a corporate sustainability rating.[78]

Expertise can be produced through academic training or from professional experience. Academic training can be further categorized as general training or training that is directly relevant to the organization's work. In order to capture these dimensions of expertise, text segments were coded as general academic training, relevant academic training, and relevant professional experience. As an example of *general academic training*, the CarbonNeutral website states that its executive vice president "holds an MBA with Distinction from the Stern School of Business at NYU and a Bachelor's degree in Psychology from UCLA." The Bird Friendly Coffee website states that the director of the organization behind the certification has a PhD in ornithology from the University of California, Berkeley, which is an example of *relevant academic training*. The website of the 100 Best Corporate Citizens provides an example of *relevant professional experience*; its director of research is described as having "more than a dozen years of experience supporting institutional investors with research and software tools for values-based investing and proxy voting."

The coding data indicates that nearly one out of five cases (18 percent) claim that at least one staff member working on the initiative has relevant professional background and expertise (i.e., substantive, full-time past work on environmental or social issues). Slightly over 10 percent claim to have staff with academic training (master's degree or above) that is relevant to environmental or social issues, while approximately 7 percent claim to have staff with academic training (master's degree or above) that does not have a clear relationship to the work of the initiative (see figure 3.3). While approximately 25 percent of the cases implemented by evaluation organizations make at least one of these claims of expertise, none of the initiatives implemented by firms make any claims of expertise. Firms therefore are significantly less likely to signal their expertise than evaluation organizations.[79]

### The Landscape of Credibility Signals

As in chapter 2, we can combine all of this information about credibility signals into a single representation of the landscape of credibility signals that these cases are sending to their audiences. As figure 3.4 shows, a major proportion of the initiatives (the 119 in the upper left-hand corner of the figure) make no claims of either expertise or independence. The eight cases in the bottom right-hand corner mention at least some level of expertise

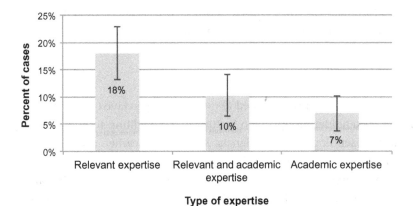

**Figure 3.3**

Types of expertise mentioned. *Notes:* Error bars indicate 95 percent confidence intervals for each sample proportion. Adapted from Bullock, "Signaling the Credibility of Private Actors as Public Agents," 201.

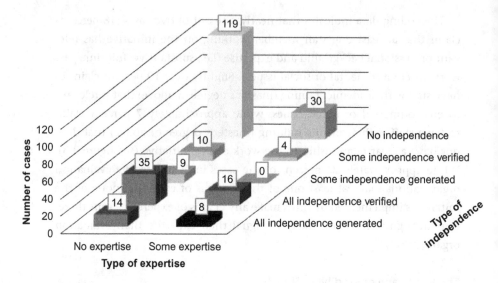

**Figure 3.4**
The landscape of credibility signals. *Note:* The 119 cases in the back left-hand corner make no claims of either expertise or independence. The eight cases in the front right-hand corner describe the expertise of their staff and claim that their data has all been independently generated.

and claim to generate all of their data. These initiatives are Rainforest Alliance Certified, Bird Friendly Coffee, Certified Best Aquaculture Practices, Certified Naturally Grown, Certified Compostable Products, Design for the Environment, AHRI Certified, and GreenGuard.

**Grants of Legitimacy**

The signals of credibility discussed above may be perceived positively by different stakeholder groups, who in turn may be willing to endorse the organization sending the right signals. As discussed earlier, such recognition is a "grant of legitimacy," and can come in many different forms. Such grants are mechanisms not only to express support for a program, but also to exert control over it. Either way, they are a useful signal themselves to other stakeholders regarding the allegiances and accountability of different information-based governance initiatives. The power resource framework developed by Warren Ilchman and Norman Uphoff at UC Berkeley provides a useful approach for identifying the key mechanisms, or "resources," that organizations and individuals use to exert power over these initiatives, which in turn become either positive or negative signals of credibility for other organizations and individuals.[80] These resources include funding, social status, authority, and information as the primary resources of power.[81] Understanding how these power resources are distributed can reveal "who claims authority over whom, and on what issues" and "who accords legitimacy to whom, on what grounds, and with what limitations."[82] More specifically, it can help explain why ratings and labels are designed the way they are and who is driving and endorsing those design decisions.

The actors who may be behind these initiatives can be divided into three general categories—organizations and individuals from the public, private, and civil sectors. The *public sector* comprises all state-owned institutions, including government agencies and nationalized industries.[83] The *private sector* "encompasses all for-profit businesses that are not owned or operated by the government."[84] The *civil sector*, or civil society, is the "sphere of institutions, organizations, and individuals located between the family, the state, and the market in which people associate voluntarily to advance common interests," and include both nonprofit and academic institutions.[85] In order to identify the extent to which the private, public, and civil sectors are using the different power resources to exert power over these cases, I coded the websites of my 245 cases for different ways in which organizations from these different sectors are involved in the cases.

For each type of involvement (e.g., funding), cases with either nonprofit or academic codes (and no other organization codes) were coded as civil sector, while cases with either retailer or supplier codes (and no others) were coded as private sector.[86] Cases with only government codes were coded as public sector. I also created codes for mixed sector involvement—public-private, private-civil, public-civil, and public-private-civil. Each type of involvement maps to the different power resources just described, and are explained in the corresponding sections that follow.

### Grants of Authority

The most obvious way an organization can exert power over an initiative is to lead it—to be the primary institution that has the authority to make its day-to-day operational and strategic decisions. Such authority can direct an initiative to focus on certain issues and methods while ignoring others, which can benefit certain actors while disadvantaging others. As panel A in figure 3.5 illustrates, initiatives that describe themselves as being implemented solely by civil sector organizations are the most common type of initiative in the dataset (33 percent). These include advocacy organizations such as Environmental Defense, certification organizations such as the Forest Stewardship Council, media organizations such as the National Geographic Society, rating organizations such as the Carbon Disclosure Project, research institutions such as the Aspen Institute, and academic institutions such as Claremont McKenna College.

Cases led solely by private sector organizations (23 percent) are the second most common type of case. Companies leading these initiatives include HP, Amazon.com, Whole Foods, and Staples. Retailers account for 72 percent of the cases led by the private sector, while 28 percent are led by suppliers (i.e., manufacturers of products being evaluated). Initiatives led solely by public sector organizations account for 6 percent of the cases, and include programs such as ENERGY STAR, Design for the Environment, and Certified Organic. Only seven cases are led by more than one sector. These include nonprofit organizations, such as the Business and Institutional Furniture Manufacturers Association, that serve as business associations for suppliers of the products being evaluated. They also include collaborations between civil and private sector organizations, such as the Climate Savers Computing Initiative. The type of implementation organization could not be identified for over a third of the cases.

These results provide a valuable snapshot of the types of organizations that are implementing these initiatives. However, authority over information-based governance strategies can be wielded in ways beyond their direct implementation. An initiative's leader may delegate or share its authority with other organizations through partnerships and coalitions, associations via advisory boards and boards of directors, and direct involvement in the design of the initiative. Such indirect authority can be used to recommend certain approaches that would cast either a positive or negative light on organizations being evaluated. Each of these three types of authority sharing were coded and combined into an aggregate metric of "organizational association" by each sector.

Panel B of figure 3.5 shows that associations with either civil (12 percent) or private (13 percent) sector organizations are most commonly mentioned on the websites, followed by associations with both private and civil organizations (9 percent). Initiatives only mentioning associations with civil

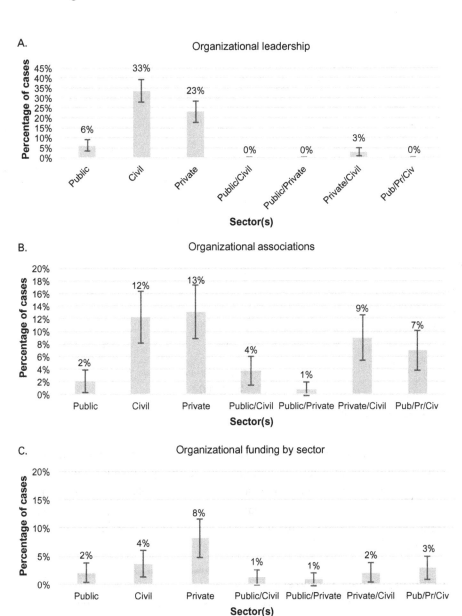

**Figure 3.5**
Power resources of the public, private, and civil sectors in environmental evaluations of products and companies. *Note:* Adapted from Bullock, "Independent Labels?," 53.

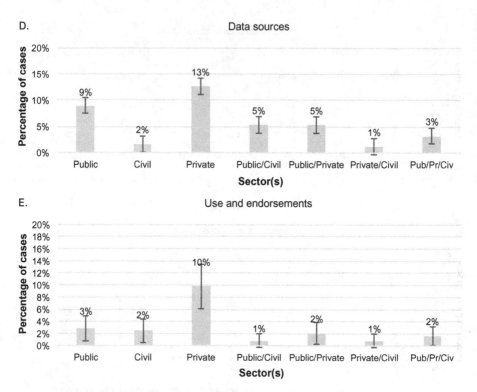

**Figure 3.5** (continued)

sector organizations include FishWise and Citizens Market, while those only mentioning associations with private sector organizations include the Corporate Responsibility Index and the Green Hotels certification. An example of an initiative with associations with both civil and private firms is the Forest Stewardship Council. Over 50 percent of the cases do not mention any such associations.

### Grants of Economic Resources

Organizations can also exert power over these initiatives by funding them. Funds can come with explicit or implicit strings attached that require an initiative to use a particular method or set of criteria that would benefit the funder. As panel C in figure 3.5 shows, cases that only mention financing from private sector organizations are the most common in the dataset (8 percent), and suppliers account for 80 percent of those organizations. One case, the Best Aquaculture Practices certification, mentions financial support from both suppliers (ten "visionary industry leaders") and retailers (Darden Restaurants). Organizations mentioning funding from only the civil sector (4 percent) and from all three sectors (3 percent) are the next most common. An example of an organization only receiving civil sector funding is the Electronic Takeback Coalition's TV Companies Report Card, and an example of an organization receiving funding from all three sectors is the Marine Stewardship Council. Approximately 7 percent of the cases receive funding from more than one sector. Notably, 79 percent of the cases do not provide any information about their funding sources.

## Grants of Information

The information used by information-based governance is a power resource itself, and can serve as a mechanism by which the sources of that information can exert power over these strategies. For example, initiatives that rely on data provided by companies are limited to what information those companies provide to them. Companies can provide false or misleading data that obscures their true environmental performance. As panel D in figure 3.5 shows, cases that only mention private sector organizations as their source of data are the most common in the dataset (13 percent), and all of these organization are suppliers (as opposed to retailers). Examples of cases that only mention the use of private sector data are ENERGY STAR, Climate Counts, and the Chemical Home. Cases that only mention the public sector as its source of data are the second most common type of case. Examples of these cases include the Auto Asthma Index and FishWise. Cumulatively, 15 percent of the cases use data from more than one sector. Across all cases (both those that use data from one type of sector and multiple sectors), the number of cases that use data from public and private sector sources is statistically equivalent (22–23 percent).

## Grants of Prestige

Organizations can also exert power over information-based initiatives by recognizing and granting prestige to them (or by withholding such recognition). Prestige can be transferred explicitly by endorsement or implicitly by their use of the initiative's information. For example, Green Home states that it has "received endorsements from throughout the environmental community, including Environmental Defense and The Earth Charter," while EPEAT lists organizations that have instituted an EPEAT certification purchasing requirement, including the United States Marine Corps, the City of San Francisco, and Yale University. Similar to funding, such endorsements can come with a quid pro quo—to earn the endorsement, an initiative must commit to a certain approach that would benefit the endorser. Panel E in figure 3.5 reveals that cases that only mention such recognition from the private sector are the most common in the dataset (10 percent), with retailers being mentioned in 62 percent of those cases and suppliers being mentioned in 50 percent of them. Some examples of cases that only mention endorsements or use by the private sector are Cradle to Cradle certification, Dolphin Safe, and the Best 50 Corporate Citizens. Approximately 5 percent of the cases mention endorsements or use by organizations from more than one sector. Nearly 80 percent of the cases do not provide any information about endorsements or use by government, nonprofit, academic, retailers, or suppliers.

## The Landscape of Legitimacy Grants

The preceding sections document how four different resources—authority, economic resources, information, and status—are distributed across a dataset of 245 information-based environmental governance initiatives. These resources are grants of legitimacy that can serve both as mechanisms of control over these initiatives and as signals of credibility (or the lack thereof) to other organizations. Figure 3.6 aggregates this data into a single snapshot of these grants of legitimacy by the public, private, and public sectors across the 245 cases. Approximately 25 percent (the sixty-two cases in the bottom right-hand corner) of these initiatives mention at least one resource from each of the three sectors, suggesting they have been deemed legitimate by at least one organization within those sectors. An additional 32 percent mention at least one resource from two of the three sectors, and another 33 percent mention at least one resource from one of the three sectors. The twenty-nine cases in the back top left-hand corner mention no resources from any of the three sectors.

## The Information Realist Perspective

As in chapter 2, observers who are optimistic about information-based governance strategies will likely interpret these results in a positive light, or at least as a glass half full. These cases are clearly sending a wide range

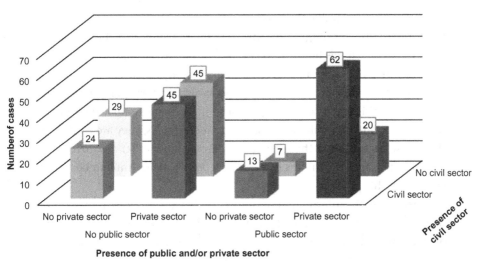

**Figure 3.6**
The landscape of legitimacy grants. *Note:* The eight cases in the back left-hand corner make no claims of connections to the private, public, or civil sectors. The seventy cases in the front right-hand corner make claims of connections to all three sectors.

of credibility signals based on both their own particular attributes—their independence and expertise—and grants of legitimacy from other organizations. The fact that more than half of the initiatives clearly describe their criteria and more than a third provide detailed and complete descriptions of their methods may be particularly encouraging for these information optimists. While they would like to see more than 40 percent verify or generate at least some of their data and more than 25 percent claim to have some type of expertise, these are still significant proportions of the dataset. They might also point out that just because they do not describe their expertise or their independence does not necessarily mean that they have none.

This is true for the data on the grants of legitimacy as well, which these information optimists will likely view as also encouraging. More than half of the cases signal that they have received at least some type of support from at least two sectors of society, and only 12 percent mention no such support from any sector. A wide range of institutions are supporting these initiatives, suggesting that they do indeed trust them. To the extent they do not explicitly trust them, the deployment of different forms of power over these programs—funding, authority, and so on—indicates these institutions have created mechanisms to keep the initiatives accountable to them. For those few that have not received many (or any) such grants of legitimacy, stakeholders can easily identify and avoid these programs, especially if their other signals of credibility are also inadequate. While these information proponents would like to see more such signals, they are confident that those that do send them are (and will continue to be) rewarded, and the collective trustworthiness of these programs will continue to be ratcheted up by this process.

Meanwhile, the information pessimists likely view this idea as incredibly naïve, and see this data quite differently. To these observers, the landscape of credibility signals in figure 3.4 looks particularly barren, with nearly half of the cases congregating in the zone of no expertise or independence. The paucity of cases that generate their own data is particularly concerning—independent verification, which relies on companies self-reporting the vast majority of their data, is a poor substitute for evaluations that are conducted truly independently. The lack of expertise (75 percent of the cases do not mention any at all) is also alarming. Not knowing the backgrounds of the people doing these assessments, how can we possibly believe they know what they are doing?

Perhaps an organization that you trust has endorsed or is running the initiative, and that is good enough for you. But the data on such endorsements would likely raise additional concerns for most information pessimists. First of all, it is not at all clear who is behind most of these initiatives. One-third of the cases do not provide any information about the type of organization behind them, over half do not provide any information about the organizations that are advising or partnering with them, and nearly two-thirds do not provide any information about their data sources. Approximately 80 percent do not mention any of their funding sources or any organizations that have endorsed them or used their services, while 12 percent do not disclose any information about any of the four resources of power discussed in this chapter. Only two cases (less than 1 percent), Fish-Wise and EPEAT, provide information about all four resources.

This opacity of the power and accountability relationships driving these initiatives may cause many observers to hesitate in trusting them. However, even the relationships that are documented will likely raise concerns for information pessimists because they often create what I call "power hybrids."[87] These power hybrids are cases in which power associated with a particular resource is shared between two or more sectors, and dominate the dataset described previously. Approximately 35 percent of the cases are "intra-resource hybrids"—multiple sectors provide them with the same power resource (e.g., receive funding from the private and public sectors). An additional 22 percent are "inter-resource hybrids"—multiple sectors provide them with different power resources (e.g., they receive funding from the private sector and data from the public sector). Information and indirect authority are the most common types of resources to be provided by multiple sectors (15 percent and 20 percent of the cases, respectively). Across all power resources, cases that use power resources from both private and civil sector organizations are the most common type of power hybrid (14 percent of cases). Tri-sector power sharing of a power resource occurs in 12 percent of cases.

These power hybrids have intricate and complex relationships with private, public, and civil society actors, and the information optimists might argue that these connections enable them to reach wider audiences and be more effective. But the information pessimists might in turn argue that they instead create significant conflicts of interests and cross pressures of accountability, which can result in what has been called "multiple accountability disorder" and the "problem of multiple masters."[88] These conflicts of interest can undermine an initiative's perceived legitimacy because they suggest conflicting accountability relationships, which can be "a critical element in the construction and contestation of legitimacy claims."[89] With so many "masters," the master that a particular stakeholder trusts may have little or no actual power in these hybrid organizations, and thus its involvement with these organizations may not be perceived as increasing their legitimacy.

Information realism recognizes these concerns about the opacity and hybridity of power in these programs, but also acknowledges that some initiatives are nevertheless doing a better job than others in communicating their trustworthiness to their audiences. For the information realist, the landscapes of credibility signals and legitimacy grants in figures 3.4 and 3.6 reveal both the challenges and opportunities of information-based governance. Unlike the information optimist, the information realist does not see an inevitable march toward progress and greater trustworthiness in this data. Nor does she view it as barren and hopeless as the information pessimist does. Instead, she sees an ecosystem of niches that programs can populate. Obviously the figures presented are simplifications of this ecosystem—many more possible constellations of expertise, independence, and organizational support are possible. Rather than simply seeing more signals of credibility and more grants of legitimacy as necessarily better, she views strategic and purposeful articulations of credibility and legitimacy as the goal for these programs.

Thus the opacity and hybridity of power discussed here is problematic if it creates confusion among stakeholders, but are not problems in their own right. Some initiatives inevitably will be opaque about some of their attributes, intentionally or otherwise, and if their audiences are aware of and can evaluate that opaqueness, then it is not a problem. Instead, it represents another way to assess their trustworthiness. Similarly, some initiatives will inevitably have conflicts of interest created by their associations with different types of organizations. If stakeholders are aware of these conflicts and how these organizations claim to manage them, then they have another data point they can use to evaluate them. Thus the management of hybridity, as opposed to hybridity itself, becomes the issue to focus on. While this perspective requires stakeholders to be thoughtful about whom they do and do not trust, it also recognizes that they likely trust organizations for different reasons. It also requires evaluation organizations and firms to be intentional about how they signal their credibility, while recognizing there is no one-size-fits-all approach to communicating trustworthiness.

The real challenge that information realism presents is building the capacity of both trustees and trustors to create trusting relationships. Fundamental to such relationships are clear signals of credibility, legitimacy and accountability, and the clarity of these signals depend on well-defined standards for different dimensions of credibility. For example, precise standards for different types of independent data verification and generation—including those that require civil, discursive, and/or consensual engagement with stakeholders—should be developed.[90] Such standards would clarify the confusion around the term "third party," and create clear categories of independence that differentiate between firm-paid vs. nonfirm-paid, governmental vs. nonprofit monitoring, and standard-setting vs. standard-checking.[91] Standards for relevant professional and academic expertise would be useful as well, particularly given the relatively low prevalence of expertise signals in the sample. The U.S. Green Building Council's LEED-related credentials for architects is one example of such a standard, and could be replicated for professionals involved in the design and operation of eco-labels and other forms of environmental evaluation.[92] Such standards would enable easier comparison among expertise claims and highlight the importance of having relevant skills. They would also enable true domain experts to credibly differentiate themselves from advocates, executives, and academics who do not possess those skills.

Audiences, however, do not magically learn about these standards or whether initiatives meet them, and so a second foundation of trusting relationships is a forum for learning and communication. In other words, in order for the market-based, ratcheting-up process that information optimists promote to work, an "information marketplace" is needed that enables direct comparisons of credibility signals across ratings and labels. Such a marketplace, which could be online and virtual, could improve the accessibility of standardized credibility signals to stakeholders, and enable them to more easily select the agents that best match their preferences. Efforts would be necessary to make this marketplace as inclusive and accessible as possible so that all actors would be able to participate in it.

The reality is that while it is probably smart to send multiple signals of credibility, limited resources may require initiatives to make tough choices if they want these signals to themselves be credible. Given that their stakeholders likely have different signal preferences, they must decide which ones they want to appear credible and accountable to, and which sources and types of legitimacy are their highest priorities. For example, if information-based programs prioritize the credibility and cognitive legitimacy associated with expertise and view themselves as most accountable to experts, then they should emphasize signals of expertise. If they prefer to enhance their trustworthiness and moral legitimacy and build support among civil society organizations, then they should emphasize signals of independence.

A similar logic may also apply to grants of legitimacy. Despite the calls for "stakeholder inclusivity" by both scholars and practitioners,[93] having support from very different organizations may send conflicting signals of credibility. An endorsement from Walmart may discredit a program among environmentalists, while an endorsement from Greenpeace may raise concerns among conservatives. Audiences may sense a "problem of multiple masters," which Barbara Romzek (University of Kansas) and Melvin Ingraham (Syracuse University) conclude often leads to managers focusing on one or two of these relationships on a daily basis with the others "being in place but underutilized, if not dormant."[94] As Oxford University political scientist Walter Mattli and Duke University political scientist Tim Büthe find in their case study of U.S. financial accounting standards, private sector agents with delegated public authority may focus more on the interests of their private principals, especially if those principals are internally cohesive and have distinct preferences from other interested parties.[95]

Thus initiatives may be smart to streamline their grants of legitimacy to improve their value as signals of credibility. As nice as stakeholder inclusivity sounds, inclusivity may not be beneficial for every initiative, and many audiences may prefer information coming from a single sector or organization. Thus the cases that only mention use of resources from a single sector accounted for 34 percent of the cases (21 percent use only private sector resources, 11 percent only civil-sector resources, and 2 percent only public sector resources) may be onto something. However, such an approach does not come without caveats and trade-offs. For example, a major caveat is that this logic applies more to NGOs, given the higher levels of trust that society has in them and the valid concern about companies evaluating their own products. In countries with highly polarized politics such as the United States, government labels may also face strong distrust from citizens who do not support the party in power.[96]

Trade-offs arise even for NGOs, however, because these single-sector cases may be losing important support from those who do not believe they are "encapsulating" their interests. Businesses may fear betrayal by NGO-driven programs, for example, and NGOs may distrust corporate initiatives. As Magnus Boström explains, it is the "combination and mutual adjustment of interests" that contribute to an image of neutrality and independence.[97] The key to working through this trade-off is to think about the broader information ecosystem and how different initiatives and organizations fit into it. In particular, it is important to understand the relative difficulty of their standards (how high they set the bar), a topic to which we will return to in the next chapter. If the only information about a product category is being provided by a multistakeholder initiative with relatively lax standards, then an opportunity exists for an NGO-led program to create

•

a program with a higher bar. Likewise, if an industry has only been evaluated by an NGO with very high standards, then a business-led program that enables a larger number of companies to do well may gain traction. We will discuss the relative effectiveness of these approaches in chapter 6, but the point here is that both types can be perceived as credible—if they clearly articulate their mission and strategy, whether it is an inclusive approach with broad participation or a more exclusive approach focused on recognizing the leaders in an industry.

## Promising and Problematic Practices

So where does this leave us in terms of promising and problematic trust-building practices? Clearly, not providing any signals of credibility is likely to be the least effective strategy. This approach will leave its audience at best feeling neutral toward it, and at worst actively distrusting it. If it is providing highly desirable and methodologically rigorous information in a highly accessible format, then it may still accomplish its goals, but with a higher risk of failure than if people perceived it as highly trustworthy as well. Without such assurance, users will likely withdraw their support if they detect any reason to mistrust the information. In this situation, rebuilding trust may be much more difficult than creating it in the first place.

By proactively sending credibility signals to their audiences, information-based governance initiatives buy themselves time to prove themselves when their trustworthiness is questioned. Having established who they are accountable to, they can quickly communicate to them why they should still be trusted and how they will fix any problems that have emerged. Chapter 6's discussion of a controversy surrounding ENERGY STAR provides an example of such a response, and demonstrates the importance of independent data generation and product testing. Generally speaking, organizations that can credibly demonstrate their independence and expertise are likely to earn and retain more trust than those that cannot. However, as discussed earlier, few programs have all of these attributes, and it is not clear which are more important than others. As this chapter explains, it likely depends on who their primary stakeholders are.

Regarding grants of legitimacy, given the greater trust placed in them generally, some association with nonprofit organizations may be more valuable than similar involvement with business and government. This dynamic, however, may depend heavily on the context and what organization, what industry, and what kind of relationships are involved. Furthermore, diversity of support across sectors may not necessarily be an asset—some audiences may trust single-sector initiatives more than multisector initiatives. The key to building trust, therefore, is not the quantity of relationships, but their quality—and how clearly that quality is communicated to key stakeholders.

### The Trustworthiness of Toilet Paper Certifications

So let us return to Carrie as she considers how to respond to the journalist's inquiry. She visits the websites of FSC and SFI to look for the signals of credibility they are sending. FSC claims that "independent certification bodies" conduct its assessments; SFI states that it is an "independent" organization whose "cornerstone" is third-party certification. Both are relatively transparent, and provide detailed accounts of their criteria and methods and where their data comes from. SFI provides bios of most of its staff, many of whom have relevant professional and/or academic expertise, while FSC only provides biographical information about its director general, who is a former WWF staff member (SFI's president and CEO is a former forestry consultant). Both have diverse sets of stakeholders associated with their organizations—through general assemblies, boards of directors, external review panels, funding support, and other mechanisms. FSC's twelve-member board of directors includes representatives from environmental NGOs and companies, as does SFI's eighteen-member board, which also includes academic and government representatives, which FSC's does not.[98]

While there are some differences, both cases are sending a wide range of credibility signals. Even someone like Carrie who knows the histories and controversies of the organizations might find them equally trustworthy. But this plethora of signals raises the "problem of many masters" discussed earlier in the chapter—with boards of twelve and eighteen members (and for FSC, a general assembly of over six hundred members), who really is calling the shots?[99] Carrie might reason that the FSC General Assembly only meets every three years, so it is unlikely its members exercise much direct influence over the day-to-day operations of the organization, and the same is probably true for these organizations' boards as well. That leaves a significant amount of control to the leaders of the organizations, and here there is a clear difference in orientation. Carrie naturally trusts the FSC's director general, with his background as a fellow environmental advocate, more than the SFI CEO, with her career as a consultant working with "government, trade organizations, and corporations."

And so she is tempted to tell the journalist that FSC is the more trustworthy organization. But she hesitates, and thinks a little more deeply about the accountability of these organizations. After the authority of the CEO and boards, funding is probably the most influential of the power resources discussed in this chapter. While FSC provides some limited information about its financial support, neither FSC nor SCI provides a full account of where its economic resources come from. The fact remains that both organizations likely rely heavily on the fees they charge companies for their certification services, which brings us back to the inherent conflict of interest highlighted by FSC-Watch that this relationship represents. Companies are motivated to hire certifiers who will indeed certify them, and certifiers are incentivized to in fact do so, with few questions asked. If they do not, companies will find someone else who will.

This logic represents a fundamental barrier to building a trust-based relationship between these organizations and their audiences. This is a challenge not only facing environmental certifications but financial auditors as well. It was at the heart of the Enron scandal in 2001, which caused the disintegration of the accounting firm Arthur Andersen. The Sarbanes-Oxley reforms of 2002 established several new mechanisms to confront this problem, including the Public Company Accounting Oversight Board (PCAOB) to "audit the auditors" and a requirement that auditors be chosen by the audit committee of the board of trustees (and not by management). While these reforms have likely helped increase the accountability of auditors, the accounting industry continues to be implicated in fraudulent accounting practices around the world; for example, in Britain (at Tesco), the United States (at HP), Japan (at Olympus), and China (at China Integrated Energy). Observers have called for a host of further reforms to the accounting industry to increase the independence and trustworthiness of external audits of companies, from expanding the required scope of audit reports to having auditors be selected by shareholder proxy votes, stock markets, or the government.[100]

All of these practices should be considered in the realm of environmental ratings and certifications as well. They can be implemented by individual initiatives or adopted by them collectively. For example, the Fair Labor Association, which focuses on the treatment of workers, already requires that companies not select their own certifiers but are assigned one by the accreditor.[101] This allows for standards and costs for certification to be standardized so that shopping around for the cheapest and least rigorous assessor can be eliminated. Another option used by Fair Trade USA is to select one organization (in this case, SCS Global Services) to conduct all certification and auditing activities.[102] This eliminates the conflict of interest problem because those being certified are no longer selecting their certifier. Governments, industries, or groups of assessment organizations could encourage or require such practices within their sphere of influence. This approach is based on the ancient principle of *nemo judex in sua cuasa*— "No person shall be a judge in his own cause"—and is encoded in professional sports, legal doctrine, and academic publishing. It is equally relevant to information-based environmental governance strategies (perhaps even more so given that the ratings and certifications they produce are credence goods), and should be applied to them as well.

Such a principle could be incorporated into efforts by organizations such as the U.S. EPA; the UK Department for Environment, Food, and Rural Affairs; the ISEAL Alliance; and Consumers Union to create standards of behavior for environmental certifications and ratings. These initiatives are laudable, and include many important criteria, some of which we will return to in the next chapter. Many of their criteria that relate to trustworthiness, however, are overly prescriptive. For example, EPA's final guidelines for its pilot assessment of standards and ecolabels to be included in its environmentally preferable purchasing recommendations to federal agencies suggest that these programs should enable broad participation by affected stakeholders, consider all relevant viewpoints, include a diversity of interests, and work toward achieving consensus in their design process. These guidelines would likely exclude single-sector initiatives that some audiences might find trustworthy and helpful.

An example from our toilet paper case will demonstrate this dynamic. In researching forest products, Carrie comes across several other certifications and ratings of toilet paper (see figure 3.7 for their logos). Two of them, Green Seal and EcoLogo, are somewhat less transparent than FSC and SFI, have similar levels of independence, and do not make any mention of their expertise. Green Seal highlights its associations with a wide

**Figure 3.7**
Organizational logos of toilet paper certifications and ratings.

range of sectors, while EcoLogo only highlights its governmental connections (to the Canadian government).[103] Nevertheless, some consumers
might trust a label associated with the Canadian government more than
the other multistakeholder initiatives whose accountability they perceive
as at best muddled.

A third option Carrie discovers is the "Shopper's Guide to Tissue Paper,"
which provides very limited information about its methods or its associations, and does not appear to have been developed through an open and
inclusive process.[104] Nevertheless, it provides specific information (buy or
avoid, percent recycled, percent post-consumer, bleaching process used)
in a relatively usable format, and was produced by the Natural Resources
Defense Council's (NRDC), a well-known environmental organization. No
controversies surround this guide, and there is no evidence that NRDC
was paid by any of the companies to produce it—unlike all of the other
certifications Carrie has reviewed. When it comes to her relative trust in
these programs, she may indeed find NRDC's assessment to be the most
trustworthy (although it may not be the most valid because it does not
appear to have been updated since 2009—an issue to which we return in
chapter 4).

So Carrie then decides to actually put her money where her mouth is and
tries to find some toilet paper online that meets her standards. She finds
products that are certified by SFI (Charmin UltraSoft), FSC (Scott Naturals),

EcoLogo (Cascades Enviro), and Green Seal (Atlas Green Heritage). She finds she is less interested in knowing about the organizations behind these certifications than what they are actually certifying (the topic of chapter 2) and how they are certifying it (the topic of chapter 4). If they are claiming to certify that the toilet paper comes from sustainably managed forests, how are they defining and measuring sustainable management? While the question of trustworthiness is important and many people use it as a cognitive shortcut in their decision making, ultimately the question of methodological validity is more important. We will discuss this topic in detail in chapter 4, and return to Carrie's quandary in that context.

# 4 Measuring Green: The Generation of the Information

## Which Is the Greenest Building of Them All?

Like Mark choosing milk products and Carrie selecting tissue paper, Lynn is faced with a decision. This decision is not for her alone, however, but for her entire institution. As I mentioned in chapter 1, Lynn is an environmental scientist at a university that is in the process of designing a new academic building. She is in the midst of a planning meeting with the architects, and they are discussing what green attributes they want the building to have. Some of her colleagues insist that the new building be LEED certified, while others dismiss LEED as arbitrary, nonsensical, and too expensive. Some are arguing for other certifications, such as the Living Building Challenge, Green Globes, and ENERGY STAR. They all turn to Lynn for her opinion, and ask her which system she thinks provides the most valid metric of sustainability performance.

Lynn has done her homework, and what she has read about LEED has left her with decidedly mixed feelings. LEED (short for Leadership in Energy and Environmental Design) was launched in 1998 by the U.S. Green Building Council (USGBC), and claims to have certified over 72,000 projects comprising over thirteen billion square feet across more than 150 countries and territories (as of February 2016).[1] The USGBC claims that eighty-eight of the Fortune 100 companies use LEED and "nearly five million people experience a LEED building every day."[2] It has trained and certified nearly two hundred thousand LEED professionals who have the authority to perform audits of buildings applying for certification.[3]

LEED's 2014 version includes a broad range of criteria across nine categories that range from indoor environmental quality to water efficiency to the sustainability of the building site, and addresses many of the public environmental concerns discussed in chapter 2. The USGBC website emphasizes the cost savings and business benefits of LEED certification as well. As an example, it cites a study that shows customers opened up more accounts and deposited more money at LEED-certified bank branches than noncertified branches.[4] It also claims to be a broad-based organization, with seventy-six chapters and 12,870 member organizations, which include "builders and environmentalists, corporations and nonprofits, teachers and students, lawmakers and citizens."[5]

Despite these impressive statistics, Lynn is also aware of criticisms of LEED's rating system. One of the most common complaints is about LEED's credit for bike racks, which critics view as either not worthy of such recognition, too easy to achieve, or not expansive enough (e.g., it doesn't include credit for giving employees bikes or locks).[6] Other critiques of LEED focus on the credits it gives for energy savings projected in computer models but not actually measured, its recognition for preferred parking for fuel-efficient cars that ultimately are used by SUVs and sports cars, and its certification of enormous homes that are located in remote, pedestrian-unfriendly neighborhoods.[7] A USA TODAY study found that building designers focus on the easiest and cheapest credits, and a study by John Scofield at Oberlin College concluded that LEED-certified buildings are no more energy efficient than comparable noncertified ones.[8] Energy efficiency expert Henry Gifford classifies the criticisms of LEED into four categories—the "Sin of Not Following Through," the "Sin of Valuing Gizmos over Appropriate Design," the "Sin of Laughably Inappropriate Use," and the "Sin of Wretched Excess."[9]

Several organizations have capitalized on these perceived weaknesses of LEED to launch and promote their own competing building rating systems. Green Globes is a project of the Green Building Initiative (GBI) that is positioning itself as a flexible, practical, innovative, and credible alternative to LEED. The GBI website emphasizes that it is a multistakeholder initiative that has developed Green Globes through "a public, collaborative, consensus-based process." Like LEED, it also highlights the private benefits of certification, stating, for example, that it helps organizations qualify for tax incentives, meet government regulations, and attract and retain employees.[10] Green Globes appears to cover a similar set of environmental categories as LEED, although critics assert that it is less rigorous and more friendly to industry.[11] For example, they point out that while LEED only recognizes FSC-certified building products, Green Globes also gives credit for the purportedly industry-friendly certifications from SFI, the American Tree Farm System, and the Canadian Standard Association.[12] Nevertheless, in 2013 the U.S. General Services Administration recommended that federal agencies should use either LEED or Green Globes as certification systems for their buildings.[13]

Rather than compete directly with LEED as a mainstream, broad-based certification, other initiatives are pursuing more focused strategies. For example, ENERGY STAR, a federal program implemented by EPA in partnership with the U.S. Department of Energy (DOE), certifies both commercial and residential buildings for their energy efficiency, but does not include any of the criteria related to indoor air quality, water efficiency, materials use, and other issues covered by both LEED and Green Globes. Thus it is not as comprehensive and may be guilty of the sin of the hidden trade-off (discussed in chapter 2), but it may be more rigorous and valid. As some commentators have pointed out, while LEED provides no guarantee that a building is actually "energy efficient," ENERGY STAR requires buildings to submit their utility bills before they are certified and on a continuing basis.[14] Other programs, including the National Green Building Standard, Passive House, and the Home Energy Rating System, have focused on energy use in the residential market.

The Living Building Challenge (LBC), on the other hand, is positioning itself as the most rigorous, holistic, and comprehensive certification system available. A project of the International Living Future Institute, the Living Building Challenge aims to help buildings "move beyond merely being less bad and to become truly regenerative." Its categories, or "Petals," are decidedly more far reaching than the other programs, and not only encompass water, materials, and energy use but also health, happiness, equity, and beauty. They consist of twenty required "Imperatives" that are based on metrics of actual performance and include requirements such as "net positive" energy and water use (e.g., 105 percent of the project's energy needs must be supplied by on-site renewable energy on a net annual basis, without any on-site combustion).[15]

In the face of this competition, the USGBC has not been sitting still, but has been actively responding to its critics. Many of the criticisms leveled against it have at least been partially addressed in its 2014 version of LEED (v4), which includes, for example, a stronger focus on building performance related to materials, indoor air quality, and water efficiency.[16] In the face of these competing claims, what should Lynn recommend to her colleagues? To some extent, this question depends on her sense of the trustworthiness of these organizations (the subject of chapter 3), as many of the criticisms of these organizations relate to their legitimacy and account- ability. But it also depends on the basic validity of their methods, and at some point we have to move beyond the trust issue and actually evaluate, compare, and decide among the underlying systems used by these initia- tives. And if Lynn does not make such an assessment, as an environmental scientist concerned about sustainability issues, then who will? But for her or anyone else to proceed, they need a framework for making such a com- plex comparative evaluation.

That is the focus of this chapter, which introduces the concepts of rep- licability, reliability, and validity as useful tools for people like Lynn to evaluate competing eco-labels and sustainability ratings. The chapter then applies these concepts in an analysis of the transparency and quality of the data and methods used in existing information-based environmental governance initiatives. The chapter explores the different types of transpar- ency that these programs should have, and highlights the importance of updating the data and methods used in their assessments, documenting the weights of the different indicators that make up their composite metrics, and applying the insights of life cycle analysis in their design.

With a few important exceptions, the data from my Environmen- tal Evaluations of Products and Companies (EEPAC) Dataset—which not only covers cases from the building sector such as LEED and ENERGY STAR but also from many other sectors as well—demonstrate the general lack of methodological replicability, reliability, and validity among exist- ing information-based environmental governance initiatives. The dataset highlights the significant challenges associated with developing robust metrics of sustainability. The chapter discusses several important trade-offs between different dimensions of information quality, and suggests several strategies for managing these trade-offs. It also identifies the most promis- ing and problematic information generation practices found in the data- set and the lessons learned from these examples. The chapter concludes by applying the insights of this analysis to Lynn's predicament and the question of which building assessment program is indeed the greenest of them all.

### An Information Quality Framework: Validity, Reliability, and Replicability

A brief metaphor will help reveal the three major questions we must ask when assessing the information quality of different ratings and certifications. When Snow White's stepmother asks her Magic Mirror who is the fairest of them all, he responds, "You, my queen, are fair; it is true. But Snow-White is a thousand times fairer than you."[17] This response begs several important questions. First, how replicable is his assessment, meaning, how repeatable is his analysis of "fairness" by others? Second, how reliable is his assessment, and how consistently does he measure "fairness?" And third, how valid is his assessment? That is, how accurately has he assessed "fairness?" If the Mirror cannot address these questions about his evaluation methods, then the Queen has every reason to doubt the veracity of his claim.

### The Statistician Should Have No Clothes: Replicability

The Magic Mirror metaphor thus introduces us to the three key dimensions of information quality discussed in this chapter—replicability, reliability, and validity. These three dimensions are presented in figure 4.1, along with several related concepts that I will discuss in more detail. *Replicability* means that results of an analysis can be reproduced. In order for this to be possible, the measurement process must be transparent so that it can be repeated (i.e., replicated). The importance of such transparency has been emphasized across a wide range of governance contexts in the last few decades, from "sunshine" laws requiring government agencies to make their meetings open to the public to calls for corporations to disclose their environmental and social performance.[18] Scholars have also emphasized the value of transparency, both in the context of their own work (e.g., by requiring the public dissemination of datasets and metadata along with the publication of articles) and the activities of industry, civil society, and government.[19] However, they have seldom distinguished between different forms of transparency.[20] In the context of information-based governance initiatives, these programs need to disclose not only their criteria and data sources, but also the data itself and the specific methods they use to analyze it. Thus if Lynn wants to assess the replicability of a certification such as Green Globes, she should identify whether it is transparent about each of these different aspects of its evaluation process.

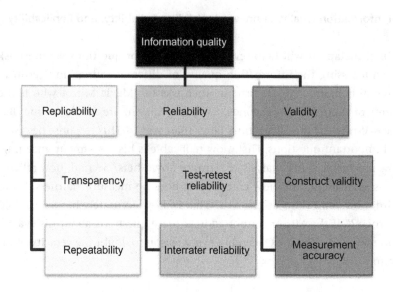

**Figure 4.1**
Dimensions of information quality.

### Consistency in Claims: Reliability

Even if a certification is transparent and replicable, however, it may still have significant methodological limitations. This brings us to our next question—how consistent are the results of the evaluation process? This is the essence of *reliability*—that the metric generates similar results each time it is used to measure something. Researchers are generally concerned about two types of such reliability. *Test-retest reliability* refers to the consistency of a metric over time, while *inter-rater reliability* relates to whether similar results are generated by a metric when it is applied by different people.[21] Thus Lynn might like to know that evaluators for the Living Building Challenge, for example, not only apply the program's criteria the same way for every building they evaluate, but also apply them the same way as each other.

### Truth in Advertising: Validity

The final and most important dimension of information quality is *validity*, which refers to the extent to which a metric actually represents a phenomenon. In other words, how well does it measure what it claims to be measuring? Validity is a central concept used across both the social and natural sciences, and different fields define it differently. For our purposes, we are primarily concerned with *construct validity*, which is a measure of the quality of a particular operationalization of a concept or behavior—the translation of something in the real world into a functional representation or metric, such as the certifications that Lynn is evaluating.[22] Such construct validity can be substantiated by either evaluating the quality of the operationalization directly or comparing it to other criteria that are related to the construct in question.[23] So Lynn might assess the validity of LEED by analyzing its own operationalization of building greenness and assessing the relevance, comprehensiveness, and specificity of the criteria it uses. Alternatively, she might compare its outcomes to other criteria she believes are related to building greenness, such as energy efficiency.

The difference between validity and reliability is subtle, but important. Lynn may find that a particular green building certification consistently applies its criteria, but those criteria may not necessarily be accurate assessments of building greenness. Likewise, she may feel that another certification has better criteria, but that they are not applied reliably over time or by different evaluators. This distinction relates to the difference between accuracy, a synonym of validity, and precision, a synonym of reliability. *Measurement accuracy* refers to the closeness of fit between a metric's estimation of a phenomenon and the phenomenon itself, while *measurement precision* refers to the level of random error associated with an estimate. Suffice to say, a high-quality metric produces both accurate and precise estimates. Lynn is therefore looking for a certification that is based on a valid and accurate measure of building sustainability, is produced reliably and with a high degree of precision, and can be replicated due to its high level of transparency.

### The Quality of the Information Generated by Information-Based Environmental Governance Strategies

The three characteristics discussed previously—validity, reliability, and replicability—are the core foundations of what we can call *information quality*. High-quality information is generated by methods that are valid (they measure what they say they are measuring), reliable (they consistently produce similar results), and reproducible (they can be repeated). Expanding beyond the specific context of building certification, this section examines the extent to which information-based environmental governance strategies are indeed generating such high-quality information. While some definitions of information quality also include dimensions relating to its relevance, comprehensiveness, and accessibility, these are the focus of other chapters (chapters 2 and 5). This section first presents data related to the replicability of these initiatives, and then summarizes the results of an analysis of their reliability and validity.

### The Key to Replicability: Transparency

As mentioned in chapter 3, transparency is not only an important dimension of information quality, but it is also a key aspect of trustworthiness and legitimacy. It is one of the most frequently mentioned subjects in the peer-reviewed literature on eco-labels and ratings,[24] and encompasses the more specific concepts of traceability and auditability.[25] University of Michigan professor of business law and ethics David Hess, for example, points out that "to have meaningful stakeholder engagement requires that we first have a robust information-based transparency policy with comparable data."[26] Likewise, Harvard scholar Archon Fung asserts that democratic transparency requires the disclosure of rich, usable, and actionable information whose availability is proportionate to the risks to which it is relevant.[27] Graeme Auld of Carleton University and Lars Gulbrandsen of the Fridtjof Nansen Institute differentiate between *procedural transparency*, which refers to the "openness of governance processes" and relates to the concept of input legitimacy, and *outcome transparency*, which "deals with the substantive ends of a given policy intervention" and can contribute to output legitimacy.[28]

The process of developing my codes for transparency revealed that it is a more complex and multidimensional concept than is usually acknowledged. In developing the coding system for the EEPAC Dataset, I identified four primary ways that initiatives can be transparent about their evaluation processes, which will be described in more detail later in this section. The coding process also distinguishes between limited and strong statements of transparency. Strong criteria transparency, for example, indicates that all of the criteria are fully explained, with at least a sentence about what is being measured and what data is being used for each cited criteria, while limited criteria transparency indicates that some but not necessarily all the criteria are listed, and they may or may not be described in any detail. Discussion of the coding results follows.

Figures 4.2 and 4.3 summarize the data for each dimension of transparency. *Criteria transparency* refers to the extent to which a case describes the criteria they use in their evaluation of either products or companies. Sixty-seven of the cases, or 27 percent, describe some but not all of the initiatives' criteria (*limited criteria transparency*), while 51 percent describe their criteria in full detail (*strong criteria transparency*). An example of a case that describes its criteria in full detail is EPA's Design for the Environment Standard for Safer Products, which documents both the product and component-level requirements for certification. An example of a case that describes some but not all of its criteria is *Fortune's* description of companies on its Green Giants list as having "gone beyond what the law requires to operate in an environmentally responsible way."

*Data transparency* refers to whether an initiative provides the actual data underlying the evaluation on its website. The content analysis of the full

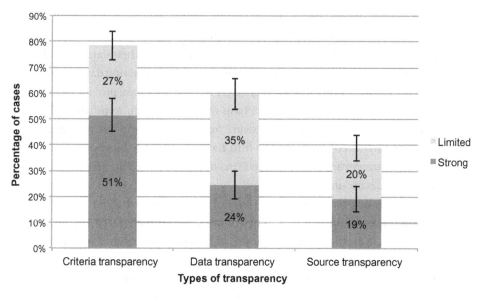

**Figure 4.2**
Types and levels of transparency (by percent of cases). *Note:* Error bars indicate 95 percent confidence intervals for each "limited" and "strong" sample proportion. Note: Adapted from Bullock, "Signaling the Credibility of Private Actors as Public Agents," 203.

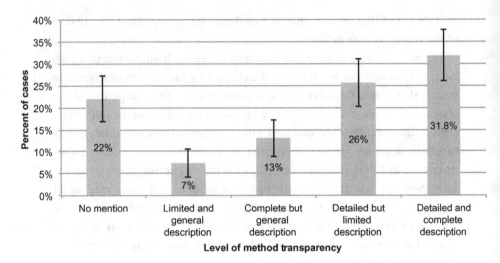

**Figure 4.3**
Levels of method transparency. *Notes:* Error bars indicate 95 percent confidence intervals for each sample proportion. Adapted from Bullock, "Signaling the Credibility of Private Actors as Public Agents," 203.

sample indicates that over 40 percent of the cases provide none of their underlying data, 35 percent provide some but not all of their data (*limited data transparency*), and 24 percent provide all of their underlying data (*strong data transparency*). *Source transparency* refers to whether a case provides a list of the sources of the data that is the basis of its evaluation. Approximately one-fifth of the cases have *limited source transparency* (some but not all of the sources are listed), and another fifth have *strong source transparency* (all of the data sources are listed). Three-fifths do not provide any information about their data sources.

Method transparency refers to the level of detail provided about how the evaluation was conducted. Given the complexity of this characteristic, four binary codes indicating increasing levels of method transparency were used to document this characteristic (see figure 4.3). These codes capture two dimensions of method transparency—the *specificity* of the information (detailed vs. general) provided about the methods used and the *completeness* of that information (complete vs. incomplete). Approximately one-third of the programs provide a *detailed and complete description* all the methods, algorithms, and processes necessary to replicate the results of their assessment, 26 percent provide most but not all of the information necessary to replicate their results (*detailed but limited description*), 13 percent provide a *complete but general description* about their evaluation process, and 7 percent provide a *limited and general description* of their methods. The remaining 22 percent provide no information on their methods at all. Cases implemented by firms were significantly less likely than cases implemented by evaluation organizations to provide detailed and complete methodological descriptions and more likely to provide limited and general descriptions.[29] No significant difference was found between cases implemented by firms and cases implemented by evaluation organizations for the other forms of transparency.

An example of a case that provides the most limited amount of methodological information is Sierra Club's Pick Your Poison Guide to Gasoline, which states their editorial interns "lump [oil companies] into three general categories, the 'bottom of the barrel' (ExxonMobil and ConocoPhillips), the 'middle of the barrel' (Royal Dutch Shell, Chevron, Valero Energy Corporation, and Citgo), and the 'top of the barrel' (BP and Sunoco)." The Green Loop is an example of a case that provides limited and general information, as it outlines a three-step evaluation process for screening products for "sustainability and aesthetics." An example of a case that provides most of the information necessary to replicate their results (detailed but limited information) is the Greener One, which outlines the specific criteria used to calculate its Green Index, but does not explain how the scores are calculated. An example of a case that provides a detailed and complete description of their methods is the University of Massachusetts Toxics 100 Air Polluters Index, which explains in detail where its data comes from and how it compiles that data into its own score.

### The Many Faces of Reliability: Missing in Action

Cases that are transparent and replicable meet a basic requirement of information quality, but they must also be reliable. If the designers of information-based initiatives are serious about confirming that their information is indeed reliable, they should cite their own studies of their metrics' reliability or cite those done by academics or other third parties. And their websites should discuss how they have determined that their metrics are being consistently applied over time and across evaluators.

Following this logic, I used the software program MaxQDA to conduct a lexical search of the text from the websites of the 245 cases in my dataset for any use of the terms of "reliability," "reliable," and "consistency." I found sixty-three references to consistency across seventeen initiatives and 130 references to either reliable or reliability across fifty-six initiatives that relate to the quality of the information they provide.[30] This analysis revealed several important insights. First of all, as figure 4.4 shows, 76 percent of the cases do not make any reference to either the consistency or reliability of their data. Approximately 2 percent only mention reliability or consistency as general ideals to work toward, and another 9 percent make general claims about the reliability or consistency of their data that are not clear about what they mean by these terms. Another 3 percent make limited claims about the reliability or consistency of their data sources, but not their own information.

Interestingly, three cases—Climate Counts, The Power Scorecard, and EcoTrotters—have legal disclaimers that explicitly state that they do not guarantee that their information is reliable. Climate Counts, for example, states that it "makes no representations about the accuracy, reliability, completeness, or timeliness of the materials or of any statements or

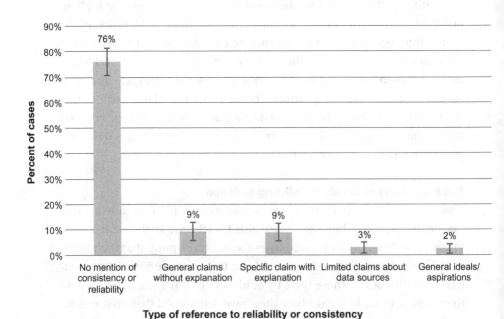

Type of reference to reliability or consistency

**Figure 4.4**
Levels of reliability/consistency. *Note:* Error bars indicate 95 percent confidence intervals for each sample proportion.

other information displayed or distributed through the site." And a fourth initiative, Covalence's EthicalQuote, explicitly states that "it does not see some sources as more reliable than others" and considers all sources equally because it "does not validate information sources" or the content of their information.

The remaining twenty-two cases (9 percent of the total) do make some specific claims about the reliability of their data. Six refer to reliability in the sense of trustworthiness, expertise, and independence, while a seventh uses it more in the sense of accuracy and validity. Beyond Grey Pinstripes explicitly states that its method of blind scoring by pairs of coders was "done to obtain inter-rater reliability," while three other cases implicitly describe their efforts to increase this type of reliability. The HERS Index, for example, describes its requirement for raters to undergo ongoing training "so that customers and the public can be assured of receiving competent and reliable services." SCS periodically splits its samples for its NutriClean Pesticide Free certification and sends them to multiple laboratories to test their consistency. While no case explicitly describes their test-retest reliability, three cases describe their data in the sense of consistency across tests. The Carbon Disclosure Leadership Index, for example, states that one of its underlying principles is for its underlying data collection processes to "use consistent methodologies to allow for meaningful comparisons of emissions over time."

Three cases—PEFC, Responsible Care, and UTZ Certified—describe the reliability of their certifications in the sense of harmonization across the different programs, countries, and companies that are implementing them. While similar to inter-rater reliability, this idea of harmonization suggests a broader consistency across not only raters but also the designers of rating systems themselves. Four other cases emphasize the importance of comparability in their discussion of reliability. For example, the National Fenestration Rating Council emphasizes that its certification enables builders and consumers to "reliably compare one product with another," while the Air-Conditioning, Heating, and Refrigeration Institute declares that its certification of HVAC equipment "provides consumers with a reliable apples-to-apples comparison of equipment they are considering purchasing." Two other cases mention specific mechanisms by which they ensure their information is reliable. GoodGuide states that it uses a "quality assurance and quality control (QA/QC) processes" to ensure the reliability of its data, while Citizens Market relies on its user-driven review process, which enable users to rate the quality of the data it uses to evaluate companies.[31]

### The Varieties of Validity: From Generalities to Specifics

Clearly, most cases in my dataset have not demonstrated their ability to reproduce their results across time and across raters, and none of them have fully met the standards of reliability outlined in this chapter. They also have limited replicability and transparency. But perhaps the programs are more effective at documenting their validity, the third dimension of information quality. This dimension is focused on the extent to which a metric is measuring what it is claiming to measure. Specifically, do these initiatives document the construct validity and measurement accuracy of their approaches to evaluating the sustainability of products and companies?

In order to address this question, I conducted lexical searches similar to the reliability analysis described earlier. I first searched the website text of the 245 cases for uses of the terms "valid," "validity," "validate," and "validation." In reviewing the 782 results generated by this search, I found that the ninety-nine initiatives using these terms were employing them in a wide variety of ways. Many were using them in the sense of trustworthiness and independence explored in chapter 3 (similar to some of the uses of reliability discussed previously), while others were using them to describe things not relevant to the quality of their metrics (e.g., this is a "valid objective of the company"). I therefore decided to narrow my search to only include uses of "validity," based on the assumption that those initiatives that were serious about assessing their validity should explicitly reference the term. Given that for many people, despite the important distinctions described in this chapter, validity and accuracy are close synonyms, I conducted a search of "accuracy" as well. This search is also relevant to assessing the extent to which the cases discuss their measurement accuracy and ability to detect and avoid errors and bias in their data collection processes.

The "validity" search yielded sixty-six results across twenty-six initiatives. Five of those cases only mention validity in a general sense. For example, the Cradle to Cradle certification states that "the certifying body will judge the validity and efficacy of each applicant's [material reutilization] program on a case-by-case basis," and does not explain what it means by validity. In contrast, the Power Scorecard provides a more detailed description of its construct validity, describing why and how it has included emission reduction credits in its scoring of electricity providers. Two additional initiatives define their validity by comparing themselves to other related criteria. Bluesign, for example, asserts that it has strong validity because its standard is oriented "to the most stringent and most commercially significant regulations and laws around the world."

Seven cases use validity in the sense of measurement accuracy. The Toxics Release Inventory, for example, describes the multistep process it employs to check for submission errors and "verify the validity" of the data companies have submitted. As a specific mechanism for ensuring their measurement accuracy, four cases discuss their validity in the context of audits they require. The Friend of the Sea certification, for example, states that the validity of its certificate is "dependent on the outcome of subsequent yearly surveillance activities." Another aspect of measurement accuracy is whether the evaluation's data and evaluation process have been updated recently. Four cases focus on their requirements for updates and recertification as indicators of their validity. GlobalG.A.P, for example, states that its certificates have "an initial validity of twelve months," while The Gold Standard notes that the validity of its versions is also time-delimited, and after a one month grace period companies must use the most recent and valid version.

How up to date both the underlying data and the criteria used to assess those data are is indeed an important aspect of measurement accuracy. For this reason, I also coded every reference to the generation and publication dates of both the data and criteria used by these initiatives, as well as any discussion of how they update and keep current these two critical components of their evaluations. Just over 13 percent state they have updated their criteria through explicit and systematic review processes, 10 percent have updated their criteria through ad hoc and limited review processes, and less than 1 percent have pending updates. Over 70 percent do not mention the age of their criteria, and over 75 percent do not mention any updating process for their criteria. Approximately 30 percent of the cases have updated their data through explicit and systematic processes, nearly 6 percent have updated their data through ad hoc and limited processes, and less than 1 percent claim their data is currently undergoing an updating process. Nearly 70 percent do not mention the age of their data, and over 60 percent do not mention any data updating process.

My lexical search for "accuracy" resulted in 241 hits across fifty-seven cases. The term was only used in a general sense by thirty-one cases. An example of such a reference is the Corporate Responsibility Index's claim that it reviews "all company submissions to ensure completeness, accuracy, and consistency." Twenty-six other cases used it in more specific ways. One initiative, Carma, uses accuracy in the sense of predictive validity, stating that it has verified that its statistical model predicts "actual emissions with high accuracy, using officially-reported emissions from thousands of power plants in the U.S., Canada, the European Union and India." Another case, HealthyStuff.org, refers to accuracy in the sense of test-retest validity, stating that it takes repeat samples "in order to evaluate the variation per product [and] to assess and verify the accuracy of [its] testing." Six cases refer to accuracy in the context of their peer review or auditing processes, while two others refer to it as dependent on the expertise of the evaluators. Rainforest Alliance goes on to emphasize that assessments "must involve individuals who are familiar with the particular region and type of forest" and "use region-specific standards." Figure 4.5 summarizes the results of these two sets of searches, and shows that over 70 percent of the cases do not mention either the accuracy or validity of their claims.

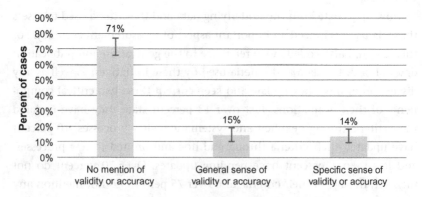

**Type of reference to validity or accuracy**

**Figure 4.5**
Levels of validity/accuracy. *Note:* Error bars indicate 95 percent confidence intervals for each sample proportion.

It is important to note that just because an initiative makes a claim about its validity or accuracy does not guarantee that it is indeed valid or accurate. While detailed assessments of the validity of all 245 cases in my dataset are beyond the scope of this chapter, I did look at two additional and more specific factors that I believe are important indicators of information quality. The first is the use of *life cycle analysis* (LCA) concepts and techniques in evaluations of products and companies. As the EPA explains:

Life cycle assessment is a "cradle-to-grave" approach for assessing industrial systems. "Cradle-to-grave" begins with the gathering of raw materials from the earth to create the product and ends at the point when all materials are returned to the earth. LCA evaluates all stages of a product's life from the perspective that they are interdependent, meaning that one operation leads to the next. LCA enables the estimation of the cumulative environmental impacts resulting from all stages in the product life cycle, often including impacts not considered in more traditional analyses (e.g., raw material extraction, material transportation, ultimate product disposal, etc.). By including the impacts throughout the product life cycle, LCA provides a comprehensive view of the environmental aspects of the product or process and a more accurate picture of the true environmental trade-offs in product and process selection. The term "life cycle" refers to the major activities in the course of the product's life span from its manufacture, use, and maintenance, to its final disposal, including the raw material acquisition required to manufacture the product.[32]

The use of LCA helps increase the comprehensiveness of sustainability evaluations and avoid the sin of the hidden trade-off, one of the key themes of chapter 2. It is also a relatively rigorous process that clearly defines processes and their boundaries, quantifies both human health and environmental impacts, and tracks uncertainties. While it has its limitations, it is currently the most sophisticated and widely accepted method available for systematically assessing the relative performance of different products and companies, and any valid environmental certification or rating system should incorporate it into its design. At a minimum, such initiatives should include criteria relating to LCA, and ideally, they should themselves be designed around the concept of the life cycle and make use of the many techniques available from the field of industrial ecology to assess the impacts associated with each phase in that cycle.

For these reasons, I also conducted a lexical search of the EEPAC Dataset for "life cycle analysis" (or "assessment") and "LCA." I found 167 relevant LCA references across twenty cases. Most of these references come from BASF's Eco-Efficiency Analysis Label, Green Format, SMaRT Certified, and the level BIFMA Sustainable Furniture Standard, all of which directly integrate life cycle analysis principles into the design of their assessment frameworks. As BASF states, "Life Cycle Inventories and Life Cycle Assessments form the basis of every Eco-Efficiency Analysis." Six other cases (CERES-ACCA Sustainability Reporting Awards, B Corporation, Pacific Sustainability Index, Climate Counts, U.S. Beverage Container Recycling Scorecard, the Sustainable Forestry Initiative, and Green Globes) include specific criteria in their assessments that relate to LCA, although it is not a primary feature of their design. For example, Green Globes Design of New Buildings certification gives points for "conducting a Life Cycle Assessment of the building assemblies and materials." Two cases mention LCA as an optional way to meet one of their criteria, while another two cases mention it in the descriptions of the companies they have assessed, implying that LCA is a characteristic they are looking for in their assessments.

The second more specific aspect of validity that I would like to highlight is the *weighting of criteria* within assessment frameworks. Any certification or rating system that assesses more than one aspect of a product or company must make decisions about the relative importance of those aspects. These decisions involve both technical and values-based considerations, and initiatives should be transparent about how they make them. This is important not only because criteria weights are an important dimension of validity and can have a significant effect on evaluation outcomes, but also because they are necessary for replicating the results of these evaluations. Initiatives that fail to discuss or at least document their weightings have significant gaps in both validity and replicability. Following this logic, I therefore conducted a lexical search for all references to "weights" or "weighting(s)," and found eighty-eight references across thirteen initiatives. Two of these cases only provide a brief mention of weightings or list the criteria weights, while the rest provide at least some description of the logic behind the prioritization process. For example, the SkinDeep Cosmetics Database explains that higher weighting factors were assigned to "categories of health concern for which studies provide evidence for effects at low doses, for permanent effects stemming from exposures during development, [and] for toxicity endpoints that tend to impact multiple biological systems in the body or to impair reproduction."

An additional theme found throughout both the validity and accuracy references are caveats and disclaimers about the information being provided. One interesting example comes from the Leonardo Academy's Cleaner and Greener certification. It highlights a trade-off between accuracy and costs, and in the case of incorporating imports and exports into its building emission footprint calculation, states that that the "possible increased accuracy to the emission factors does not justify the additional workload necessary." Eighteen other initiatives make more straightforward disclaimers (often on their Terms and Conditions pages) about the accuracy or validity of the information they are providing. For example, SCS, in its description of its Indoor Air Quality Certification Program, states that it "does not make any warranty (express or implied) or assume any liability or responsibility to the user, reader or other third party, for the accuracy, completeness or use of, or reliance on, any information contained within this program."

## The Information Quality Landscape

The data presented above on the replicability, reliability, and validity of the 245 cases in the EEPAC Dataset can be brought together to provide a snapshot of the information quality landscape associated with information-based environmental governance strategies. Figure 4.6 presents such a snapshot. The diagonal axis measures the cases' level of transparency, combining data from figures 4.2 and 4.3. If a case has at least limited levels of all four types of transparency (criteria, source, data, and method transparency) discussed earlier, it would receive a score of 4 and be in the bottom row of cases in figure 4.6. If it is not transparent on any of these dimensions, then it would receive a score of 0 and be in the top row. The horizontal axis, on the other hand, combines the data from figures 4.4 and 4.5 on the reliability and validity of the cases. The first, left-most column includes cases that have no references to either their reliability or validity, while the fourth, right-most column includes cases that discuss both their reliability and validity. The second column from the left includes cases that provide some mention of their validity but none of their reliability, while the third column includes cases that provide some discussion of their reliability but none of their reliability.

Figure 4.6 shows that the extent of transparency and discussions of information quality vary widely across this landscape. The seven cases in the upper left corner—the World's Most Sustainable and Ethical Oil Companies, the Big Green Purse, EcoChoices, EcoMall, Green Culture, Green

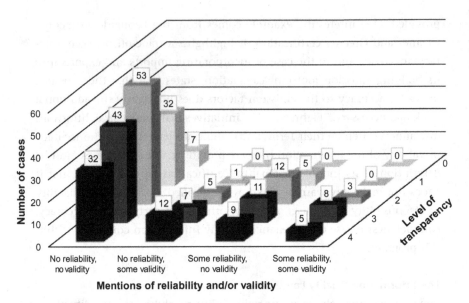

**Figure 4.6**

The information quality landscape. *Note:* Each column represents the number of cases that fall into the specified category. Thus, the seven cases in the upper right corner neither discuss their validity or reliability nor are transparent about any of the four methodological factors coded for—their criteria, sources, data, or methods. Likewise, the five cases found in the lower left corner discuss some aspect of both their reliability and validity and have at least limited transparency across all four of these methodological factors. The "level of transparency" axis is a count of these four types of transparency found in each case.

Options, and EnergyGuide—neither discuss their validity or reliability nor are transparent about any of the methodological factors that I coded for. On the other hand, the five cases found in the lower right corner—the GreenSpec Directory, the College Sustainability Report Card, the Sustainable Forestry Initiative, GoodGuide, and Seafood Watch—do describe their validity and reliability and are at least somewhat transparent about their criteria, methods, sources, and data.

Overall, figure 4.6 shows that the majority of initiatives (nearly 70 percent) do not directly discuss either their reliability or validity, and have only limited transparency into their methods, criteria, data, and sources. While these concepts are inherently difficult to measure and this graphic is a simplification of the data presented earlier in the chapter, it nevertheless reveals that these programs generally lack a sophisticated approach to information quality assurance. Even the cases that do mention their validity or reliability usually do so in limited and unspecific ways.

## The Information Realism Perspective

For commentators who are skeptical of information-based governance strategies, this landscape of information quality will likely reinforce their concerns expressed in earlier chapters. Not only are the majority of these initiatives overly narrow, lacking in independence, unsupported by appropriate expertise, and opaque about their accountability relationships, they also lack fundamental quality assurance and control mechanisms. They generally are not replicable—60 percent of the cases do not reveal the sources of their information and 75 percent do not provide any information whatsoever about their assessment methods. If we do not know where the initiatives are getting their data from or how they are conducting their analyses, why should we trust their results? Likewise, 78 percent do not discuss their reliability, and 83 percent do not discuss their validity. Over three quarters of these cases therefore are not providing any assurance that their metrics of sustainability performance are measuring what they claim to be measuring and that they generate similar results each time they are used to measure something.

More specifically, the vast majority of these cases do not mention the age of their data and criteria or any process for updating them. Only 8 percent incorporate principles or methods based on the most sophisticated method for evaluating the environmental impacts of products and companies—life cycle analysis—into their criteria. Only 5 percent describe how they weight the different criteria in their evaluations. Information pessimists might reasonably suggest that such lack of methodological rigor is summed up by eighteen initiatives' telling statements disclaiming any responsibility for the accuracy, reliability, or validity of the information they provide. If these initiatives will not stand behind their results, why should the rest of us pay any attention to them? How do we know what they are pedaling is not misinformation at best and disinformation at worst?

Observers who are more optimistic about the potential of information-based governance strategies will likely respond very differently to the data presented in this chapter. They might view the transparency data as generally encouraging, given that 78 percent of the cases provide at least limited information about their criteria, and 59 percent make at least some of their underlying data available to the public. The fact that one-third of the initiatives provide detailed and complete descriptions of their methods demonstrates that a large proportion of these ratings and certifications are likely to be replicable.

These information optimists might reasonably question whether lexical searches are themselves valid metrics of validity and reliability, given that they might miss cases that do not use the particular keywords searched for. They might dismiss the disclaimers about accuracy as legalese required by lawyers afraid of costly lawsuits that do not reflect the confidence of the designers in their evaluations. They might point to the limitations of life cycle analysis as a method of assessment—LCAs are expensive, require data that are not always available, and are not appropriate for every issue and topic of analysis. They might complain that the focus on different forms of reliability and validity reflects a disciplinary bias toward quantitative and statistical analyses, and privileges organizations with the resources and backgrounds that can conduct those analyses.

For the information realist, there is much to both agree with and contest on both sides of this debate. To the information optimist's argument about the quality of the data presented in this chapter, validity, reliability, and replicability are inherently complex and multidimensional concepts that are indeed difficult to measure, and any measure of them will likely have a large degree of error. However, it is important to realize that such uncertainty means that the estimates provided in this chapter may be underestimating OR overestimating how valid, reliable, and replicable these cases are. On the one hand, more cases may be transparent than suggested in this analysis, but on the other hand, those cases coded as being transparent may not actually be that replicable. Likewise, I did not assess the quality of the reliability or validity claims that the initiatives make on their websites, only whether they make them at all. So more cases may discuss these aspects of their methods without explicitly using the search terms, but alternatively, fewer cases may demonstrate high levels of actual reliability or validity. Given that there is no clear reason to believe that the results are skewed in one way or another, it is reasonable to assume that they are relatively accurate estimates of these characteristics.

As with the subjects of the earlier chapters, these estimates demonstrate to the information realist both the challenges and opportunities facing information-based governance strategies. On the one hand, their generally low levels of information quality and methodological sophistication raise important concerns about their usefulness to the public. On the other hand, some initiatives have demonstrated that it is possible for them to at least discuss the quality of the information they are providing with some level of sophistication. They show that it is realistic for these programs to engage questions relating to their replicability, reliability, and validity, and begin making use of more sophisticated forms of analysis and quality control. Regarding the complaint that such an emphasis reflects a quantitative or statistical bias, the quantitative bias is already embedded in many of these rating and certification programs, as they often involve numerical data in their assessments. Until critics or designers propose some other means to assure the public of the methodological rigor of these programs, it makes sense to use the concepts and standards of quality that are most commonly utilized to assess this kind of data.

As for the contention that data may not be available for certain types of life cycle analyses and that such analyses may not always be appropriate, this may indeed be the case. But given that LCA is the most accepted method for assessing sustainability performance, the onus is on evaluation organizations to explain why they are not using it. Just as engineers should have to justify why they are not using the most accurate and valid methods for building a bridge, so must the designers of these programs justify why they are not using the most accurate and valid methods for evaluating products and companies' environmental impacts. Once they do so, then their audiences can determine if these are valid reasons for using an alternative evaluation method.

The argument that LCA (and other methods that increase information quality) is too expensive raises the important trade-off of cost and accuracy, which was also mentioned by the Cleaner and Greener certification. While no guarantee, the availability of both time and money can enable the use of more replicable, reliable, and valid evaluation methods. As the old adage says, "you get what you pay for." This places some of the responsibility on the public and particular stakeholder groups to demand and be willing to invest in—either as policymakers, consumers, executives, philanthropists, or taxpayers—high-quality information. Just as we demand high-quality bridges, we should demand high-quality sustainability information. But the evaluation organizations also need to offer and market such information to these groups, which this chapter clearly shows that the vast majority have not done.

Nevertheless, this result does not necessarily warrant giving up entirely on these initiatives, as the information pessimists might suggest. The cases that do discuss and defend the quality of their information and make the effort to update their data and criteria, document their weightings, and incorporate LCA into their processes can and should be recognized and rewarded for doing so. They provide guideposts for other initiatives that are committed to increasing the replicability, reliability, and validity of their evaluations, and a stark contrast to those that refuse to address their methodological deficiencies. The next section highlights further sources of inspiration and promising practices from both the academic literature and other sectors of society that have struggled with these same challenges.

In critiquing the quality of the information being produced by these information-based governance strategies, it is important to remember that these issues plague traditional governance approaches as well. Replicable, reliable, and valid measures of the effectiveness of environmental regulations are also few and far between, for example. Even the academic community, which we would assume should be the paragon of methodological virtue, has struggled with ensuring the quality of the data and conclusions that it produces. The information realist acknowledges these limitations of existing data across all of these domains, but rather than giving up on them recognizes that some of these data are nevertheless better than others. If we believe their underlying enterprise to be important and have value (conclusions that are discussed further in both chapter 2 and chapter 6), then the key is to incentivize them to improve by rewarding those that are using the best available methodological practices—and penalizing those that are not.

## Promising and Problematic Practices

This chapter has suggested a number of important practices that evaluation organizations can follow to demonstrate to their audiences that they are providing high-quality information. These are also practices that people like Lynn can be looking for when they are evaluating competing certifications and ratings. Lacking such practices constitutes a problematic practice itself, and should raise concerns for Lynn as she tries to identify the most valid, reliable, and replicable program for her institution. This section summarizes those practices and provides additional examples from both the academic literature and other domains.

Before diving into the specific areas of replicability, reliability, and validity, I have three general recommendations for certification and rating designers. The first relates to standards. Just as these organizations have been encouraging companies to report their corporate social responsibility performance in standardized formats and topics (through common sets of topics and questions such as the Global Reporting Initiative framework), they too should be reporting their own performance in a similar manner. Key components of that performance are measures of their metrics' replicability, reliability, and validity, which are fundamental to any research report. As William Trochim explains, such reports should "include a brief description of your constructs and all measures that will be used to operationalize them. ... For all [measures], you should briefly state how you will determine reliability and validity. ... For reliability, you must describe the methods you used and report results. A brief discussion of how you have addressed construct validity is essential." Specific measures of information quality, such as intra-class correlation and Kappa coefficients, should be reported whenever possible.

The second general suggestion relates to language and the use of the terms discussed in this chapter. Given that they are used in different ways by researchers and the public, it is important to clearly signal when you are discussing the reliability and validity of your data in statistical terms, and not just in the general sense of being trustworthy and credible. "Reliability," for example, is often used in two distinct ways—one is in the statistical sense of consistency described earlier and one in the more colloquial sense of trustworthiness and credibility described in chapter 3.[33] The general use of these terms is technically correct and can be found on many initiative websites as well as in the academic literature. For example, in their analysis of the "reliability" of product eco-labels as instruments of ensuring agricultural biodiversity, VU Utrecht University researcher Mariëtte Van Amstel and her colleagues focus on the capacity and independence of the evaluation organization and not specifically on the consistency of their results.[34] Ideally, however, we should reserve these terms for their use in the statistical sense, and use other words—like "trustworthy" and "credible"—for other uses. While such a distinction may seem tedious, it is a valuable signal to stakeholders and the public that a program has thought systematically about the quality of its information, which can serve to differentiate it from its competitors.

A third general recommendation relates to who is conducting all of these analyses. As discussed in chapter 3, expertise is a key signal of credibility, and it is reasonable to expect that someone who has experience and training in statistical analyses be the person behind the numbers being used in these sustainability evaluations. Evaluation organizations should therefore report not only how they are assuring their information's quality but also who is doing that quality assurance. Given the complexity of these analyses, ideally such quality assurance is a multistep process involving multiple experts, some who are actually crunching the numbers and others who are reviewing them for errors. Such a process might involve both inside and outside reviewers; regardless, it should be clearly articulated to the public.

### Replicability Practices

Specifically regarding the replicability of their results, environmental certifications and ratings should clearly disclose their data, data sources, evaluation criteria, and the methods by which they use these criteria to analyze their data and generate their information. Michael Sadowski and his colleagues at the consulting firm SustainAbility assert that this process of opening up their methodological "black boxes" builds trust and can increase (rather than undermine) the acceptance and use of sustainability ratings.[35] An excellent example of such transparency comes from the world of academic publishing, which is increasingly requiring researchers to publicly disclose their data and methods when their articles are published. Nature journals, for example, require that their authors "make materials, data, code, and associated protocols promptly available to readers without undue qualifications," preferably through public repositories.[36] The *American Journal of Political Science* has a similar requirement, but has also contracted with the University of North Carolina's Odum Institute for Research in Social Science to verify that the submitted replication materials do indeed produce the reported results.[37]

These developments have aptly been described as the "new scientific revolution,"[38] and are finding a large number of false positives in the literature.[39] But as David Boockman and Joshua Kalla from UC Berkeley Political Science Department argue, the discovery of these errors "does not suggest scientists are especially prone to making mistakes. Rather, it shows that scientific errors are increasingly likely to be detected and corrected instead of being swept under the rug."[40] Similar protocols can and should be established in the realm of information-based environmental governance strategies. Initiatives could use a standard checklist or form to disclose how they have verified the replicability, reliability, and validity of their claims. Policymakers or other stakeholders could develop information quality indices that take into account the different forms of transparency, reliability, and validity discussed in this chapter. Ideally, they would take into account the results of efforts to replicate the results of these initiatives. These "metrics of metrics" would assist the public in evaluating the quality of the information these programs provide.

### Reliability Practices

Reliability is an equally important attribute of environmental ratings and certifications. These programs should follow the same principles recommended by the Global Reporting Initiative (GRI) for corporate sustainability reports, which include comparability, accuracy, timeliness, clarity, and reliability. Specifically regarding reliability, the GRI specifies that organizations "should gather, record, compile, analyze and disclose information and processes used in the preparation of a report in a way that they can be subject to examination and that establishes the quality and materiality of the information."[41] Their methodological disclosures should have both general summaries for the public as well as more technical information for experts.

Organizations should also show how they have confirmed that their information is indeed reliable. Eugene Szwajkowski from the University of Illinois Chicago and Raymond Figlewicz from the University of Michigan Dearborn provide a good example of a check of test-retest reliability in their study using the SOCRATES database developed by the socially responsible investing firm KLD Research & Analytics (KLD), and find that it does have an acceptable level of this form of reliability.[42] Judith Walls, a professor at Nanyang Business School, led a study that provides a similar check of the inter-rater reliability of a new metric of environmental strategy by having a second researcher code a random selection of reports and calculating the related Cronbach's alpha.[43] The goal of these tests of reliability are to quantify the level of precision—and conversely, the uncertainty—of the information that these initiatives are providing.

Along these lines, designers should also keep track of the sources of uncertainty in their analyses, and communicate their estimates of uncertainty and levels of confidence in their final evaluations. An excellent example of such a practice is the National Research Council's assessment of research doctoral programs in the United States. Acknowledging the high levels of uncertainty in such an endeavor, the assessment includes two parallel rankings, one based on faculty members' stated preferences for different characteristics of high-quality programs and one based on faculty members' revealed preferences for such characteristics by their ranking of a sample of programs. Both rankings are presented as ranges of values—as opposed to specific ranks—that represent the middle 90 percent of a large number of ratings, and thus take into account the variability in the underlying data.[44] Thus instead of being ranked #3 or #11 in the country, Princeton's Economics Department rank is estimated as being between #4 and #11 or between #3 or #5, depending on which metric is used.[45] Underlying this approach is an appreciation that the weights placed on different criteria can create major changes in rating results. In their analysis of three different methods to rank NFL teams, a team of researchers led by Tim Chartier, a professor of mathematics at Davidson College, demonstrated that this effect can occur even in rankings based on well-established linear algebra models.[46] Thus it is critical to assess the sensitivity and uncertainty of rating systems based on weightings that are often much less systematically derived.

As the research in this chapter shows, none of the cases employ all of these promising practices, and thus all of their data has limited reliability. Nevertheless, it is important to reiterate that this is a problem that plagues many other domains as well. As AccountAbility staff members Nicole Dando and Tracey Swift explain, financial auditors, even after decades of experience, still remain "unable to guarantee the robustness and reliability of financial accounting and reporting and to impart public confidence."[47] Social, ethical, and environmental accounting is in its infancy and it is not surprising that it is facing similar challenges.

**Validity Practices**

These challenges extend to the area of initiatives and organizations ensuring the validity of their data. However, numerous opportunities exist for them to improve their performance in this area as well. One way for these initiatives to increase their validity is to not make unqualified general claims or exaggerate what they are measuring. As the Federal Trade Commission (FTC) states in its Green Guides, "marketers should not make unqualified general environmental benefit claims ... [because they] are difficult to interpret and ... likely convey that the product, package, or service has specific and far-reaching environmental benefits" that marketers will not be able to substantiate. Such unqualified general claims are easy for consumers, competitors, and the FTC to identify as lacking validity. Claims that are based on methods and criteria that are out of date also can be quickly dismissed as having limited validity.

It is also important for evaluation organizations to be as comprehensive in their assessments as possible. Following the logic of the values discussion in chapter 2 and the principles of life cycle analysis discussed in this chapter, ratings and labels that are unduly narrow risk obscuring trade-offs among performance criteria. In a study of higher-education sustainability ratings, Nick Wilder and I show that these initiatives have significant gaps in coverage of key environmental and social issues.[48] Nevertheless, two cases—STARS and the Pacific Sustainability Index—cover a broader range of criteria than the others and have greater validity as a result. An example from academia further demonstrates the problem of narrowly-defined rating systems. Joel Baum, a management professor at the University of Toronto, shows that the design of the Impact Factor rating of journal article citations enables both articles and journals to "free-ride" on a few highly cited articles, and concludes that the metric has "little credibility" as a measure of publication quality, despite its widespread use. The use of food miles as a metric of food sustainability is another example—a report for the UK's Department of Environment, Food, and Rural Affairs concluded that "a single indicator based on total food kilometres is an inadequate indicator of sustainability [because] the impacts of food transport are complex, and involve many trade-offs between different factors."[49] In order to have high validity, product and company sustainability assessments must take a holistic approach to evaluate all stages in the life cycle, from their raw materials to their end-of-life. And following the preceding discussion, those stages and their associated criteria need to be weighted appropriately.

The validity of these metrics not only depends on their construction but also on how they are measured. First of all, they must be based on relatively *up-to-date data sources*. Second, they should avoid possible *threshold effects* in the data collection process. Lori Bennear at Duke University shows that the exemptions from reporting requirements for companies that use a chemical below a certain threshold level can account for up to 40 percent of observed declines in toxic release emissions. Furthermore, rankings of the highest- and lowest-emitting facilities can change significantly due to this threshold effect.[50] Third, they should consider the *accuracy of user-generated data,* which is a key element of many of the cases in my dataset. Beyond questions about whether these users have the relevant expertise to evaluate the sustainability of products and companies, Tim Chartier and his colleagues reveal another limitation of ratings based on such data.[51] Ratings based on a small number of user contributions can be misleading and unstable, as each user can have a large impact on the average score. Programs that make use of such user-generated data should explore linear algebra approaches (such as the Colley method) that compensate for this effect.[52]

Other studies demonstrate how the validity of sustainability ratings and certifications can be assessed by comparing them to other related metrics. For example, Noushi Rahman at Pace University and Corinne Post at Lehigh University compare KLD's environmental ratings to their own proposed measure of environmental corporate social responsibility to confirm the validity of their new metric.[53] In earlier work, Aaron Chatterji and his colleagues test both the retrospective and predictive validity of KLD's environmental ratings by comparing them to corporate data on environmental fines and toxic releases.[54] They find that companies with high environmental concern scores generally had poor environmental performance in the past and end up producing more pollution and having more compliance violations in the future. However, no significant relationship exists between KLD's environmental strength score and future environmental performance data.

As I discuss in chapter 6, showing that information-based environmental strategies have a positive effect on such environmental outcomes is one of the key measures of their effectiveness. When trying to show such an effect, it is important to establish that the relationship is causal and not just a random correlation. In an innovative study using propensity score matching to demonstrate such causality, Resources for the Future researchers Allen Blackman and Maria Naranjo test the validity of organic coffee certifications by comparing the environmental impacts of Costa Rican farms that received organic certification with similar farms that did not.[55] Given that the only characteristic that systematically differs between the two sets of farms is their organic certification, they find that the certification does indeed improve the environmental performance of coffee production, both in terms of reducing chemical inputs and increasing use of lower-impact management practices.

Programs can also use less-statistical mechanisms to enhance both the validity and reliability of their information, such as peer reviews and audits. They can encourage these processes to incorporate local expertise and have some level of regional specificity to them, as Rainforest Alliance does. However, this point raises an important trade-off for these metrics. The more locally—and regionally—oriented a certification or rating is, the more it risks losing its global comparability and reliability. Other trade-offs exist as well—between cost and quality, between depth and breadth, between validity and reliability, among the mix of input, output, and outcome metrics, between the need to update criteria and the importance of maintaining consistent metrics over time, and between the value of transparency and the risk of companies attempting to game the system. The validity-reliability trade-off is particularly important—a battery of questions that a company can easily answer may be highly reliable, but the questions may not get at the more complex but critical aspects of sustainability.[56]

These trade-offs are admittedly challenging to manage. The first step is to identify and be honest about them, both internally and externally. The next step is evaluate them in the context of the evaluation taking place—what does the current landscape of similar ratings look like? Is a more replicable, valid, or reliable metric necessary? Can a strong case for the need for higher-quality information be made? Who are the key stakeholders, and what are they demanding? What are they willing to pay for? The final step is to decide how to manage those trade-offs, and develop the highest-quality metrics that the particular context calls for and allows. While CoValence's approach of treating all of its data sources as equally reliable counts as one of the most problematic practices in terms of its basic validity, CoValence at least is being open about how it has managed this particular reliability/validity trade-off. For CoValence, the validity benefits of evaluating the reliability of its sources do not warrant the associated costs and reliability concerns with such source reliability assessments. It is then up to the public and stakeholder groups to evaluate how these ratings have managed these trade-offs, and provide grants of legitimacy—as discussed in chapter 3—to those they evaluate as providing the level of information quality they demand.

## The Quality of Green Building Information

So how does this help Lynn in her evaluation of green building certifications? Armed with the framework presented in this chapter, she can use it to compare the different programs. First, she can assess their replicability by analyzing their transparency. All four of the main green certification programs for commercial buildings in the United States—LEED, Green Globes, Living Building Challenge, and ENERGY STAR—are relatively transparent about their criteria, methods, and data sources. However, the transparency of their data is more limited and varied. None of the programs publicly reveal the underlying data upon which the certifications of individual buildings is based. LEED provides a scorecard of all of the individual points that each of its certified buildings has earned, while ENERGY STAR publicizes the 1–100 ENERGY STAR score that each of its certified buildings has received.[57] Green Globes only provides the date of certification and the number of globes earned by each of its certified buildings, while the Living Building Challenge only provides its buildings' certification status and type (NetZero Energy, Petal, Full Living, and Living Community Challenge).[58]

The replicability of all of these programs is therefore limited, although they have made a commendable effort to publicize their methods and criteria. Their reliability and validity, however, are much more uncertain. The Living Building Challenge employs a technical director who is responsible for overseeing "certification and technical consistency,"[59] and LEED has Technical Advisory Groups that review the consistency and technical rigor of credits and prerequisites.[60] Reliability and validity are mentioned in an ad hoc fashion in many of LEED's point interpretation sections. ENERGY STAR emphasizes the importance of consistency in energy benchmarking exercises and its requirement that all types and amounts of energy used by the property must be documented for twelve consecutive months.[61] Green Globes asserts that it "sets the standard for accuracy, consistency, and credibility," and claims its lack of prerequisites, recognition of nonapplicable criteria, and incorporation of partial credit "results in the highest possible accuracy of the final Green Globes score and rating."[62] However, the organization provides no evidence on its website to support these claims; indeed, none of these initiatives provide any evidence of the overall validity or reliability of their certifications.

The four programs differ significantly in their practices regarding data updates. Green Globes has no requirements for recertification, while only one of the LEED certifications (for Operations and Maintenance) requires buildings be recertified every five years.[63] The Living Building Challenge conducts its final audit at least twelve months after construction has been completed, but has no further updating requirements.[64] ENERGY STAR requires annual recertification, although the logo displayed on certified buildings does not indicate the years certified.[65] In contrast to the failure of these programs to provide up-to-date data about the buildings they have certified, all of the programs have updated their evaluation criteria relatively recently—LEED 4.0 in 2013, ENERGY STAR for New Homes in 2010, Living Building Challenge 3.0 in 2014, and Green Globes New Construction v.2 in 2013.[66]

### Insights from Academic, Government, and Other Sources

Regardless of these recent updates, however, these programs are not providing much, if any, persuasive evidence that they are providing high-quality information about the sustainability of the buildings they are certifying. While recognizing the updating requirement of ENERGY STAR and the data transparency of both ENERGY STAR and LEED, Lynn turns to other sources that have attempted to evaluate and compare the validity of the different programs as metrics of building sustainability. Several online articles compare Green Globes and LEED, and conclude that the former is more user friendly, while the latter is more stringent.[67] BuildingGreen's Tristan Roberts and Paula Melton find that LEED is stronger on site sustainability, energy performance and renewables, water efficiency, materials and resources, ventilation, daylighting, product emissions, regional bonus points, and rewards for innovation.[68] In a review of all four programs, Sustainable Performance Solutions' Lawrence Clark concludes that the Living

Building Challenge is "the built environment's most rigorous performance standard," and concurs with others that LEED and Green Globes are about doing less harm while the LBC is about doing good.[69]

A few scholars have assessed competing green building certifications with regard to their coverage of a particular environmental performance area, such as indoor air quality or energy efficiency. For example, Wenjuan Wei and her colleagues at the University of Paris examine the extent to which different aspects of indoor air quality (IAQ) are measured by thirty-one different green building certifications, and find that IAQ metrics make up between 8.2 percent and 9.1 percent of LEED's rating system.[70] Other researchers have focused on establishing the validity of LEED in particular. Oberlin College's John Scofield found that a sample of LEED-certified buildings in New York City had similar greenhouse gas emissions and energy costs to comparable noncertified buildings.[71] More specifically, gold-certified buildings had lower emissions and costs while silver-certified and certified buildings had higher emissions than comparable buildings. UC Santa Barbara's Sangwon Suh and his colleagues found that the difference in life-cycle environmental impacts of LEED-certified vs. noncertified buildings can range from 0 to 25 percent, depending on which LEED points are earned.[72] They estimate that the most significant impact reductions are in the categories of acidification and human respiratory health, and the occupancy phase of buildings accounts for the vast majority of the environmental impacts in ten of the twelve impact categories.

While these energy and life cycle studies are revealing, they do not compare the different certifications and thus have limited value in evaluating their relative validity. The data on energy use by LEED-certified buildings are concerning, but I could find no similar studies for Green Globes and so that certification program could face a similar challenge to its validity. This is less of an issue for the Living Building Challenge, which requires net-positive renewable energy use and thus should by definition have improved energy performance, and ENERGY STAR, which requires certified buildings to demonstrate a reduced amount of energy use. However, the lack of environmental performance tracking over time by LEED, LBC, and Green Globes also reduces their validity as metrics of environmental impact.

In its regular review of green building certifications, the U.S. General Services Administration (GSA) systematically compares the criteria and methods of Green Globes, LEED, and the Living Building Challenge against the minimum sustainability requirements for federal buildings. The Energy Independence and Security Act of 2007 (EISA) requires that the GSA identify a green building certification system that it "deems to be most likely to encourage a comprehensive and environmentally sound approach to certification of green buildings."[73] The GSA's 2012 report finds that none of the three systems cover all of the federal requirements: Green Globes aligns with the most requirements at some level (twenty-five out of twenty-seven) while the Living Building Challenge meets the most outright (twelve out of twenty-seven).[74] In 2013, the GSA recommended that agencies use either LEED or Green Globes, but not the LBC.[75]

The GSA's technical advisory group, while recognizing LBC's laudable focus on sustainability performance as opposed to technical standards, had concluded that it "did not align well with Federal requirements [because] it does not specify how to meet its performance requirements."[76] This decision not to recommend the Living Building Challenge reflects the command-and-control nature of the federal standards. If a building demonstrably achieves net-zero energy use, for example, who cares if it did so following the federal government's precise requirements? Granted that LBC does not cover some substantive federal criteria (as LEED and Green Globes also do not), but those it does address are covered more deeply than any of the other systems.

The federal evaluation criteria also focus on whether the certification was developed through an inclusive, moderated, and consensus-based process.[77] The Living Building Challenge was developed by the International Living Building Institute to define "the most advanced measure of sustainability in the built environment possible" and it did not systematically include all stakeholders in this process.[78] As discussed in chapter 3, this focus on inclusivity may not be important to all audiences (and may actually undermine the development of cutting-edge initiatives), and it is inappropriate for the federal government to use it to evaluate validity and exclude a particular certification system.

While the federal guidelines are designed to identify comprehensive certifications, they do not focus on life cycle analysis as a means to encourage such comprehensiveness. Green Globes, LEED, and LBC in this context are more valid than the federal guidelines, as they each incorporate LCA into their design and criteria. While further progress can be made, these efforts represent the most innovative practices in this area for ratings and certifications more generally. LEED grants three points for building projects that "conduct a life cycle assessment of the project's structure and enclosure that demonstrates a minimum of 10 percent reduction, compared with a baseline building, in at least three of the six impact categories listed ... , one of which must be global warming potential."[79] Green Globes grants 33 points for building projects that use an approved LCA impact estimation method "to evaluate a minimum of two different core and shell designs" that results "in selection of the building core and shell with the least anticipated environmental impact."[80]

LEED also grants up to six points and Green Globes grants up to twenty points to projects that use products with environmental product declarations that incorporate life cycle assessments. Furthermore, LEED used life cycle analysis in determining the weights of the different criteria and the allocation of the 100 points in the system.[81] Knowing this, Lynn may be less concerned about architects going for the easiest points and LEED's credits for building components such as bike racks and alternative vehicle parking spaces. If points are weighted appropriately, then there is nothing wrong with architects choosing the most appropriate credits for their situation (just as football coaches must choose between field goals and touchdowns). It is unclear from its website how Green Globes determined its criteria weights, while the Living Building Challenge is an "all or nothing" approach that does not use weighted criteria.[82] Two comparative analyses show that Green Globes places more weight on energy-related criteria, while LEED has a more even distribution of weights across different impact categories. Nevertheless, both studies conclude that their category weights are generally similar.[83]

### Connecting Information Quality Considerations with Institutional Interests

Thus from this review of the literature, Lynn decides that all four programs leave something to be desired in terms of their information quality. None of them provide any evidence of their reliability or validity. However, they meet at least a basic standard of transparency, and compared to certifications and ratings in other domains, the quality of their documentation and explanation of their methods is relatively high. Given this relative parity in information quality, what should Lynn recommend? Should her university fall back on cost as the determining factor, as some of her colleagues are suggesting? Cost is indeed an important factor, and as mentioned earlier, may be inversely related to the quality and validity of certifications. The relative costs of these certifications are complex and difficult to compare, as they involve the direct application costs as well as the indirect costs associated with consultants and necessary architectural changes.

In an effort to take this complexity into account, Jeffrey Beard concludes that Green Globes is significantly faster and less expensive than LEED.[84] Similarly, a study of two comparable dormitories built on the University of North Carolina, Charlotte campus—one certified by LEED and the other by Green Globes—reveals that LEED's direct costs were $640 less than those for Green Globes.[85] However, the architecture and engineering service costs required for LEED were four times as much as those for Green Globes, making the total cost of LEED certification nearly $42,360 more expensive.

It is important to distinguish between the potential causes of this price difference. It may be due to differences in validity, reliability, or operational efficiency, but it may also be due to differences in the rigor and difficulty of the different standards. Two certifications may be equally valid, but one may have a higher standard than the other. This distinction relates to the relationship between standards and effectiveness. A label that has a relatively low standard that certifies 50 percent of a market may create more environmental benefits than one that has a relatively high standard that certifies only 10 percent of a market. This relationship is a complex one that we will return to in chapter 6, but for now the point is to distinguish between the quality of a certification's metrics and the level of performance required on those metrics. This distinction points to a broader point about the potential value of having ratings that recognize different levels of performance, and this is the logic underlying the multi-tiered approaches of LEED and Green Globes (i.e., silver to platinum, 1–4 Globes). It is also a justification for the existence of multiple and competing certification systems.

So what certification Lynn recommends will depend as much on her evaluation of their quality as her sense of the importance of her university reaching a high level of sustainability performance. This relates back to chapter 2 and to what extent these certification programs relate to and activate Lynn's values and the values of her institution. It will also depend on its available resources; budget constraints may limit its choices even if a strong commitment to sustainability exists. In this sense, Lynn's answer may be a series of questions directed back to her colleagues and the donors funding the new building's construction about these different factors.

Given their answers to these questions, if Lynn's university has a broad-based and deep commitment to environmental performance and confidence they can raise the money to pay for the extra up-front costs (some if not all of those costs will be recovered over time), she might recommend they go with the Living Building Challenge. If her university highly values methodological rigor and clarity, financial savings, and greenhouse gas and air pollution reductions but has more limited resources and a less holistic commitment to sustainability, then she might suggest ENERGY STAR. If it has an intermediate level of commitment to holistic environmental performance and a moderate level of available resources, LEED may be the option she should propose. However, if its resources are more limited and its commitment to sustainability is growing beyond energy concerns but is still nascent, then she might recommend Green Globes, particularly if the alternative is no certification at all. Some of her colleagues have proposed that they avoid the costs of certification altogether and do their own sustainability branding for the building. Equipped with this chapter's framework, Lynn should remind them that they will need to establish the replicability, reliability, and validity of these self-reported claims of sustainability, just as these different certifications should be doing.

On that note, the bottom line for these building certification programs—and information-based governance strategies in other domains as well—is that they need to establish the level of information quality they are providing. The challenges they face are embodied in the only general statement about reliability and accuracy found on the LEED website, which similar to the other cases cited earlier, is a disclaimer of responsibility.[86] These initiatives need to move beyond such legalese and take responsibility for the quality of the information they are providing. These four cases are doing relatively well in terms of their transparency, but can improve significantly in other areas. Their audiences need to recognize the progress they have made, which has likely been motivated by their competition with each other, but also push them to continue to make improvements. Quality is unlike the "fairness" that the Queen and her Magic Mirror are so focused on, as it can be improved with intentionality and persistence. At whatever level of performance they focus on, these programs can improve and better document the accuracy and consistency of their claims. They should conduct, cite, or fund studies that demonstrate their validity, reliability, and replicability, and use their results in their marketing and outreach as a point of competitive advantage.

So after examining their options closely and thoughtfully deliberating about their priorities, Lynn and her colleagues decided that as representatives of an institution that prides itself on being on the cutting edge of innovation and performance, they wanted to lead by example and design the new building be certified by the Green Building Challenge. The value of having a building that is certified as having "net positive" energy and water use and is actually environmentally regenerative was particularly compelling for them. They were also attracted to the strong validity of LBC's life cycle focus and its requirement that the building's environmental performance be measured twelve months after construction. While they acknowledged the strengths of the other programs, they felt that LBC was the most holistic and comprehensive of the competing options and best matched their organization's values and sustainability aspirations. A certification from the LBC, in their eyes, represents the most valid and highest-quality information an institution can provide to the public about the environmental performance of its buildings.

With all of this focus on quality, it is important not to miss another key insight from the world of building certifications. One of the key advantages that commentators have emphasized about Green Globes is its greater ease of use. A certification can be the most trustworthy, salient, and valid program in the world, but if it is too difficult for anyone to figure out and appropriately utilize, it will inevitably fail. LEED has been responding to its critics in this regard, and its LEED Dynamic Plaque is one of the more exciting and engaging forms of certification to be developed in recent years.[87] Designers of information-based governance strategies must make the information they are producing and disseminating attractive, intelligible, and usable for their different audiences, from consumers to architects to policymakers. At the end of the day and even after Lynn's in-depth analysis, the certification she and her colleagues selected was likely the one they most understood and were the most excited about using. For this reason, the usability of the information provided by these initiatives is the subject of the next chapter.

# 5 Delivering Green: The Communication of the Information

## Decisions about Directions, Choices about Messages

Anu was recently hired as the chief marketing officer for a major consumer products company. One of the many items on the task list given to her by the firm's CEO is to improve perceptions of the company's corporate social responsibility, particularly in the area of climate change. In recent months, the company has received the lowest grade among its peers from the rating organization Climate Counts, and it has been labeled a "Climate Laggard" by nonprofit organizations Greenpeace and Ceres.[1] The firm is increasingly getting hammered by negative comments on social media, and the CEO is beginning to worry that this negative spotlight may be having an effect on sales. In fact, a consumer survey commissioned by Anu's predecessor revealed that consumer loyalty to several of the company's most important brands has shown a marked decline over the past year.

Anu's highly talented staff has no shortage of ideas to reverse this trend. Some want to double down on the successful green innovations the product development department has been rolling out, from bio-based packaging to nontoxic cleaners. Others are focused on getting their products certified by respected third-party organizations, arguing that claims of technical advances mean nothing to skeptical external audiences without such recognition. Still others are interested in highlighting the company-wide sustainability policies that the CEO has recently put in place. They are particularly proud of a sustainability award from *Working Mother* magazine, which named the firm as one of the Best Green Companies for America's Children.

Meanwhile, a strong contingent of "cautious skeptics" are resistant to over-emphasizing the green message, and advise Anu to pay closest attention to what most consumers want, which they assert is high-quality products at low prices. The concerns of these skeptics about consumer perceptions of eco-labeled products as confusing, overpriced, and not even that green overlap with those of the information pessimists discussed in earlier chapters. However, they are also based on a belief that many consumers and other stakeholders are still skeptical of climate change and the project of sustainability more generally. And then there are a few who Anu has deemed the "practical idealists" on her staff, who are less interested in advertising specific products or policies and more focused on communicating the company's broader contributions to solving the many global environmental challenges facing society. To what extent has the company actually helped tackle these critical challenges?

Anu has had a successful career at several large corporations and knows when she sees a train wreck. She also realizes she needs to develop a clear strategy for her team to implement. Her sense is that the past approach has been ad hoc and opportunistic, with little thought to a broader vision of how sustainability really fits in with the company's sales, marketing, and product development processes. The company has a corporate social responsibility report with a vision, objectives, and data, but it reads like a mashup of ideas rather than a thoughtfully developed strategy. And regardless, very few people actually read the report anyway. Her sense is that the company needs to both improve its sustainability performance and its communication of that performance. Her CEO is genuinely committed to doing so and wants the company not only to be seen as environmentally responsible but also to make a real difference. In light of this mandate, Anu knows she needs to find a better way to incentivize sustainability innovations within the company as well as communicate its accomplishments and aspirations to consumers and its other stakeholders, and she needs to do it soon. Anu's overarching goal is therefore to deliver a new green vision to both the firm's internal and external audiences strategically, holistically, and efficiently.

To sift through all the different options her staff is proposing, she divides them into three teams. One team will focus on developing strategies to effectively communicate sustainability information about the company's products, another will focus on information about the company's overall sustainability performance, and the third will focus on information about global challenges and the firm's contributions to solving them. Anu purposefully divides the cautious skeptics across the teams and asks them to get on board with the assignment, but encourages them to actively offer constructive criticism to their teams that is informed by their skepticism. Her staff is working on their assignments now, and Anu needs a framework to both evaluate their proposals and integrate them into a seamless communication strategy for the company. It would also be helpful to have some benchmarks of promising and problematic practices for communicating this type of information that can guide her in this process.

This chapter aims to provide such a framework and set of benchmarks for people like Anu. This framework is built around the concepts of determinance, importance, and salience, and is relevant not only to corporate leaders, but also to activists, policymakers, and other designers of sustainability information. The challenge that Anu faces is not unique to companies but common in these sectors as well. Nonprofit organizations and government agencies both need to effectively deliver their messages to their members and the broader public. They might do all the right things to build the value, trustworthiness, and quality of their information that were discussed in the previous three chapters, but if they fail to deliver that information effectively, all that work will have been a waste.

### Understanding the Nature of Determinance

The theoretical framework presented in this chapter integrates concepts from a variety of disciplines and will help designers of information delivery systems avoid such a waste. As figure 5.1 shows, it begins with the concept of *determinance*, which refers to the extent to which a particular characteristic of an item actually drives—or "determines"—a decision about that item.[2] Determinance is driven by two main factors—the (1) salience and (2) importance of the characteristics of the item in question (also shown in figure 5.1). These two factors capture the fact that an information delivery system must effectively take into account how the content and form of information influence each other. On the one hand, *salience* refers to whether a characteristic stands out or is "top of mind," and is influenced by both the prominence and intelligibility of that characteristic—do people see it and do they understand what they see?[3] Prominence refers to factors associated with usability, while intelligibility is determined by both the

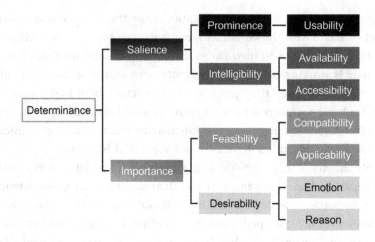

**Figure 5.1**

A framework for understanding the drivers of content relevance. *Note:* This figure maps out some of the key relationships between these different concepts, and does not capture all of their possible interconnections. The chapter discusses each of these concepts in more detail and explores some of the most relevant linkages between them. These linkages may not always be in a linear direction but may involve complex feedback loops.

availability and accessibility of the characteristic as an existing concept in people's minds.[4]

*Importance,* on the other hand, represents the significance or import of a characteristic, and is determined by two additional factors—the desirability of that characteristic (how attractive or meaningful it is) and the feasibility of actually acquiring it. *Desirability* is a more abstract concept that can be influenced by both emotional and rational processes, while *feasibility* is a more concrete idea that centers on the compatibility and applicability of the characteristic to the current context, including the capacity of the individual involved.[5] The relationships between different concepts in this theoretical framework are mapped out in figure 5.1, and will be explained in more detail in the following sections.

While they may overlap, these are all distinct factors. Returning to Mark from chapter 2, he may be able to clearly see the USDA Organic label on the milk carton (it is prominent to him) and he may have heard of it before (it is available to him), but he may not be very familiar with it (it is not accessible to him). Thus the label is not very intelligible and therefore only has limited salience for Mark. Nevertheless, upon figuring out what "organic" actually means, he may decide that it is indeed a desirable characteristic as he both cognitively and emotionally grasps its relevance to his values. But its higher price makes it less feasible to obtain, and therefore its overall importance remains limited to him. Thus while the presence of the organic label is an objective fact that may differentiate it from other products, many subjective factors that depend on the background and mental state of the individual consumer can influence its overall determinance in specific decision contexts.

Chapter 2 explores many of the factors that influence the desirability, or value, of ratings and certifications to different audiences, from their evaluation focus and coverage of public and private benefits to their evaluation scope and coverage of different sectors, countries, and levels of corporate activity. The chapter concludes that broad coverage of these different elements can enhance the perceived importance and desirability of information-based governance strategies. This chapter focuses instead on the three other factors that determine the uptake of these strategies—their prominence, intelligibility, and feasibility. These factors relate to important cognitive processes and are critical for information designers such as Anu to understand as they work to create sustainability information that will actually be useful to people like Mark.

### Prominence: Can Audiences See It?

As figure 5.1 shows, the salience of information and the likelihood of it being noticed and remembered is determined by two key factors, the first of which is its prominence. Following this logic, Anu knows that her first challenge for her team is to get their intended audiences' attention. These audiences need to notice whatever information that the team develops about the company and its products. No matter how great that information is, if people do not see it, it is useless to them. But beyond being seen, it is also helpful for that information to be viewed in a positive light—as superior and distinctive. The concept of *prominence* captures both of these factors, as it represents both a "state of being conspicuous" and a "quality of notoriety, eminence ... and superiority."[6] A study of business school reputations led by Violina Rindova, a professor of business at the University of Texas at Austin, finds that these two dimensions of prominence can have different antecedents. The study concludes that while prominence as quality is predicted

by input metrics (e.g., student GMAT scores), prominence as conspicuousness is associated with certifications from institutional intermediaries
(e.g., media rankings, peer-reviewed faculty publications) and affiliations
with high-status actors (e.g., faculty members with prestigious degrees).[7]
But it also reports that organizations perceived as producing high-quality
products are more likely to be prominent—as in conspicuous—in people's
minds. Those prominent organizations are in turn more likely to command
a price premium due to their strong reputation.

The concept of usability can help us further understand the factors that
determine the prominence of a particular organization or form of information, and how it can contribute to such price premiums. *Usability* is a widely
used term in the information design and management field, and can make
something both more conspicuous and eminent. The International Organization of Standards (ISO) provides a practical definition of "usability" as
"the extent to which a product can be used by specified users to achieve
specified goals with effectiveness, efficiency and satisfaction in a specified
context of use."[8] Usability has become an important focus in the design of
websites and online experiences, and experts on the topic have articulated
a wide range of principles to guide such design.[9] While usability is also
relevant to the other dimensions of determinance discussed later in the
chapter, a few of these design principles are worth noting here as they can
enhance both dimensions of prominence discussed previously.

Usability expert Jakob Nielsen discusses the importance of "aesthetic
and minimalist design" and suggests that sites "should not contain information that is irrelevant or rarely needed."[10] Steve Krug, another usability
specialist, also points out that "people tend to spend very little time reading
most Web pages" but instead "scan (or skim) them."[11] This insight is reinforced by research by Patrick Lynch and Sarah Horton, who have had leadership roles in information technology at Yale University and Dartmouth
College, respectively. They focus on perceptibility, simplicity, and flexibility
as universal design principles, and conclude that highly usable information
is *perceptible* regardless of ambient conditions and sensory abilities, *easy to
use* regardless of a user's background, and *flexible* across a wide range of
contexts.

Douglas Van Duyne (MarketShare), James Landay (Stanford Computer Science Department), and Jason Hong (Carnegie Mellon School of
Computer Science) also discuss the value of "organizing information in a

hierarchy of categories [that] can help customers find things," even though creating such a hierarchy is challenging because different users think differently.[12] While hierarchies can increase efficiency and flexibility of use, they often result in the structuring of websites into multiple layers of pages that become increasingly inaccessible to users from the homepage. One potential metric of a website's usability therefore is the distance (e.g., number of clicks) that important information is from the homepage—in other words, how accessible that information is to the user and how well the hierarchy actually reflects the importance of that information. Another simple metric is the extent to which information is provided in PDF files, which Jakob Nielsen describes as "unfit for human consumption" in the online context because they often provide a jarring user experience, cause software problems, deliver an undifferentiated wall of content, and are generally hated by users.[13]

While the field of usability has yet to develop a standard set of design principles, the discussion in this section suggests that information interfaces that have high perceptibility, simplicity, and flexibility are more likely to be prominent and salient to a broad range of audiences. In designing her company's new sustainability engagement strategy, Anu should ensure that the information it provides is quickly noticeable, easily accessible, and highly flexible. Regardless of their situation, people should be able to see and explore it without much difficulty or exertion. As we will discuss further, however, she will also need to manage important trade-offs between such simplicity and ease of use with other needs and desires of her audiences.

### Intelligibility: Can Audiences Understand It?

The second dimension influencing an information interface's salience is its *intelligibility*. The Oxford English Dictionary defines "intelligibility" as the "capability of being understood."[14] So once someone notices a message from Anu's team about their company's corporate sustainability, they have to be able to make sense of it. Research on cognition and framing has identified three primary determinants of this capability of being understood— availability, accessibility, and interactivity. First, the provided information must refer to concepts that are *available* in the viewer's mind and memory.[15] At the most basic level, it needs to be in a language the person can understand; next, it must use words in that language that they comprehend.

Then those words must refer to images, objects, or ideas that the viewer has a memory or conception of. Without such availability, the intelligibility and salience of the provided information are likely to be very low.

However, in the event that the person has no such conception available (e.g., they have never heard the term "corporate sustainability"), all is not lost. In this context, *learnability* is critical—any introduced concepts or objects need to be easily learned and incorporated into an individual's lexicon and memory. Scholars in the fields of linguistics, cognitive psychology, education studies, and machine learning have studied learnability across a wide range of contexts and found that it can build on a variety of learning pathways.[16] The machine learning literature, for example, distinguishes between rote learning, learning by being told, learning by analogy, learning from examples, and learning from observation.[17] One approach developed and widely used in the field of education has identified four distinct learning styles—visual (through graphics or figures), aural/auditory (through hearing or speaking), textual (through reading and writing), and kinesthetic/tactile (through activity and movement).[18]

Information-based strategies can explain relevant concepts like sustainability, organic, or the product life cycle using any combination of these approaches. For example, they can provide examples and images of product life cycles, have a narrator compare these life cycles to the efficiency of biological systems, or ask audiences to identify ways to reduce waste after watching a video about a particular product life cycle. While it is unclear whether selectively presenting information visually, orally, textually, or through kinesthetic experiences is more helpful to different groups of learners,[19] it has been shown that repeated exposure to information does indeed increase the *accessibility* of that information.[20] Accessibility refers to whether a concept is accessible and retrievable from long-term memory at any given moment in time.[21] Providing information in multiple formats—images, videos, text, or tangible objects—may activate different modes of learning and thereby increase such accessibility.

Heightening the *memorability* of the information can also increase information accessibility. Steve Krug suggests that memorable websites, for example, are likely to be more usable ones because people are less likely to have to relearn how to use them.[22] Such memorability has two dimensions—the ability of something to stay in one's mind and the ability to remember how to use something. Many strategies exist for making information stay in one's mind, from using bright colors to employing surprising language or images. A study by Scott Bateman and his colleagues at the University of Saskatchewan, for example, found that individuals shown charts embellished with elaborate and detailed imagery were more likely to remember the information two to three weeks after being exposed to it than those exposed to a simple, unadorned chart.[23] This effect may in turn lead to an improved ability to access and make use of that information.

The third determinant of intelligibility is *interactivity*, which in our context can be understood as the degree to which the user and the information mutually influence each other.[24] Studies of interactivity can be found in the education, computer science, and marketing literatures, and generally highlight the positive effects of interactivity on users' online experience, learning, attitudes, emotions, and behavioral intentions.[25] In one study, for example, students using web-based instruction systems with a high level of interactivity (e.g., discussion forms, chat room, instant messaging) had higher levels of satisfaction and perceived learning than those using systems with low levels of interactivity.[26] Another study shows that consumers visiting a highly interactive version of a coffee brand's website (e.g., with a recommendation option, online ordering, store locator, 3D product demo) are more likely to experience a sense of flow and total absorption, which in turn was associated with both stronger cognitive engagement and more positive emotional engagement with the brand and the website.[27] A third study demonstrates that instructional videos provided in an interactive digital environment (with a synchronized online note-taking tool, supplemental resources, and practice questions) resulted in higher recall test scores than noninteractive videos.[28] Thus strong evidence exists that interactive features increase the usability, learnability, memorability, and ultimately the intelligibility of the information provided.

### Feasibility: Can Audiences Use It?

Anu realizes, however, that even if her team adds such features and is able to make its new and improved sustainability information both prominent and intelligible, it needs to be feasible for their customers and stakeholders to understand and make use of that information. Along with the desirability of the information, feasibility is a key determinant of the perceived importance of information-based initiatives, and refers to the degree to which particular actions can be easily completed.[29] In Anu's context, those actions include consumers noticing, processing, and acting on information about her company's sustainability practices. The literature on decision making has identified two factors that can increase the likelihood that an individual will make use of a particular piece of information. They include its *applicability*, which refers to how relevant the information is to the current context, and its *compatibility*, which refers to how compatible the information is with that context. A third factor is the cost of using the information, which not only includes any transaction costs associated with such use but also any price premiums associated with the products or companies recommended by the information. Transaction costs will be examined in more detail in this chapter, while price premiums are discussed further in chapter 2.

The concept of applicability has been most extensively used in research on framing and priming effects. Scholars have found that efforts to influence decisions by priming individuals with potentially biasing information beforehand are constrained by the applicability of that information to the particular decision at hand.[30] Applicability thus reflects the degree of match between the features of the context and the features of the provided information.[31] Research has shown that the accessibility of knowledge in a person's memory is moderated by such applicability. In other words, the closer the match between the context and the provided information, the more likely individuals will remember and utilize their available knowledge about the decision in question.[32] So for it to be feasible for Anu's customers to utilize the sustainability information her team is developing, it needs to be applicable to their situation. Customers may see and understand the information, but if they do not make the connection between that information and their current purchasing decision, then it is unlikely to influence their behavior.

The second prerequisite for audiences to feasibly make use of information is its compatibility with the situation at hand. Even if consumers recognize the applicability of the information to a particular situation, it must be compatible with how they make decisions in that type of situation. As Archon Fung, Mary Graham, and David Weil at the Harvard Kennedy School explain, "people have settled routines and habits for making choices. Some carefully compare the price-per-pound labels for different brands; ... others don't bother. Some browse reviews of products and services ... others shop on impulse. ... [Information] can become embedded [in their decision-making processes] only if it is compatible with these settled routines."[33] In particular, it is important that the information is provided at a time and place when and where people make decisions that are relevant to that information. The authors assert that Los Angeles restaurant hygiene grades and fuel economy ratings on new car stickers are particularly well designed in this regard—they are made available when and where "users are accustomed to making decisions."[34]

Certainly, these programs are better designed than the counter-example that the authors discuss (technical government reports and dispersed public databases), but what if no other restaurants nearby have better grades? Or the car dealer only sells cars with poor fuel economy? In those contexts, the information may not be that useful. These scenarios highlight the importance of information flexibility to the feasibility of its use. Mentioned earlier as a key aspect of usability, *information flexibility* implies that information can be used and is useful across a wide range of contexts. In-window restaurant grades and on-car fuel economy ratings are examples of information being provided at the *point of sale*, which may be a useful place for many consumers.[35] Indeed, as much as 70 percent of brand selections are made at stores.[36] However, it may not always be the *point of decision*— the time and place where individuals make a decision to buy a particular product or engage in a particular activity.[37] Some consumers may decide on a course of action not at the point of sale, but at the *point of research*—the time and place when and where individuals learn about and investigate different options relevant to their current needs and desires.

Information that is flexible would be available at as many potential points of decision as possible, including both points of sale and points of research. It would be available in stores, on windows, on shelves, and on products, as well as on company websites and comparison websites that are accessible on computers, tablets, and mobile phones. This flexibility makes it more feasible for individuals to integrate the information into their personal routines and contexts. Of course, these different information interfaces must be highly usable in these different contexts. This point reinforces the importance of several additional design principles from the usability literature. For example, Jakob Nielsen identifies a need for "flexibility and efficiency of use,"[38] while usability consultant Bruce Tognazzini emphasizes the principles of "efficiency of the user" and "explorable interfaces."[39] Likewise, Steve Krug lists effectiveness ("does it get the job done?")

and efficiency ("does it do it with a reasonable amount of time and effort?") as important aspects of usability.[40]

To summarize, in order for audiences to feasibly make use of any particular form of information, it must not only be applicable to the decision at hand but also compatible with that decision. Such compatibility requires flexibility, efficiency, and effectiveness in the delivery of the information to those audiences. This point relates to Herbert Simon's concept of "bounded rationality," which posits that humans are only capable of rationally acting upon a limited amount of information.[41] It also connects with work on the shortcuts that humans use to deal with information overload and reduce the transaction and search costs associated with that information.[42] Humans have limited time, resources, and cognitive capacity, and therefore efforts to inform them about their decisions must be designed with these limitations in mind. In order to overcome these limitations, become integrated into these bounded cognitive processes, and become a shortcut itself, the information provided by these initiatives must be prominent enough to perceive, intelligible enough to understand, and feasible enough to utilize.

## Interface Attributes

The framework described in the previous sections can help executives such as Anu by helping them focus on the prominence, intelligibility, and feasibility of the information interfaces they are developing. The framework organizes a disparate set of concepts, from accessibility to usability, into a structured approach to designing such interfaces. But to supplement this theoretical perspective, it would be helpful for Anu to know what the specific options are that she has at her disposal to increase the salience, importance, and determinance of the sustainability information her team is developing. Understanding the range of current practices among existing environmental certifications and ratings would also help Anu formulate her own information delivery strategy.

The sections that follow present data from my Environmental Evaluations of Products and Companies (EEPAC) Dataset that address these two needs of executives such as Anu. As explained in chapter 1, this dataset consists of the website text from 245 cases of information-based environmental governance initiatives, which were coded for a wide range of characteristics and attributes. The data presented in this chapter focuses on three specific information attributes that represent particularly important interface decisions. These attributes include the *form of the information provided*, which range from the simple to the complex and from having a positive to a negative orientation. They include the *pathways by which the information is provided*, which include both physical and digital options. And they include the *information's architecture*, specifically how it is organized when it is presented online. These attributes directly relate to three of the key concepts—simplicity, flexibility, and efficiency—presented in the framework earlier in this chapter, and allow us to further explore how they can be incorporated into information interfaces. The other concepts also discussed earlier, including perceptibility, memorability, and learnability, are explored further in a later section.

### Forms of Information

The final component of the information value chain presented in chapter 1 is the interface by which the information is delivered to its audience. The core of this interface is the underlying form of the information that has been developed by the information value chain. There are eight basic forms of information that I have identified in the EEPAC Dataset—awards, ratings, databases, certifications, rated certifications, reviews, boycotts/watch lists, and rankings. Table 5.1 presents descriptions and examples of each of these forms of information.

These forms vary along several dimensions, each of which reveals important insights about the design of "simple" information interfaces. As discussed, simplicity is a key driver of information determinance, as it can make information more prominent, usable, and salient. However, simplicity is difficult to achieve, as information can be simplified in a variety of ways, and each involves important trade-offs. Such simplification can include the exclusion of positive, neutral, or negative information; it can also mean the reduction of information granularity, the combination of criteria, or the elimination of qualitative descriptions. All of these forms of simplification remove information that may be useful and desirable to at least some audiences.

A key distinction among these different forms of information is whether they provide positive, negative, or neutral information. Certifications and awards have a clear positive orientation, boycotts and watch lists are clearly

**Table 5.1**

Forms of information in information-based strategies

| | |
|---|---|
| Database | Provides basic data on performance, with no attempt to rate, rank, award, or shame using that data (e.g., EPA's Toxics Release Inventory). |
| Review | Evaluates performance qualitatively, either in absolute or relative terms, with no direct comparative analysis that rates, ranks, or certifies relative performance (e.g., Green America's Responsible Shopper). |
| Award | Recognizes exemplary performance relative to a peer group, with no differentiation in different levels of performance (e.g., Innovest's 100 Most Sustainable Companies). |
| Certification | Recognizes exemplary performance for meeting certain absolute standards, with no differentiation in levels of performance (e.g., ENERGY STAR). |
| Rated Certification | Recognizes exemplary performance meeting certain absolute standards, with more than one level of performance specified (e.g., gold, silver). Negative performance is not assessed (e.g., LEED certified). |
| Ranking | Ordinally ranks companies or products in terms of their absolute or relative performance on one or more criteria (e.g., UMass Amherst's Toxic 100 Ranking). |
| Rating | Rates using numbers, words, or letters the performance of a company or product based on either an absolute or relative scale that provides more than one level of performance recognition. Both negative and positive performance is assessed (e.g., Greenpeace's Greener Electronics Guide). |
| Boycott | Recognizes poor performance relative to a peer group, with no differentiation in different levels of performance (e.g., Ceres' Climate Watch List). |
| Hybrid Systems | Some combination of the above types. |

negatively oriented, and databases, ratings, rankings, and reviews can provide a mix of all three types of information. Awards, for example, highlight the strongest relative performers in a sample (the strategy behind Innovest's 100 Most Sustainable Companies), while boycotts point out the weakest (the strategy behind UMass Amherst's Toxic 100 Ranking). Archon Fung from the Harvard Kennedy School and Dara O'Rourke from UC Berkeley assert that the success of the Toxics Release Inventory (TRI) is due to the use of negative information and "blacklists" by advocacy groups that target the worst polluters because they are simple to understand and can mobilize diverse audiences.[43]

A study led by Gunne Grankvist from Sweden's University of Trollhät-
tan/Uddevalla shows experimentally that consumers with "intermediate
interest in environment issues" are indeed more responsive to negative
eco-labels, while consumers with a strong interest in the environment are
equally affected by both negative and positive information (and consumers
with limited or no interest were unaffected by both kinds of information).[44]
Another study by Wageningen University marketing scholars Ynte Van
Dam and Janneke De Jonge shows that negative eco-labels have a stronger effect on people who are oriented toward avoiding negative outcomes
in their lives, while positive eco-labels have a stronger effect on people
who are focused on achieving positive outcomes.[45] Thus the exclusion of
positive, negative, or neutral information represents an important simplification decision that may affect who engages with the information.

Another mechanism to simplify the information is to reduce its granularity. Certifications such as ENERGY STAR aim to recognize all performers
who meet a certain absolute standard, and do not differentiate between
those they certify. Ratings, rankings, and rated certifications such as LEED,
Greenpeace's Greener Electronics Guide, and Timberland's Green Index
seek to differentiate performance along a more granular spectrum. In this
sense, ENERGY STAR provides binary information, while the others provide
nonbinary scales of information. The binary certification may be simpler,
more prominent, and more salient to consumers, but it leaves out information that some users may value.

These forms of information are not only different ways to deliver information but also reflect different emphases in the information development
process. Databases, for example, focus on creating quantitative information, while reviews rely more on qualitative assessments. Likewise, both
databases and reviews are generally more descriptive forms of information, in the sense of attempting to just "state the facts."[46] Awards, certifications, rankings, ratings, and boycotts are inherently more evaluative forms
of information, as they are "closely connected with choice, decision, and
action" and include prescriptive and normative meaning.[47] Simplification
efforts that eliminate the descriptive elements of the information and their
focus on facts and attempt to boil everything down to a simple recommendation can increase the information's salience. However, this may increase
concerns about bias and reduce the perceived trustworthiness of the simplified information.

Another simplification strategy is "bundling," which combines similar information goods into one package.[48] Multi-attribute programs—such as the Green Index, LEED, and Greenpeace's Greener Electronics ratings—are examples of information bundling. They combine multiple criteria into a single index and set of results for users to focus on. By reducing the amount of information audiences have to interpret, such a strategy can reduce the transaction costs associated with using sustainability certifications and help overcome a sense of information overload.[49] Again, however, such simplification has an important trade-off—it hides information that some users may feel is important and want to utilize in their decision-making process. By combining two or more of these forms of information, hybrid systems, such as the Greenwashing Index's use of both ratings and reviews, are designed to avoid these trade-offs. They provide a blend of outputs that have both simplified information for the general audience as well as more granular data for those with particular interests.

The 245 cases in the EEPAC Dataset were coded for each of these information forms, and the data are presented in figure 5.2. Certifications are the most common (41 percent), followed by awards (30 percent) and ratings (23 percent). Least common are boycott/watch lists (6 percent) and rankings (4 percent). Two-thirds of the cases provide only one form of information, 20 percent provide two forms of information, and just over 12 percent provide three forms of information. The granularity and valence (whether they provide positive or negative information) of the cases were also coded. Just over half of the cases only provide positive forms of information (awards, certifications, and rated certifications), while very few (2 percent) provide only negative information (boycotts). The remainder provides a mix of both positive and negative information. Nearly half of the cases only provide simple binary information (awards, certifications, boycotts), while just under 30 percent only provide more complex nonbinary information (ratings, rankings, certifications, databases, reviews). The remaining cases provide both simple and complex information to their audiences.

These different forms of information can be further differentiated by the extent to which they analyze and simplify information as opposed to the extent to which they collect and compile it. Awards, boycott lists, ratings, rankings, and certifications tend to be more oriented toward information analysis and simplification, while databases and reviews are generally

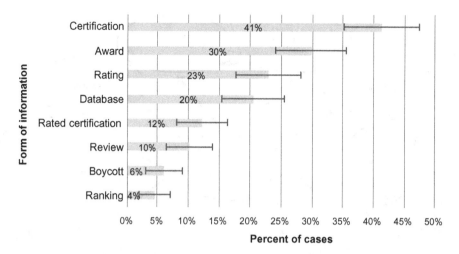

**Figure 5.2**
Form of information provided. *Note:* Error bars indicate 95 percent confidence intervals for each sample proportion.

designed as important mechanisms of information collection, production, and dissemination. As figure 5.2 shows, most of the programs in the EEPAC Dataset serve as providers of information analysis, signaling, and shortcuts for their audiences, while a smaller number are sources of more extensive and less-processed sets of information (such as the Toxics Release Inventory).

## Pathways of Information

Once Anu and her team have decided on the form or forms that their information will take, they must decide how they are going to actually deliver that information to their key audiences. This design choice revolves around selecting the appropriate pathways for their information to flow through, and the most relevant principles to this choice that we have discussed are flexibility, simplicity, and efficiency. As figure 5.3 illustrates, such pathways can be either physical or electronic. Physical pathways include on-product labels, paper newsletters, wallet guides, books, and magazines, while electronic pathways encompass company websites, social media outlets, phone and tablet apps, shopping sites, and email newsletters. This distinction is important for several reasons—first, different audiences may be more familiar and comfortable with particular pathways. For example, older,

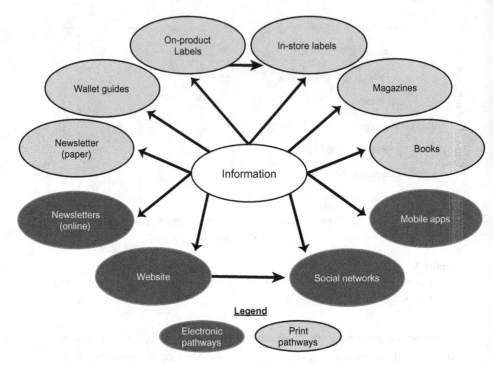

**Figure 5.3**
Electronic and print pathways of information. *Note:* Cases were coded for a variety of specific website pathways, including individual pages for rated items, search engines, pulldown menus, lists of all rated items, PDF reports, and online shopping options. Cases were also coded for two specific mobile app pathways (iOS and Android apps) and four specific social network pathways (Facebook, Twitter, LinkedIn, and Instagram). Arrows only indicate primary flows of information—additional pathways are possible but not shown in this figure.

less educated, and lower-income audiences generally are less likely to be exposed to electronics pathways. According to a 2013 study by the Pew Research Center, over 40 percent of adults 65 years old or older and adults without a high school diploma do not use the Internet (compared to 2 percent of 18–29-year-olds and 3 percent of college-educated adults), while one out of four adults who make less than $30,000 per year do not surf the web (compared to 3 percent of those who make over 75,000 per year).[50] Thus information provided via electronics pathway is less likely to reach these groups.

In order to investigate how existing information-based environmental governance initiatives are delivering their information, my research assistant and I coded the websites of the 245 cases in the EEPAC Dataset for these different electronic and physical information pathways. Given the speed at which information technologies have been changing in recent years, we manually recoded all 245 cases in January–March 2016 for these pathways (for more on the coding process, see appendix I). Interestingly, the websites of forty-seven of the initiatives included in the original dataset are now defunct, suggesting that these initiatives are currently nonoperational. This is an important insight to which we will return in chapter 7. For now, the data presented in figure 5.4 and discussed as follows are from the 198 cases that still maintain functional websites.

As a requirement for being included in this dataset, all of these cases use at least one electronic pathway, a functioning website. Just over 40 percent use at least one physical pathway to deliver its information, and over 85 percent of these pathways are on-product labels, such as USDA Organic or ENERGY STAR. The primary exception is paper pathways. Corporate Knights and CR, for example, produce magazines in which their corporate sustainability rankings are published. The Monterrey Bay Aquarium provides a wallet guide called "Seafood Watch" that evaluates the sustainability of different fisheries, while the Better World Shopper's ratings comes in the form of a book (as well as a website). Another exception is labels that are provided by a retailer in-store but may not be physically on products. An example of this relatively rare pathway is Home Depot's Eco-Options label.

While all of the cases have websites, the electronic pathways by which they deliver their information online vary significantly. Just over half of the initiatives provide lists of their product or company evaluations on a single page, or on a series of pages when the volume of ratings is high. Such lists make these program's basic information relatively accessible and easy to find. Just under half provide evaluation information about individual companies or products on a separate page or pop-up screen, enabling them to provide more details about their certification or rating without overloading a single page. Nearly one in three cases provide both comprehensive lists and individual pages.

One in six cases provide their information in downloadable PDF files, but usually this information is supplemental to information provided in

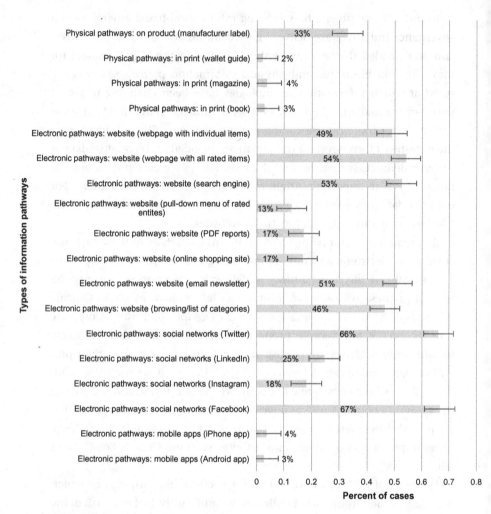

**Figure 5.4**
Pathways of information (by percent of cases). *Note:* Error bars represent the standard errors for each sample proportion.

html format. Two of the few exceptions include Greenpeace's Chemical Home Report and CR's 100 Best Corporate Citizens rankings, which are only published as PDFs. Just over half of the cases provide the ability to search directly through their ratings and certifications; another 30 percent provide general search functionality of their website, while 20 percent provide no ability to search their sites at all. Search engines enable users to find specific products or companies that they are interested in more quickly.

Nearly half enable users to browse the ratings with product category links, while one in eight sites have pull-down menus that provide quicker access to their rating and certification results.

Each of these functionalities are designed to make the information on these sites more accessible. But first they have to get audiences to visit the site in the first place. Otherwise, even the most user-friendly site is worthless. An increasingly common way to increase the salience and prominence of these websites is through social media sites, which have grown dramatically in popularity in recent years. For example, a 2015 survey found that over 62 percent of American adults use Facebook, the most popular such site, while 24 percent use Instagram, 22 percent use LinkedIn, and 20 percent use Twitter.[51] Slightly more women use Facebook and Instagram, and slightly more men use Twitter.[52] High school and college graduates use Facebook in equal proportions, 18–29-year-olds are the most common users of Instagram and Twitter, and LinkedIn is particularly popular among the college educated and the employed.[53] As of March 2016, these four sites are among the most widely used social networking sites in the United States.[54]

These social media sites therefore provide a range of pathways for information-based governance initiatives to reach a diverse set of audiences. We coded their websites for whether they have links to their own pages on these social media sites (and not just links to share a page through them), and found that two-thirds of them had such a link to either Facebook or Twitter pages (or both). Just under 25 percent have a link to LinkedIn, and just over 18 percent have a link to Instagram. Approximately one-third have no links to any social media pages, while only 5 percent have links to all four sites. Another electronic pathway that these initiatives can utilize is email newsletters, which users can sign up for on their websites and forward on to friends and family. Just over half of the cases give visitors to their websites the opportunity to sign up for such newsletters or action alerts, which may be directly related to the case itself or more generally about the organization implementing the case.

Users can also access information from these initiatives through mobile phone and tablet apps. A 2015 study found that nearly two-thirds of Americans own smartphones (phones that can access the Internet), and 25 percent report they have limited online access beyond their phone (10 percent have no other access at home).[55] Younger, less well-off, less-educated, and nonwhite American adults are disproportionately dependent on their phones for Internet access.[56] The two dominant platforms for these devices are iOS (for Apple products) and Android (for products made by Samsung, LG, and other manufacturers). According to 2015 data from Kantar Worldwide, Android smartphones make up 59 percent of the market while iPhones make up 40 percent. iPhones in the United States are more popular among women, adults with a graduate degree and incomes over $125,000,[57] and iPhone users are more likely to visit e-commerce sites from their phones.[58] Despite the growing importance of cell phones for Internet access, only seven cases in the EEPAC Dataset have apps designed for either iOS or Android. Seafood Watch and B Corporation, for example, have apps for both platforms, while GoodGuide and Climate Counts only have iOS apps available.

These initiatives might also add the ability to purchase rated or certified products via their websites. Such functionality would represent an additional pathway of information for users to follow. It would also increase the applicability of the information and the feasibility of actually using it at their points of research, decision, and sale. My analysis found that 51 percent of the cases provide such shopping functionality, either directly on their site or as a direct link to purchase specific products on external sites.

### Architectures of Information

The third major set of user experience choices Anu and her team have to make relate to the structure—or architecture—of the information delivery system itself. If that system is an eco-label on a product, how will that eco-label be designed? What will it look like? If that system is a database, how will users access the information through it? What fields will they be able to search? Given that the EEPAC Dataset focuses on websites, in this section I focus on the architecture of those websites. How are they structured, and how easy is it for people to find relevant information on them? Specifically, how many clicks (and how much time and effort) does it take for them to find important information about these programs on their websites?

The average number of pages on these websites is 9.4, with 38 percent having ten or more pages, 42 percent having between five and nine pages, and 20 percent having fewer than five pages. Less than 2 percent have only one primary page and approximately 25 percent have only primary and secondary pages (i.e., pages that are one click away from the homepage). The remaining 73 percent on average have five tertiary pages (i.e., pages more than one click away from the homepage); the highest number of tertiary pages per case in the dataset is fifty-six (Rainforest Alliance). As figure 5.5 highlights, approximately one-fifth of all the codes were found on the primary homepages of the sites, half on secondary pages, and one-third on tertiary pages (pages two or more clicks away from the homepage). Approximately 20 percent of the cases have a majority of their codes on their homepages, 47 percent have a majority of their codes on their secondary pages, and 20 percent have a majority of their codes on tertiary pages (codes were more evenly spread across the page levels in 15 percent of the cases). Over 40 percent provide information in at least one PDF file on their website, which is higher than the statistic noted previously because this includes all PDFs, not only those that include product or company evaluation data.

## The Landscape of Information Delivery

As in previous chapters, I have created a snapshot of the landscape of information delivery systems being used by the cases in the EEPAC Dataset.

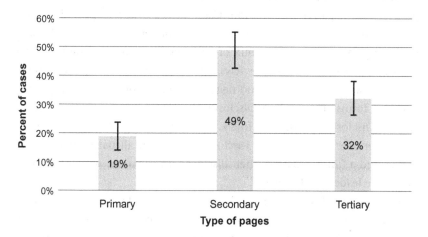

**Figure 5.5**
Coded information location in website structure (by percent of codes). *Note:* Figure shows the proportion of coded text segments that were found on the primary webpage, secondary webpages (pages one click from the primary webpage), and tertiary webpages (pages more than one click away from the primary webpage). Error bars indicate 95 percent confidence intervals for each sample proportion.

Figure 5.6 presents this landscape, which combines and simplifies a subset of the data presented earlier. The diagonal axis aggregates the data on the different forms of information previously discussed, dividing the cases into seven different types. The top three rows, where 60 percent of the cases are congregated, include initiatives that only provide positive information (awards, certifications, or rated certifications). These cases are further categorized by the complexity of their information—the cases in the first row only provide simple binary forms of information (awards like the Ceres-ACCA Sustainability Reporting Awards and certifications like USDA Organic), and the cases in the second row only provide more complex nonbinary information (rated certifications like EPEAT). The cases in the third row provide both binary and nonbinary information, such as the Corporate Responsibility Index, which recognize the top 100 companies it rates as "Companies That Count" (a form of a binary award) while also rating them in four bands—platinum, gold, silver, and bronze.

Figure 5.6 highlights the dearth of cases that only provide negative information—no cases only provide complex negative information, and only two (CERES's Climate Watch and ECRA's list of Current Consumer Boycotts) provide simple negative information. The rest of the cases (nearly 40 percent) are concentrated in the final three rows, which provide both positive and negative information. Cases in the fifth row provide simple forms of this information (awards, boycotts, or certifications), while cases in the sixth row provide this mix of information in more complex forms (ratings, rankings, databases, reviews, and rated certifications). Cases in the bottom row provide positive and negative information in both simple and complex forms. The Carbon Disclosure Project, for example, publishes a Carbon Disclosure Leadership Index, which recognizes the top performers in different sectors, as well as disclosure scores for all participating companies and a full database of their disclosed information, which lists both disclosure leaders and laggards.

The horizontal axis further categorizes the cases by some of the pathways of the information they utilize. Cases in the columns to the far left (20 percent of the total) do not provide any information directly on products or via the four social media sites discussed (Facebook, LinkedIn, Twitter, and Instagram). The second column from the left includes cases that have social media sites but no information on products (45 percent of the total), while the third column from the left includes cases that provide information on

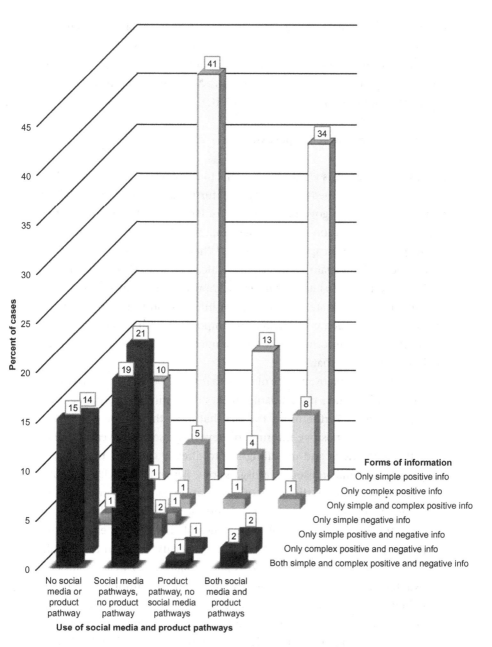

**Figure 5.6**

The landscape of information delivery systems. *Note:* The ten cases in the back left-hand corner use no social media or product pathways and only provide simple positive information. The two cases in the front right-hand corner utilize both social media and product pathways and provide both simple and complex forms of positive and negative information.

products but no social media sites (10 percent of the total). The remaining 24 percent of the cases in the fourth column from the left have both information on products and social media sites. Only two cases use both social media and product pathways and provîde a mix of both simple and complex positive and negative information.

**The Information Realist Perspective**

Information pessimists would likely point to the landscape of information delivery systems outlined in this chapter as further evidence of the depravity of information-based governance strategies. First of all, the range of different forms of information shown in figure 5.2 highlights how confusing and challenging this landscape is to navigate. What is the difference between a rating and ranking again? Between a certification and rated certification? How are ordinary consumers, and even knowledgeable experts, supposed to compare and evaluate such a range of information forms? In the view of these skeptics, such comparisons are impossible to make because these different forms of information are incommensurable, and no amount of analysis will change that.

The almost complete absence of cases that solely provide negative information about products or companies makes this form of governance even more problematic. Furthermore, the fact that more than 50 percent of the cases only provide positive information reveals these initiatives' complicity with corporations and unwillingness to directly confront them about their environmental and social transgressions. Beyond this lack of courage to speak truth to power, the range of information complexity discussed in this chapter highlights an even deeper and more fundamental limitation of these strategies. As legal scholars Omri Ben-Shahar (from the University of Chicago) and Carl Schneider (from Yale University) assert in their book, *More Than You Wanted to Know: The Failure of Mandated Disclosure*, disclosure strategies face an unresolvable dilemma—complex information is too overwhelming and ultimately counter-productive for users, while simplified information is too limited and ultimately misleading for users. Providing both simplified and complex information represents not a happy medium but even more information for users to sift through. Given the time and knowledge constraints of the vast majority of people who might use information disclosed through these strategies, they are at best a fool's errand and at worst actively harmful to society, as they waste time, distract from more effective strategies, and lead to complacency and a false sense of progress.

The pathways and structural data raise further concerns for information pessimists. The general scarcity of paper pathways makes many of these programs inaccessible to people without access to the Internet, and the provision of their data in PDF files undermines the prominence and intelligibility of their product and corporate evaluations, even for those with online access. The broad lack of well-designed search engines, rating lists, browsing abilities, and pulldown menus further reduces the flexibility and ease of use of the cases. This is particularly concerning given the amount of important information these initiatives provide about themselves that is not available on the homepages or even secondary pages of these websites. The number of clicks necessary to reach much of this information reduces its salience and prominence, and raises concerns that the website designers may be strategically hiding some information deep within their site hierarchies.

Furthermore, the fact that nearly half of the cases do not have a clear social media presence, two-thirds do not provide any information directly on product labels, and nearly 85 percent do not provide shopping functionality on their sites (and one in seven do none of the above) demonstrates the limited feasibility for consumers and other stakeholders to actually make use of the information in their purchasing decisions. The lack of detailed pages about individual rated companies or products on more than half of the cases also likely reduces the learnability and memorability of the information, not to mention its credibility and replicability. Providing details about individual products allows for more interaction with the data, and can provide context for the rating or certification provided, making it more cognitively and emotionally accessible in the future.

And so on the one hand the pessimists see nothing to assuage their pessimism. Information optimists, on the other hand, would likely see this landscape not as confusing, but as exciting, and as evidence of the innovativeness and dynamism of information-based governance. From social media to email newsletters, these initiatives are trying out new ways to engage their audiences and make their information salient to them. How many government regulatory programs have mobile apps? The lack of homogeneity in this landscape reflects the efforts of designers to use tools that help them reach specific audiences. To the extent there is homogeneity, such as the lack of purely negative information, that more likely reflects a belief that positive information is a more effective approach to engage their audiences than evidence of corporate complicity or a lack of courage. And in any case, 40 percent of the cases provide both positive and negative information in the form of ratings, rankings, databases, and reviews.

Regarding the "unresolvable dilemma" mentioned earlier, optimists would likely disagree in several ways. The first would be to contest the notion that this dilemma is unresolvable—just because it has not been resolved yet does not mean that it cannot be. The second would be to contend that some initiatives have done a relatively good job of balancing the competing needs for simplicity and complexity. And the third would be to point out that the dilemma may be more resolvable for certain audiences and in certain contexts, and a blanket fatalism is unsupported, inappropriate, and unhelpful.

Optimists would also likely be more positive about the pathways and structural data presented. First of all, given the challenges of getting information on products, setting up online shopping opportunities, and producing print publications, the fact that over 80 percent of the cases engage their audiences through at least one of these pathways is quite impressive. Second, as discussed earlier, not all purchasing decisions are made at the point of sale—many are made at the point of research, and so sites that provide information that consumers use to research products and companies can still be influential.

Third, not all pathways are used by all audiences, and following the logic of chapter 3, initiatives may be strategically emphasizing certain pathways and certain information to reach certain audiences. For example, designers focused on reaching younger audiences may prioritize having a Twitter and Instagram presence, and see less need for a LinkedIn site. Similarly, initiatives may be delivering their information in more extensive PDF reports because they are trying to influence policymakers and executives, who may value the more detailed, contained and finished product that a PDF document allows for. Some consumers may also find PDF files easier to review and navigate than a sprawling website. Indeed, given that a huge amount of information is necessary to establish the relevance, validity, and trustworthiness of these programs, it is natural for them to organize and prioritize that information in some kind of digital hierarchy, which might include at least a few PDF files. For the most credible and rigorous programs to include all of its information on its homepage would be truly overwhelming. Indeed, only the most vacuous programs could include all of their information on a single page.

And finally, while on-product certifications are great, manufacturers are not always amenable to them. It is therefore important to have alternative information pathways available, such as online comparison websites, and not dismiss them as "inaccessible." Pessimists cannot have it both ways— if they want independent information that has no conflicts of interest, they have to be open to information that is not directly available on products.

In light of these opposing arguments, the information realist must articulate a clear perspective on the quality of existing information delivery systems that provide sustainability information about products and companies. She will first acknowledge the limitations of both the usability theories and data described in this book and elsewhere to evaluate these delivery systems. The reality of user experiences is much too complex for any single theory or dataset to fully capture the quality of those experiences. Therefore, some humility is necessary on both sides of the debate, particularly given the nascent and rapidly changing nature of relevant technologies. Nevertheless, the information realist can use some of the concepts and data presented in this chapter as starting points for a discussion on how to approach the design of environmental information delivery systems (and information delivery systems more generally).

Regarding the first debate over the commensurability of the diverse forms of information provided by the EEPAC cases, it is true that awards and certifications are inherently incommensurable. While greater methodological transparency can increase the comparability of their results, at heart they are intrinsically different given that an award is a relative evaluation (only a few programs can receive the award regardless of their level of performance) while a certification is an absolute one (any program attaining a certain level of performance can earn it). However, commensurability is not the only attribute we value and it is only relevant when we have to choose between two different forms of information. We are also interested in the effectiveness of these initiatives in their own right, as well as how they might complement rather than compete with other forms of information. We will return to these questions in chapter 6, but for now we will focus on thinking about these different forms of information as components of stand-alone delivery systems.

The question of the unresolvable dilemma raised by Ben-Shahar Schneider is an important one, as a trade-off between simplicity and complexity does exist. However, the pessimist's view is too defeatist—if we were to adopt this attitude more generally, we would give up on every effort to educate and inform the public, and perhaps on democracy itself. If it is hopeless to get the balance of information right, why fund public schools, or even attempt to educate our own children? Why entrust any decisions to the public, including voting for their representatives, that depend on them being informed about particular issues? However, the pessimists are correct to say that many, if not most, of these initiatives, fail to find the optimal "not too much, not too little" amount of information. Whether it is worth continuing the pursuit of this optimum depends on the likelihood that alternative modes of governance will have greater chances of success, which is another topic that we will return to in chapters 6 and 7.

The debate about the value of stand-alone negative information can be informed by psychological research on the effects of fear-based appeals. One insight from this research is that such appeals can be effective, but only if they also show how to avoid the threat and audiences believe they can indeed do so (i.e., they have high self-efficacy and response efficacy).[59] In other words, fear works as a motivator only if people are provided with mechanisms to escape that fear. Such research can explain the value of hybrid approaches that combine both positive and negative information. Boycotts and poor ratings can make an issue salient and important for an audience, while awards, certifications, and excellent ratings can point them toward alternative companies and products that perform better than the worst offenders. If implemented effectively, such a strategy can increase both the prominence of an information-based initiative as well as the feasibility of utilizing the information it provides. Thus the optimists may be right that such hybrid approaches may be based on a well-informed strategy. However, it must be implemented well, with a clear connection drawn between the negative and positive information. This connection is rarely made, by either the hybrid initiatives or the initiatives that only provide positive information.

Different audiences may also prefer certain types of information more than others, which relates to the multitude of pathways documented in the results discussed here. The information realist's perspective on these pathways is that some may in fact be more attractive to particular audiences, and it makes sense for programs to tailor their information delivery systems to the audiences they are focused on, both aesthetically and functionally. However, once they have made their information salient and feasible for their primary audience, it also makes sense to make your information accessible to as many other audiences as possible. Indeed, many of the website features described previously, from search engines to category hyperlinks, have become relatively standard, and can be helpful to a broad range of users.

Nevertheless, it is still important to have a realistic and intentional strategy, particularly given the fast-changing nature of the technology landscape. For example, mobile apps were deemed a requirement for engaging cell phone users in 2014,[60] but in 2016 commentators called for a greater focus on having mobile-friendly websites.[61] Others are suggesting that both mobile browsers and apps are important components of a comprehensive mobile engagement strategy—the former to reach a broad audience of light users and the latter to retain a more narrow audience of heavy (and loyal) users.[62] This example highlights the still-early and evolving state of modern web-based information technologies, and the ubiquitous presence of new innovations and pathways to reach users. Designers should be bold but focused in this age of experimentation—try one thing or try many things, but move on to other options quickly if initial pursuits are not working out.

Regarding the data on the architecture of the cases' websites, it is certainly necessary to prioritize the information and structure that information into an intuitive hierarchy of webpages. But mechanisms exist that can make this hierarchy flexible, prominent, and accessible to users. Site maps in the footer or sidebar and pulldown menus in the header of the homepage can reveal the structure of the entire website, and enable users to reach any page within it, or at least any section, with one click. Tabs can enable more content to be embedded on the homepage, while reducing the need to click on to any secondary pages (and eliminating the page-loading time associated with such clicks). A standard of having most if not all key facts relating to the content, methods, trustworthiness, and effectiveness of these initiatives within one click of their homepages is a realistic and sensible one that they should all be aspiring toward.

## Promising and Problematic Practices

In light of these debates, what is Anu to do? Should she join the pessimists and give up on her entire enterprise? Or should she proceed with the unbridled enthusiasm of the information optimists? What would a third path of information realism look like for her? Her three teams are coming back soon—what have we learned in this chapter that can help inform her next decisions?

As a general rule to guide her work as an information realist, Anu should add attributes to her information delivery system that have high determinance and are likely to increase the likelihood that her audiences will actually use the information she is providing to them. Such characteristics should make her information more prominent, intelligible, and feasible to use. Prominent attributes will increase the conspicuousness, perceptibility, and usability of the information. Intelligible characteristics will make the information more available, learnable, accessible, memorable, and interactive. And attributes that increase the information's feasibility will make it flexible, efficient to use, and applicable to and compatible across a range of decision contexts.

The cases in the EEPAC Dataset provide valuable exemplars of these attributes. Blue Ocean's partnership with Whole Foods, for example, enables the grocery store chain to place informative labels on seafood in all of its stores, making information about fishery sustainability prominent, conspicuous, and compatible with consumer's regular shopping routines.[63] The Wildlife Habitat Council's Corporate Lands for Learning Program enables companies to place large signs on their certified properties that highlight their certification status and inform visitors about the company's conservation efforts at these sites. While on-product labels can also increase the prominence of these programs' information, the less-common print publications, such as Gardens Alive's catalog and Monterey Bay Aquarium's wallet cards, can also increase the conspicuousness and usability of this information, particularly for consumers who do not often use online information sources. Catalogues provide a fast way to quickly learn about and compare a wide range of products, while wallet cards provide quick shortcuts to the best and worst choices when consumers are in stores or at restaurants.

The relative usability and intelligibility of these initiatives is difficult to precisely measure, but several online services are available that provide systematic assessments of websites. For example, WooRank, PowerMapper, and Google Webmaster Tools provide a wealth of both free and fee-based website tests and statistics that can be used to improve the usability and navigability of sites. Pingdom.com assesses the speed and performance of websites and their individual pages and recommends ways to improve sites free of charge. WebpageFX and Juicy Studio provide free online evaluations of the intelligibility of websites that include detailed text statistics and readability indices.[64] High scores on these indices suggests that designers should consider simplifying the language, words, and sentences used on their sites, particularly if they are trying to reach a broad audience.

Several cases effectively use video and graphics that increase the accessibility and learnability of the information they are providing. For example, Rainforest Alliance and the Environment Protection Agency's Safer Choices Program have engaging videos embedded directly on the front page of their sites that explain the value and purpose of their certification programs, providing specific examples and statistics. The EPA video features EPA Administrator Gina McCarthy with her dog as she introduces the Safer Choice Program to viewers, while the Rainforest Alliance program uses an engaging cartoon-drawing video that tells viewers more about its certified products and increases their salience and memorability. It also provides an interactive map that shows all of the farms that it works with around the world, with information about their acreage and crops grown. The Marine Stewardship Council (MSC) and B Corporation websites have links to YouTube playlists of videos about their certifications that include inspirational program overviews, engaging interviews and presentations by thought leaders endorsing the program, and focused testimonials from specific companies and fisheries that have been certified. The MSC also has a companion site, Fish and Kids, that teaches children about the importance of marine sustainability through a variety of interactive games, films, activities, and "fishy fact files."

All of these mechanisms can increase the salience and intelligibility of these initiatives. They enable stakeholders to learn what environmental certifications do and why they are important, and help them remember to look for their labels when they are shopping. As discussed earlier, other programs make this process even easier by providing links directly to certified products that they can buy online. Blue Dolphin, for example, labels itself as the "first and only ALL-green general store" and enables visitors to its site to find and buy certified products that it has identified directly on its site. GoodGuide provides direct links to Amazon.com for products it has rated on its website and mobile app. It also experimented with a "Transparency Toolbar" (embedded as a Firefox or Chrome add-on) that would appear at the bottom of the browser whenever users searched for a product on Amazon. This toolbar would show social, environmental, and health ratings of the product searched for as well as higher-scoring alternatives.[65] Knowmore.org developed a similar Firefox add-on that provided sustainability ratings of companies in Google search results and as a toolbar alert when users visited company websites.[66] CSRHub created a widget that can be embedded into any website and enables users to quickly look up CSRHub's corporate sustainability ratings directly from that website.[67] These innovations have either been discontinued or not recently updated, but similar efforts to increase the prominence and applicability of sustainability information and layer it into the online browsing experience could build on these experiments.

These are just a few of the innovative practices that initiatives have employed to make their information more accessible and usable for its audiences. A few examples of characteristics that likely decrease the usability of this information are worth mentioning as well. The most obvious among them is the erection of paywalls that limit the information to paying customers. Eight of the cases in the EEPAC Dataset have such paywalls, and they clearly reduce the usability of the information for those who cannot or will not pay for access to it. Of course, such payments can have clear benefits by increasing the financial sustainability of the initiative and perhaps its trustworthiness as well. Initiatives with paywalls are often more oriented toward corporate audiences with greater resources, and their payments can enable the initiative to increase the usability of the data for those who do pay for access to it. From the perspective of most stakeholders, however, this practice essentially assures that the information will not be used by the vast majority of the public, and so this trade-off should be carefully considered. Consumer-oriented initiatives that have paywalls, such as Consumer Reports, must invest in building their brand, ensuring their credibility, and persuading the public that their information is worth paying for.

A related practice by sites such as Ethical Consumer and the Princeton Review requires email signups or user sign-ins before being able to view the information (or more detailed information), which can turn away a huge number of potential visitors—without the benefit of payments. Likewise, splash pages (such as those used on the Environmental Working Group's SkinDeep website) that pop up requesting email addresses or newsletter signups before entering the main website can be a deterrent and frustration for users as well. Organizations should resist these and other practices that create friction for their users unless they are absolutely necessary.

## The Delivery of Corporate Sustainability Information

Anu is now listening to her team's presentations, as they explain their ideas for improving perceptions of their company's sustainability performance. They have some great proposals, such as committing to third-party product and company certifications, moving away from unverified green claims, and tracking their contributions to solving three key environmental challenges—global water shortages, product toxicity, and climate change. From the life cycle analyses they conducted on a sample of their own products, they identified these challenges as the ones most relevant to the company's supply chains and the ones they could make the most impact on.

More specifically, each team proposes ambitious goals over both the short and long term, and recommend specific certifications to pursue. They recommend that these goals be adopted and integrated across the company's management, finance, marketing, and product development divisions to create a broader corporate culture around them. As a step in this direction, the team focusing on corporate sustainability would like to become the first Fortune 500 company to be certified as a B Corporation, which encourages such integration by recognizing companies that meet the "meet the highest standards of verified, overall social and environmental performance, public transparency, and legal accountability."[68]

Also along these lines, the team focusing on product sustainability wants to have all of their products certified to the multi-attribute Cradle to Cradle Platinum standard, which sets minimum requirements for the safety of the materials used in the products, the safe reutilization of those materials, the use of renewable energy during the manufacturing process, and the cleanliness of the water used during that process.[69] Furthermore, they want the packaging of all of their products to be certified by the U.S. Department of Agriculture's BioPreferred Program, which sets standards for products derived from plants and other renewable materials—and not from petro-leum.[70] Likewise, they want all of the palm oil they use in their products to be certified by the Roundtable on Responsible Palm Oil.[71]

The team focusing on global impacts suggests partnering with nongov-ernment organizations to track progress and implement programs related to the sustainable use of energy, water, and chemicals, both within and beyond the company's own operations. They have drafted five- and ten-year objectives for the company to achieve in each of these three areas. They have incorporated goals into their proposal that, if achieved, should ensure that the company scores well on corporate sustainability ratings such as Climate Counts and is removed from boycott lists such as Green-peace's "Climate Laggards."

Anu is impressed by both the visionary and grounded nature of her teams' work. She sees how her staff has responded to the poor evaluations of several information-based environmental governance initiatives by thoughtfully incorporating the certifications from several other initiatives into their own homegrown and hybridized information-based governance strategies. Getting the cautious skeptics and practical idealists to work side by side seems to have paid off, as the ambitious goals are coupled with clear connections to the company's customers. The skeptics pushed each team to articulate the relevance of each recommendation to people who buy their products, and they did so—often using the language of trustworthi-ness, validity, and values, concepts discussed at length in earlier chapters. Nevertheless, and perhaps not surprisingly, Anu senses a lack of integration and linkages both among the teams' work and between their work and the company's diverse stakeholders.

Fortunately, Anu has been busy working on forging this broader inte-gration as the teams have been developing their specific strategies. At the heart of this strategy is a new holistic website that is fully immersive and interactive. The site has two modes—one designed to be deeply engaging and the other to be extremely accessible. For the first mode, she has hired a renowned game designer to make the site into an online gaming experi-ence, simulating classic roleplaying games like World of Warcraft or the Sims. It starts with an engaging video trailer explaining the company's sus-tainability strategy and the online game, which enables users to dynami-cally explore and learn about the company's supply chains. Users choose a character (male or female, adult or teenager) who begins in a simulated grocery store, where they can choose to learn about a variety of the company's products. Once a user chooses a particular product, their character is able to fully investigate its supply chain, working their way from the grocery aisle back to the delivery and manufacturing processes, and then all the way to the farms and forests where the product's ingredients come from. Along the way, the character can earn points and coupons by meeting real workers in the supply chain and learning about both the real-life challenges and opportunities facing the company's efforts to become more sustainable.

While the first mode is focused on making the company's efforts more salient, learnable, and memorable, the second mode is focused on information accessibility and usability. It provides easy-to-use pulldown menus of product categories, an interactive search engine, a comprehensive site map, and highly readable text that puts all the information about the company's sustainability efforts at the user's fingertips. It is structured in a relatively flat and easy-to-navigate hierarchy, with each page uncluttered and simple to use. Importantly, it is designed to be mobile-friendly so users can access it easily in stores. This is critical because Anu envisions customers scanning the barcodes of their products and instantly accessing sustainability information about them. This information is organized in a nested fashion, from quick summaries on the first page to detailed statistics on other pages for those who want to dive deeper. This way, customers who have never heard of the Cradle to Cradle label, for example, can quickly learn more about it.

Through this nested and dual-mode approach, Anu hopes to solve the "unresolvable dilemma" discussed earlier. By providing two tracks, she can create a fully immersive and engaging experience for stakeholders who want to really understand her company's strategy, while also providing a more efficient and flexible information delivery system for those who just need straightforward information to decide among different products. She is realistic about how many people will choose the immersive option, but also recognizes the value of providing that option for those who are willing to take the time to engage with it. Without such engagement, they are likely to remain cynical about and critical of her company's sustainability efforts; with it, they may become less cynical, and even positive, about those efforts. It may even lead them to become advocates for the company's efforts and share what they have learned throughout their social networks. Anu is also hopeful the game-like atmosphere of the site will be particularly engaging for Millennials, who have grown up with such online experiences.

At the end of the day, however, Anu also knows that the quality of her company's sustainability efforts—and the hard work of her staff—will not be measured by how interactive or intelligible its website is, or even how well designed their full information delivery system is. This system may make her company's sustainability information prominent, usable, and compatible for a range of different stakeholders, and that information may be as salient, trustworthy, and methodologically valid as possible. But if the use of that information by those stakeholders does not result in tangible benefits to the environment and society, no one will care. Indeed, the information pessimists will be justified in telling her, "I told you so." Ultimately, regardless of whether they are developed by corporations, nonprofit organizations, or government agencies, it is the effects and effectiveness of these information-based governance strategies that really matters. And so it is these topics we turn to in chapter 6.

# 6 Being Green: The Effects of the Information

## Evaluating Electronics

Vernon is a senior official at a state-level government agency, and has been tasked with identifying which electronics certifications and ratings his agency should use to identify preferable products in its procurement process. The agency has to make a large order of new computer monitors, and its staff members have a mandate for purchasing environmentally friendly products. The administrator of Vernon's agency has expressed a strong preference for certifications that show they actually "make a difference." The administrator agrees with Vernon that issue salience, organizational trustworthiness, methodological validity, and interface usability (the topics of chapters 2–5) are also very important. But at the end of the day he wants to be able to demonstrate specific and tangible benefits from these purchases to the people of their state, and wants his agency to use a certification or rating that ensures those benefits are being delivered.

The materials Vernon has reviewed suggest that there are three front-runners to focus on in the electronics space—ENERGY STAR, EPEAT, and TCO. Following the advice of the earlier chapters of this book, Vernon has already analyzed the usability, validity, trustworthiness, and salience of these three programs. But he is stuck on this issue of "making a difference." Holding all of these other variables constant, how can he evaluate and differentiate between the programs' ability to demonstrate their environmental benefits? He recognizes that while they may be easy to use, developed by trustworthy organizations, focused on important issues, and based on robust methods, they may still not deliver the tangible and exemplary results that his boss is looking for. But what should these results look like, and how extensive and specific should they be?

He finds that certifications in the electronics sector are indeed making claims about their impacts. ENERGY STAR, for example, states that its displays are 25 percent more energy efficient than standard options. It also highlights that its certified office equipment, which includes computers, monitors, and printers, has saved more than 500 terawatt-hours (TWh) of energy and more than $50 billion in energy-bill costs in the United States.[1] EPEAT, which not only includes criteria related to energy efficiency but also toxic substances, recycled materials, and other environmental concerns, states that its 757 million registered electronics products that were purchased between 2006 and 2013 have reduced 528,000 metric tons of hazardous waste, eliminated enough mercury to fill 4.6 million fever thermometers, and decreased the equivalent amount of waste produced by 248,000 U.S. households annually.[2] It also claims that its certified products in 2013 alone will reduce 20 million kilograms of water pollutant emissions and 2.2 million metric tons of greenhouse gas emissions.[3] TCO's criteria are more comprehensive than those of ENERGY STAR and EPEAT, and include a broad range of metrics related to environmental performance, socially responsible manufacturing, labor rights, conflict minerals, and product ergonomics and usability.[4] However, it makes no similar claims about the specific impacts of its certification on its website.

So which program is the most impressive? Which should Vernon recommend to his agency's administrator? Vernon likes the breadth of TCO, but is concerned by its lack of information about its effectiveness. While he appreciates their efforts to be explicit about their impacts, he finds it difficult to compare the claims of ENERGY STAR and EPEAT. He realizes he needs a framework for thinking about and evaluating these claims. This chapter provides such a framework, and is designed to be helpful for policymakers such as Vernon—or activists such as Cathy, corporate executives such as Anu, academics such as Lynn, or consumers such as Mark—who are trying to evaluate the effectiveness of competing sources of sustainability information. For at the end of the day, Vernon's boss is right—it is indeed the impact of these programs that really matters. All of the aspects of the information value chain that I have discussed in previous chapters (the issues covered, the methods used, the organizations involved, and the interfaces provided) ultimately must be oriented toward creating tangible environmental benefits. Otherwise these certifications—and the entire enterprise of information-based governance—is at best a waste of every-one's time and at worst actively causing harm.

This chapter begins with a discussion of several important concepts and distinctions that will be useful to us as we think about the impacts of these programs. They include audience responsiveness, effects, effectiveness, and the differences between inputs, processes, outputs, and outcomes. The chapter then summarizes the insights from sixty-eight interviews I conducted with a broad range of individuals about their perceptions of information-based environmental governance strategies. These interviews reveal a wide variety of ways in which the effects and effectiveness of these programs are framed and perceived. To explore what types of claims environmental certifications and ratings are making about their own effectiveness, I then introduce data on the outcome transparency of the cases in my Environmental Evaluations of Products and Companies (EEPAC) Dataset. The results demonstrate that these programs have not focused on publicizing their effectiveness to the public, and several types of outcomes are barely mentioned at all. They also show that EPEAT and ENERGY STAR are actually leaders in providing such transparency, and I discuss the most promising practices that they and other programs have employed.

The chapter continues with a further discussion of the tension between information pessimists and optimists, and how greater clarity about the goals of these initiatives and greater intentionality about measuring their progress toward those goals may help reduce this tension. Such clarity and intentionality are hallmarks of information realism, which are further elucidated at the end of the chapter. The chapter ends by returning to Vernon's conundrum, and analyzes his options in light of the data and conceptual framework introduced in the following sections.

## An Information Effectiveness Framework: Audience Responsiveness

This book's previous chapters illustrate the complexity of the development process behind information-based governance strategies, from identifying issues to cover and data to use to building institutional relationships and creating interfaces to connect with users. But what are the effects and consequences of these strategies once they are implemented and begin releasing information to the public? If they are environmental governance strategies, through what mechanisms do they actually have an impact on the environment, if at all? This chapter discusses the nature of information from this more consequentialist perspective, building on the philosophical position that the normative qualities of an action "depend only on its consequences."[5]

### The Importance of Audience Responsiveness

Because of the basic voluntary nature of information-based governance,[6] audience responsiveness to these programs is the primary mechanism through which they act and have consequences. If audiences respond positively to the information, they may then pursue complementary forms of governance based on regulations, market dynamics, technological development, moral arguments, or additional information-based strategies. They then become, if they are not already, stakeholders in the initiative, which can be understood as "any group or individual who can affect or is affected by the achievement of the organization's objectives."[7] In order to be effective, information-based governance strategies need to target specific audiences and convert them into stakeholders who will positively contribute to their success. Consumers and institutions can change their purchasing behavior, manufacturers can introduce new technologies, government agencies can enact new regulations, and advocacy organizations can begin new campaigns, all in response to the information provided by these information-based strategies. The environmental performance related to the original focus of each particular information-based governance strategy can then be improved and a public or common good created.

The effects of these programs are therefore strongly mediated by the responsiveness of different audiences. Figure 6.1 illustrates this dynamic, and shows the potential for a wide range of potential actions that audiences can take in response to the information provided. While certain actors are more associated with particular responses (the solid lines), they may support and pursue other strategies as well (the dotted lines). The important point is that information-based strategies are not necessarily dependent on one audience (e.g., consumers) to be effective, but can stimulate a range of collective actions by several types of audiences to create public and common goods. Indeed, David Vogel, a professor of political science and business ethics at UC Berkeley, emphasizes the limitations of consumer-focused voluntary programs, given many consumers' unwillingness to "internalize

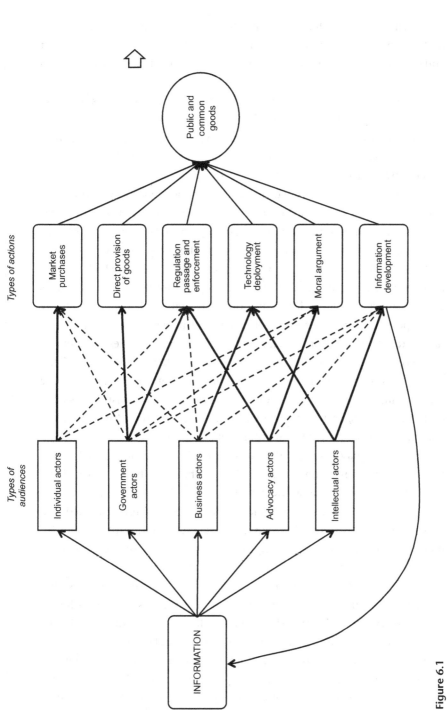

**Figure 6.1**

Information-based governance effect pathways. *Note:* Solid lines indicate responses most likely pursued by indicated actors, while dashed lines indicate other important actions each actor might likely undertake.

the environmental externalities of what they consume," and concludes that the real value of these programs is when they leverage more stringent and effective government regulation.[8] This is a theme that we will return to later in this chapter.

Regardless of the type of action stimulated, the responsiveness of these audiences is likely to be strongly influenced by the salience, trustworthiness, credibility, and usability of the information provided, as described in earlier chapters. In addition to these factors, audiences are also likely to be affected by their perceptions of the effectiveness of these strategies, which in turn are likely to be defined in terms of the audience's own interests.[9] Different audiences are furthermore likely to have quite divergent interests. Government policymakers, for example, most likely view information-based initiatives from the perspective of whether they enhance their own authority and create a race-to-the-top "California effect" in which companies go beyond regulatory requirements, replicating the modeling effects of California's regulatory leadership on other states.[10] They are less likely to support programs that diminish their own power and create an alternative race-to-the-bottom "Delaware effect" in which states compete to create the least-regulated business environment (often deemed to be Delaware).[11] Research on consumer motivations to buy green products, on the other hand, indicate it is the product's performance, symbolism and status, cost effectiveness, and credible environmental claims that drive consumer willingness to pay for environmentally labeled products.[12]

Companies may conclude that an information-based initiative is effective if they view it as improving their corporate profits and stock prices through anticipated marketing benefits, attraction of new, more affluent customers, or increased satisfaction of existing customers.[13] Reduced production costs, improved employee morale, preemption of regulations, increased costs for rivals, and improved opportunities for industry cooperation may also motivate companies to support these initiatives.[14] Corporate responsiveness to information may also depend on factors that scholars Neil Gunningham (Australian National University), Robert Kagan (UC Berkeley), and Dorothy Thornton (UC Berkeley) identify as influencing corporate environmental compliance with regulations, such as community and activist pressures and companies' environmental management styles.[15] Alternatively, an effective initiative for a civil society activist ultimately needs to drive consumer pressure, government regulation, or direct corporate action that results in improved corporate environmental performance and improved environmental quality.[16] While overlap in their interests often exists, each of these actors likely has different underlying perceptions of the effectiveness of information-based governance strategies, which are summarized in table 6.1.

It should be highlighted that if an audience perceives a strategy as antithetical to its interests, it may decide to pursue strategies that aim to undermine it. Benjamin Cashore, Graeme Auld, and Deanna Newsom's work at Yale University shows how the Forest Stewardship Council (FSC) label drove many U.S. foresters to support the alternative Sustainable Forestry Initiative label, which ultimately forced FSC to revise its standards.[17] Thus it is important to pay attention to both the positive and negative feedback effects among different initiatives and different types of governance. The success of a voluntary certification, for example, may encourage the adoption of stronger industry standards and regulations, but it may also defer such standards and regulations as well. Different governance types should therefore not only be analyzed individually but also in terms of how they interact and either complement or undermine each other.

### The Effects and Effectiveness of Information Regimes

An important question about the effects of these programs relates to the level of intentionality behind them. A results-oriented perspective would argue that intention is irrelevant, as long as positive results are achieved (however "positive" may be defined). A goal-oriented perspective would instead focus on the objectives of these programs and whether they are

**Table 6.1**

Potential factors driving perceptions of effectiveness

| Actor | Potential factors driving perceptions of effectiveness |
| --- | --- |
| Government | Perceived policy complementarity, legal mandate support, and environmental protection |
| Consumer | Perceived improvements in product quality, cost effectiveness, health and safety, and environmental protection |
| Corporations | Perceived improvements in efficiency, employee morale, customer satisfaction, competitor costs, policy preemption, industry cooperation, and environmental protection |
| Activist | Perceived environmental benefits through consumer pressure, government regulation, or direct corporate action |

achieved. From this perspective, other benefits are fine, but irrelevant if the explicit goals of the initiative are not achieved. These different perspectives echo the distinction that Harvard scholars Archon Fung, Mary Graham, and David Weil make between effects and effectiveness: "A policy has *effects* when the information it produces enters the calculus of users and they consequently change their actions. Further effects may follow when information disclosers notice and respond to user actions. A system is *effective*, however, only when discloser responses significantly advance policy aims."[18]

An effective information-based program therefore has its intended effect, but it may have other effects (both positive and negative) as well. This distinction, however, raises the question of whose policy aims and intended effects are we referring to? As the comments from the interviewees that follow demonstrate, stakeholders have quite diverse opinions on what the goals of these initiatives should be. Both results and goal-oriented perspectives are therefore needed to evaluate information-based strategies. It is important to evaluate both the effectiveness of these strategies in achieving their stated objectives and their broader effects on society, individuals, and the environment. In some cases, they may fail to achieve their objectives, but still have important effects. In other cases, they may achieve their stated objectives, but have limited effects. This disconnect between effects and effectiveness may in part be due to poor, limited, or nonexistent goal setting by these programs.

One of the challenges of setting strong goals is the variety of ways to measure a program's effects. As political scientists William Gormley at Georgetown University and David Weimer at the University of Rochester point out, these effects can be related to the resources they use (a program's "inputs"), the means by which those resources are used (a program's "processes"), the direct products of those processes (a program's "outputs"), and the valued consequences of those outputs (a program's "outcomes").[19] Ideally these effects are mutually and positively reinforcing, such that the use of high-salience and high-quality inputs (e.g., program criteria and data, the subjects of chapters 2 and 4, respectively) is associated with the use of valid participatory and analytical processes (e.g., governance incorporating relevant stakeholder groups and methods using life cycle analysis, the subjects of chapters 3 and 4, respectively). These processes are then in turn associated with meaningful outputs (e.g., certified products sold) and outcomes (e.g., acres of forest preserved). However, such associations cannot be assumed, as these dimensions of performance may in fact lack linkages or even conflict with one another. Economical but shortsighted use of inputs, for example, may result in significantly weaker outcomes.

As discussed in chapter 3 and reinforced by work by Sébastien Mena and Guido Palazzo, stakeholders may evaluate the legitimacy of organizations on the basis of any of these dimensions.[20] While Gormley and Weimer define an organization's performance as "its impact on outcomes," they also acknowledge that this conception of performance is difficult to measure, often because the organization is far removed from its ultimate outcomes or it is impossible to measure those outcomes comprehensively or weight them precisely.[21] This explains why many metrics of effectiveness are based instead on input, output, and process variables—or on less than ideal outcome variables. Each of these options have their strengths and limitations, and stakeholders may have different preferences for them. Funders may focus more on efficiency and inputs, while communities that believe their voices have not been heard may emphasize participation and processes. Meanwhile, managers who prefer specific production targets to achieve may favor output-based metrics, while advocacy groups may emphasize the need for specific environmental and health outcomes.

These different aspects of performance highlight the fact that effectiveness is ultimately in the eye of the beholder. How effective a program is perceived to be may depend a great deal on who you ask. As emphasized earlier, the fact that information-based governance strategies are essentially voluntary in nature makes audience perceptions of these strategies even more important. If audiences perceive the strategy as being effective, they may be more likely to respond to the information it provides, which in turn can further increase its effectiveness. The logic of these feedback effects is supported empirically by experiments by Boston University marketing professor Sankar Sen and his colleagues Zeynep Gürhan-Canli (University of Michigan) and Vicki Morwitz (New York University). Their work shows that consumers are more likely to participate in a boycott if they view it as effective.[22] Thus it is valuable to understand how different audiences perceive these programs and what their more general effects are. It is also important to identify how these programs themselves define their effectiveness and what factors they believe may be driving that effectiveness. Through this process, the roles of different mechanisms by which certifications and ratings may be contributing to the creation of public goods and governance efforts can be better understood. It can also allow us to explore the possibility of identifying a unifying concept and measure of effectiveness across multiple contexts and sectors for these programs.

### Perceptions of "Green" Effectiveness: Interviews with Stakeholders

This section presents insights from sixty-eight interviews with consumers and representatives from companies, nonprofit organizations, government agencies, academic institutions, and organizations behind several different ratings and eco-labels. I consider all of these individuals to be important stakeholders of these programs because they have the capacity to both affect and be affected by their outcomes. In the sections that follow, I first describe the methods used to select the interview participants and to conduct the interviews, and then discuss the interviewees' views on the effects and effectiveness of product eco-labels and corporate green ratings. This research identifies a wide range of both effects and measures of effectiveness articulated by these participants. While clear environmental outcomes were the most commonly cited metric of eco-label effectiveness, respondents did not agree on any single overarching definition of effectiveness for these types of programs. These interviews provide a relatively comprehensive view of how different audiences—consumers, activists, regulators, executives, academics, and raters themselves—perceive the dynamics and consequences of eco-labels and sustainability ratings.

### Interview Methods

For these interviews, I selected a stratified sample of consumers and representatives from nonprofit organizations, companies, government agencies, academic institutions, and evaluation organizations. In total, I interviewed sixty-eight individuals for approximately one hour each. The interviews with organizational representatives focused on understanding their perspectives on the effects and effectiveness of product eco-labels and corporate environmental ratings that they were already knowledgeable about, while the consumer interviews presented information about several labels and ratings to them and then explored their impressions of them. I chose to interview representatives from each of these groups in order to hear from a wide range of individuals with different backgrounds and to better understand the similarities and differences in their views of eco-labels and green ratings. Quotes included in the sections that follow are only identified by the interviewee's type of organization (company, nonprofit, etc.) and thus do not identify specific individuals. Before discussing their specific comments, I will first summarize the backgrounds of the participants in my interviews—for a full description of my sampling methods, see appendix II.

In selecting the company representatives, I limited the sample to staff working at companies in the consumer electronics sector. The consumer electronics industry has a long history and wide range of eco-labels and green ratings, and provides a rich case study from which we can learn. I interviewed representatives from nine companies, including the #1 seller of music products (Apple), the #1 seller of personal computers (Dell), and the #1 seller of audio-visual equipment (Sony).[23] I also interviewed individuals working at nine organizations who are implementing eco-label or green rating initiatives related to the electronics sector. They included individuals involved with the implementation of 80Plus, ENERGY STAR, EPEAT, Greener Electronics Guide (by Greenpeace), and TCO Certified.

For the other stakeholder groups, I did not limit my sampling to the electronics sector, primarily because not as many individuals in these groups are exclusively focused on electronics. For government stakeholders, I was able to conduct interviews with sixteen individuals representing one congressional agency (the Government Accountability Office) and three executive agencies (the Environmental Protection Agency [EPA], Federal Trade Commission [FTC], and Department of Energy [DOA]). I also interviewed representatives from ten nonprofit organizations, including Rainforest Alliance, World Resources Institute, Consumer Federation of America, Union of Concerned Scientists, and EarthJustice. And I conducted interviews with twelve academic researchers with expertise in either electronics or eco-labels and with backgrounds in economics, political science, public policy, marketing, or engineering. They come from a range of institutions, such as Arizona State University, Ohio State University, Harvard University, and the Georgia Institute of Technology. I also interviewed a diverse sample of twelve consumers that included six men and six women, seven age 40 or older individuals and five under 40, and three high school-educated, two in college, four college-educated, and three graduate school-educated.

The consumer participants also have a range of annual household incomes, from less than \$25,000 to more than \$100,000. The final sample of interviewees is presented in table 6.2. The sections that follow summarize their perspectives, beginning with the perceived effects of information-based environmental governance strategies.

### Perceived Effects of Eco-Labels and Green Ratings

After going over their knowledge and impressions of existing product eco-labels and corporate environmental ratings, I asked all of the interviewees an open-ended question about what they thought the effects of these kinds of programs have been. I then followed up with more specific questions about their effects on the policies and behavior of companies, government agencies, nonprofit organizations, and consumers. I also asked whether they believed these programs have undermined or complemented other environmental policy initiatives, such as regulations. While the sections that follow highlight the range of perspectives among representatives from different stakeholder groups, I did not detect any systematic differences across those groups.

**Company Effects**   The main effect on companies that corporate representatives cited was the role of eco-labels and ratings as a "motivational tool." One manufacturer representative stated that these programs are "definitely driving design decisions," and are highly influential in manufacturing processes. Another noted that there is an "absolute need for [such] aspirational

**Table 6.2**
Interview sample summary

| Interviewee background | Total | % of total |
|---|---|---|
| Nonprofit Organization | 10 | 15% |
| Consumer | 12 | 18% |
| Academic Expert | 12 | 18% |
| Company | 9 | 13% |
| Government Agency | 16 | 24% |
| Evaluation Organization | 9 | 13% |
| Total | 68 | - |
| Average | - | 17% |

standards." One retailer representative stated that he believed these initiatives have motivated manufacturers to perform better, and have allowed retailers to effectively promote the environmental and energy efficiency benefits of certain products. Another noted that they are "geared to many different audiences"—some, such as ECMA 370, are oriented toward businesses and procurement officers who know what they are looking for, and are not designed for the general consumer.

Other stakeholder representatives expressed similar sentiments. One government representative stated he believed that these programs have encouraged companies to make greener products, while a second said he thought one of their biggest effects was "innovation stimulation." A third asserted the specific effects were that they "taught companies not to be afraid [of sustainability efforts]" and "how to make money from [greener products]." A fourth government official, however, expressed a more skeptical view—that the actual results of these programs are mixed and that while many make companies feel good and give them a "green badge of courage," in reality they do "squat." A fifth stated that while they have done some good, their contribution has been very limited in the broader context of environmental policy.

Representatives from nonprofit organizations expressed similar caveats about these programs, but in general were positive about their effects on companies. They stated that companies "take them seriously," "pay attention and are motivated by them," and are incentivized by them to improve their performance. One of the consumers interviewed thought that corporate leadership is an important mediator of these effects—"I think that in general companies are being pressured in trends in political consciousness to create a rating system ... and based on who runs the company [and] who is associated with it, that's going to [determine] how effective it is."

**Consumer Effects** Nonprofit organization representatives were also relatively positive about the effects of these initiatives on consumers. For some, the best eco-labels and ratings are "quick tools" that "empower consumers" and provide "information resources" to consumers. Others asserted that eco-labels have made the issues they cover, from climate change to deforestation, more familiar to consumers. Several of the academic experts on consumer behavior interviewed expressed similar attitudes—one stated that these initiatives have done a "decent job matching consumers and producers," and another cited a specific example from his own research that showed product sales increasing after an eco-label was introduced.

Nevertheless, some participants expressed reservations about the effects of these programs on consumers. One government official said that even one of the most successful eco-labels, ENERGY STAR, still did not cover much of the market (in actuality, ENERGY STAR's market penetration varies widely, from 0 percent for small scale servers to 100 percent for cable boxes, and depends on a variety of factors).[24] Another asked rhetorically, are these programs "a drop in the bucket or a huge success?" and answered his own question, "Hard to say." Several representatives from the companies, evaluation organizations, and other groups expressed concerns about the effects of "eco-label proliferation" on consumers. Such proliferation, in their eyes, might be causing confusion, disillusionment, and "green fatigue" among shoppers. However, others did not see a problem with this expansion, and believed that this phenomenon is still in its infancy and only covers a fraction of what it should be covering.

As evidence against an enduring overload effect, one respondent cited the example of nutrition labels—when they are first introduced or when people first encounter them, they may seem overwhelming, but once people become familiar with them they are able to "filter out" the extraneous information and focus on what is important to them (vitamin A vs. calories vs. sugar content). Another respondent, however, used nutrition labels as an example of how providing lots of detailed information has been overwhelming and has not had the intended effect—despite the introduction of these labels, obesity levels have increased over the last twenty years.

What do consumers themselves say? Those that I interviewed expressed a range of views, but in general were positive about these programs. When asked whether they would make use of the eco-labels they learned about in the interview, one said, "I think I would take them into account, but I wouldn't go to the ratings as my first stop … I would probably narrow it down to a few washing machines, and then I might see if they are on a list of labeled or ratings products." Another said she thought "there should be more of them—they should be standards for what we buy," and another concluded, "I would want to use a combination of them, as none of them covered what I wanted. I felt they were incomplete, but now that I know I would definitely want to look at them." But a fourth participant remarked, "I might compare one or two but not all of them, TMI [too much information]!"

**Government Effects**  The most commonly cited effect on government was the use of eco-labels as procurement standards. In 1999, for example, President Clinton issued an executive order mandating all federal agencies to select ENERGY STAR-labeled products.[25] In 2007, President Bush issued a similar order requiring federal agencies to buy EPEAT-registered products for at least 95 percent of their needs.[26] More recently, President Obama signed Executive Order 13693 in 2015 that requires federal agencies to "promote sustainable acquisition and procurement" by purchasing products whenever practicable that are certified not only by ENERGY STAR and EPEAT but also by the BioPreferred, WaterSense, Safer Choice, and SmartWay Programs (as well as other products identified by the EPA or DOE as "energy and water efficient").[27] These orders have forced government agencies to be leaders in procuring certified products, and have created an important market for them.

Other interviewees mentioned the greater efficiency that these voluntary initiatives have over traditional regulatory processes—they are much more informal and enable conversations with industry, nonprofit organizations, and even other countries that do not normally happen in the more adversarial and bureaucratic regulatory process. One government official stated, however, that she believed these programs actually are less efficient and more expensive than traditional regulatory processes, because they take a lot of time and money to collaborate with industry and other groups to jointly develop their standards. The length of time that it can take to complete a regulation or a voluntary standard can vary significantly, and interviewees disagreed as to which takes less time on average. Another official thought these voluntary programs can often be a distraction from the mission of the EPA, which is to "protect human health and the environment." This relates to the more general issue of whether these initiatives complement or undermine regulatory efforts, which I will return to later in this chapter.

**Nonprofit Effects**  Several participants noted that eco-labels often create divisions within the advocacy community, where some are positive and optimistic about them and others are more skeptical and pessimistic. This dynamic leads the former to be more engaged in these efforts, while others remain critical and focus on other strategies. One advocacy organization representative noted that even though his organization has been involved in creating a green rating program, it "was not in isolation from other ongoing projects, [such as] pushing for state laws, working with purchasers, etc."

Another said that one criticism of these initiatives is that "NGOs are often outgunned and outweighed in their development processes," as it is the companies who have the resources and staff to participate in ongoing meetings and workshops around standard-setting and criteria development. For some of the government participants, this issue underscores the importance of using weighted voting procedures during the standard development process. Such procedures limit the voting power of any one particular stakeholder group to 33 percent, for example, or no greater than 50 percent. Such rules underscore the importance of defining the stakeholder groups in a manner that balances the different interests appropriately. These participants also highlighted the value of practices employed by organizations such as NSF International that pay for the travel and meeting expenses of nongovernmental organizations to attend their standards development meetings.

**General Effects**   Several other, more general effects of these programs were cited as well. Citing consumer surveys commissioned by his agency, one government official asserted that general claims of environmental friendliness or greenness create confusion and skepticism among consumers, and therefore specific claims about environmental attributes are more appropriate and helpful. Several interviewees, and in particular two academic experts, expressed concerns about the unintended consequences of these programs and their potentially negative effects on environmental protection efforts in the long term. As an example, one interviewee said LEED's point system may encourage tearing down buildings, which may not be the best environmental outcome.

I also asked every interviewee about another potential general effect of these programs, which is whether they complement or undermine other forms of environmental governance, and in particular environmental regulations. The majority of the respondents believe that eco-labels and green ratings complement regulatory efforts, although there were some strong minority opinions. On the complementary side of the argument, one government interviewee described an important downside of regulation that voluntary programs can address. In the building industry, for example, regulations create "perverse incentives" that encourage "builders to treat building codes as the maximum they are supposed to do." Their goal becomes minimizing their efforts at compliance, and therefore performance and enforcement greatly depend on the diligence of the inspector. Voluntary ratings and labels attempt to change this dynamic and create competition among builders in going beyond compliance. In this way, the regulatory code can become the floor of performance, rather than the ceiling.

This logic was echoed by many other interviewees, although some emphasized that the extent to which labels work in this manner depends on important contextual factors, such as the expense and difficulty of meeting the voluntary standards, the threat of further regulatory action, and the culture of the industry. One participant involved in the electronics sector stated, for example, that the competitive culture of his industry had made it more amenable to competing on environmental criteria, which may not necessarily occur in other sectors. Most interviewees therefore emphasized the complementary relationship between voluntary and regulatory programs, and that both are needed to improve environmental performance. Several argued that the key is to ensure that the voluntary standards that begin as goals are ultimately transformed into expectations for the entire industry.

A few respondents, however, expressed skepticism about the extent to which this occurs, and cited the opposite phenomenon as being just as likely—"successful" voluntary programs providing an excuse for not passing and implementing more extensive regulations. One respondent cited ENERGY STAR as an example of this dynamic. Even though it has certified a large proportion of products in a range of different product categories and has raised its standards for many of those categories, there are still many products on the market that do not meet even the original ENERGY STAR standards. This government official claimed that the program is nevertheless seen as successful, and is used as a strong argument against further regulation: for example, "Why are government standards for these appliances needed when we have ENERGY STAR?" Other participants asserted that the lack of full market penetration of programs like ENERGY STAR demonstrates the need for mandatory standards (set by government agencies such as the DOE) to set a performance "floor" for all products, certified or not.

## Definitions of Effectiveness

A factor driving these debates may be differences in how these participants define "successful" or "effective." Indeed, differing perspectives on the nature of effectiveness may explain why some interviewees emphasize particular effects of eco-labels and de-emphasize or ignore others. I therefore also asked all of the participants in my interviews to explain how they themselves define effectiveness, and what it means to them in the context of environmental certifications and ratings. The sections that follow summarize their responses.

**Environmental Outcomes**  The most common definition used in the interviews focused on the environmental outcomes of the program. Some participants answered in the form of questions, such as "Does it improve the environment?" or "Are they solving some specific problem?" Others said they must be evaluated in terms of their "observable environmental improvements," "overall benefits," "physical benefits," or "making an impact." Several participants put effectiveness in the context of the goals of the program, asking "What are the environmental impacts they are trying to reduce?" and "Does it achieve their objective—whatever they set out to accomplish?" Others cited specific metrics of performance, such as a "net reduction in $CO_2$ emissions." One academic expert stated that the standards need to "be strict enough that their impacts are significant," while another emphasized that they must focus on present and past performance, not future expectations. An advocacy organization representative emphasized, however, that the standards should take into account the goals of companies as well as their past performance, but need to penalize them when they retreat from those goals. He also emphasized that their standards must be "beyond what is required by law," and result in "transformative change."

**Consumer Behavior Outcomes**  The second most commonly cited definition of effectiveness relates to changes in consumer behavior. Common phrases included: "Does it change consumer behavior?" "Has it caused a shift in consumer demand?" "Do consumers recognize it?" "Do they motivate purchasers to change their decisions?" "Do they help consumers identify a recognizable brand message?" Others mentioned more specific metrics, such as the share of a market that an eco-label has certified. One company representative said effective initiatives must "actually result in sales of products that are better for the environment," explicitly linking this focus on consumer behavior to the environmental outcomes discussed in the previous section. A nonprofit representative emphasized the ability of these programs to "resonate with consumers," implying that eco-labels must be salient and relevant to consumers in order to be effective. One participant emphasized the difference between product eco-labels and corporate "scorecards," which she asserted differ in their audience orientation. Eco-label effectiveness should be measured by the labels' market penetration because that is their orientation, but scorecards are less consumer oriented and should be evaluated differently.

**Company Behavior Outcomes** Along these lines, other interviewees emphasized that these information-based governance strategies can also be effective by eliciting changes in company behavior directly. As one academic representative asked, "Does it change company behavior?" The main point here is that rather than operating indirectly through consumers and markets, these programs can influence companies themselves as "effective campaign tools" that allow advocacy groups to "go after individual companies," as one nonprofit representative explained. Another nonprofit representative said that people should realize that scorecards and ratings used in this context are "designed to be opinionated and subjective" and are used to make a point about society's values. They are not meant to be a full and final scientific assessment of a company's environmental impact.

Other interviewees emphasized that effectiveness can also be defined in terms of specific changes in company behavior, such as being more transparent about their product's manufacturing processes or ingredients. Encouraging companies to "really do innovation" and bring new green products to market was also mentioned as a dimension of effectiveness, as was a broader effect of "promoting competition" among companies on green attributes. Others mentioned that some company-supported labels and rating systems are more internally oriented in order to motivate and organize a company's environmental management efforts. Other programs are focused on enhancing communication and collaboration among companies so that lessons learned are shared and a sense of industry momentum is created. Another aspect of changed company behavior discussed was procurement—programs that are regularly used by corporate procurement officers may also be viewed as effective.

**Public Policy Outcomes**   Rather than focusing on consumer or company behavior, several interviewees noted the importance of changes in public policy as a measure of effectiveness. As one NGO interviewee asked, "How does it influence policy?" Another interviewee who has been involved in producing an environmental ranking of companies stated that the goal of that effort was "not to affect consumers but to impact public policy." These interviewees emphasized that such ratings and rankings can raise awareness of the issue in question, and create demand for stronger regulations. In this context, one of these participants highlighted the importance of having results that are interesting to the media, which can then raise the profile of the initiative and attract attention from policymakers.

**Awareness and Education Outcomes**   Some interviewees also mentioned a more general measure of effectiveness that was unconnected to any specific audience. This measure was increased "awareness and education" about the issue in question and the environment more generally. Does it educate consumers, policymakers, or executives about corporate or product environmental performance? Does it increase awareness about the importance of the environmental impacts of consumption and production? Such a definition of "effectiveness" implies a longer-term and more indirect mechanism of social change and environmental progress—through learning and sensitization over time. An emphasis on this type of outcome might justify easier-to-achieve standards if it raises awareness among key stakeholder groups and introduces key issues to a significant portion of the public.

**Knowledge and Information Outcomes**   Similarly but more specifically, other participants emphasized the intrinsic importance of the accuracy of the information provided by the initiative. As one interviewee stated, it "must be credible"; or another, "it must be verifiable"; or another, it must have "quality control." On the surface, such statements may appear to be more descriptions of drivers of effectiveness than definitions of effectiveness itself, and indeed some interviewees did appear to conflate the proximate drivers and the ultimate goals or definitions of effectiveness. On a deeper level, however, increasing society's knowledge and the quality of information about the environmental performance of products and companies may indeed be a goal in itself. With such information, policymakers and citizens can make better decisions about whether the environmental impacts of a product's performance are significant and which areas of performance are most important to address.

Another participant asserted that programs must be able to differentiate between companies and products on their environmental performance so that audiences can effectively choose between them. Others mentioned the importance of creating "simple," "clear," and "easy to understand" information. These are slightly different goals than information accuracy or quality, as in some cases slight differences found between two products or simplified data presentations may not be statistically significant, defensible from a scientific perspective, or important relative to other aspects of environmental performance. But it is additional information that may still be considered useful by some audiences and may incentivize further efforts to improve performance.

**Process Outcomes** Several interviewees also mentioned specific attributes relating to the processes by which eco-labels and ratings are created as metrics of effectiveness. Again, these may be interpreted as drivers and not definitions of effectiveness, but they also can be seen as ends in themselves. The first such attribute mentioned was related to trust—that "people know it and trust it." Building a trustworthy eco-label, and a trustworthy process behind it, can not only build the public's confidence in claims about particular products but also about environmental issues more generally. How such trust is built is of course another question, although another participant also emphasized the importance of democratic processes, which perhaps is one factor that can contribute to building such trust. But the fact that an eco-label was created with input from a wide range of voices in a democratic manner may also be an explicit goal as well—democratic decisions are often seen as more valid and legitimate, regardless of their content and outcome. And finally, one participant defined effectiveness in terms of the long-term "durability" of the program. "Is it built to last over time?" Will these programs be around in forty or fifty years? Such durability can obviously contribute to other dimensions of effectiveness, but creating an institution that persists over time can also have independent value, as it becomes an established source of benefits and sustained progress for society.

## Outcome Transparency of Information-Based Environmental Governance Strategies

Clearly the stakeholder representatives I interviewed described a wide range of real and potential outcomes of information-based governance strategies. The importance of these different forms of effectiveness depends on not only your particular interests and background, but also whether you value more direct and tangible outcomes that may be more limited in scope, or more indirect and intangible outcomes that may have a broader scope. Other trade-offs exist as well. For example, specific environmental outcomes ultimately may be preferable to some groups, but also more difficult to measure than other types of outcomes—such as consumer awareness or purchases. These trade-offs and the many different forms of effectiveness mentioned raise the question of what types of outcomes are existing initiatives focusing on in their communications to the public? To what extent are they claiming to have produced the different types of effects highlighted by the interviewees in the preceding sections?

In order to address these questions, my research assistant and I conducted a content analysis of the website text of the 245 cases found in the EEPAC Dataset. Through this analysis, we identified text that mentioned either existing or potential outcomes associated with the case in question. Two levels of such outcome transparency were identified. The first was *limited outcome transparency,* which includes any general claims regarding the potential social or environmental benefits of the initiative (e.g., a computer with this label has a 30 percent smaller carbon footprint). *Strong outcome transparency,* on the other hand, includes specific claims regarding the actual benefits from the initiative (e.g., number of trees saved through an eco-label). Slightly more than 10 percent of the initiatives make specific claims regarding the actual benefits they create, while nearly 20 percent make general claims about their potential social or environmental benefits but do not discuss actual outcomes. Over 70 percent do not mention either real or potential outcomes of their program.

I then further coded the identified text for the more specific types of outcomes discussed earlier. As figure 6.2 shows, the most commonly mentioned type was environmental outcomes (22 percent of all cases). Approximately one-third of these claims were about specific outcomes that the initiative had itself created. The next most common type was company outcomes (19 percent of all cases). Nearly two-thirds of these claims were more limited and general statements about potential benefits. Only 7 percent of the cases directly mentioned any consumer outcomes, and only a handful discussed outcomes related to awareness and education (five cases), knowledge and information (six cases), public policy (two cases), or process (one case) outcomes. None of these latter four types of claims were coded as having strong outcome transparency—all were limited and general in

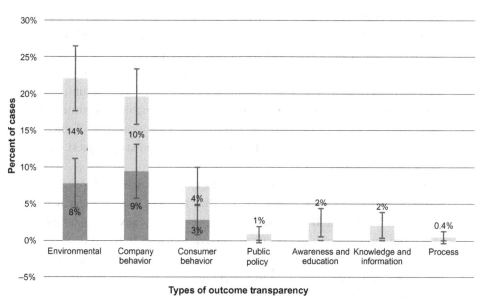

**Figure 6.2**

Types and levels of outcome transparency. *Note:* Figure shows seven types of outcomes coded for in the website texts of the cases. Light shading indicates cases having limited transparency about these outcomes, dark shading indicates cases having a high level of strong transparency about these outcomes. Error bars indicate 95 percent confidence intervals for each sample proportion.

nature. Any outcome transparency can increase the output legitimacy of these initiatives that was discussed in chapter 3, but the strong form is more likely to do so. However, it is important to remember that neither are measures of actual effectiveness, only the extent to which these organizations are making claims of effectiveness.

Some examples of these different claims of effectiveness highlight their diversity. As an example of a strong environmental outcome claim, the Forest Stewardship Council (FSC), mentioned in chapter 3, provides a four-page literature review of independent research on its "Impact in the Forest." The review cites, for example, studies showing that deforestation rates were twenty times lower in FSC certified areas than noncertified areas in Guatemala. An example of a limited environmental outcome claim was found in Earth Advantage's discussion of its home certification program. It states that every Earth Advantage home is designed to improve energy efficiency by 15 percent, but does not provide any estimates of actual energy savings or pollution reduction. Strong claims of company behavior outcomes were made by the Rainforest Alliance certification and Smithsonian's Bird Friendly Coffee programs—they both provide specific statistics on how many farms and hectares of land they have certified. The Corporate Responsibility Index provides an example of a limited claim of company behavior change—it states that its feedback reports enable companies to "identify areas for improvement and ensure efforts focus on areas of maximum impact." Sounds good, but no evidence is provided to suggest that this actually happens.

In terms of consumer behavior outcomes, EPA's WaterSense program makes a relatively strong claim, stating that it "helped consumers realize more than $55 million in water and sewer bill savings." An example of a more limited claim of consumer behavior change is Food Alliance's statement that its certification results in "positive customer feedback," "increased customer loyalty," and "sales increases," but no specific evidence of these effects is provided.

An example of a limited public policy outcome claim comes from ISO 14001, which states that its standards "provide the technological and scientific bases underpinning health, safety and environmental legislation" and "are the technical means by which political trade agreements can be put into practice." An example of a limited awareness and education outcome comes from the Corporate Lands for Learning certification, which claims that it fosters "a clear understanding of the interdependence of ecology, economics, and social structures in both urban and rural areas" in both children and adults. The 100 Best Corporate Citizens provide an example of a limited knowledge and information outcome with its quote from Intel's Director of Corporate Responsibility stating that the initiative has had "a huge impact internally" at Intel and his colleagues view its scores and rankings as a significant "learning opportunity."

Responsible Travel provides an example of a limited process outcome claim in its annual Responsibility Report, which details its vision, targets, outcomes, and next steps across nine major areas, from its membership and customers to its local community and the broader tourism industry. It states that it initiated a debate about the Sustainable Tourism Stewardship Council's (STSC) plans to develop Global Sustainable Tourism Criteria, and recruited more than eighty individuals to sign a petition demanding more transparency in the process. While not citing any specific outcomes, this action represents an effort to make the development of a new tourism standard more open and democratic. No instances of strong transparency claims about processes, public policy, knowledge and information, or awareness and education were found in the dataset.

As has been done in previous chapters, figure 6.3 maps out the landscape of outcome transparency across all 245 cases in the EEPAC Dataset. Due to their collective low occurrence, it groups the public policy, knowledge and information, awareness and education, and process outcomes in one general category of "indirect" forms of effectiveness, and plots it with the consumer, company, and environmental forms discussed earlier. Limited and strong forms of claims are grouped together in order to make the figure easier to read. It shows the 70 percent of the cases (172 in total) in the upper left-hand corner that do not make any claims about their effectiveness. The fourteen other cases in the top row make no claims about indirect outcomes or environmental outcomes while the fifteen cases in the far-left column make no claims about consumer or company outcomes. The second highest pillar represents twenty-five cases that make no assertions about consumer or indirect outcomes but do make some assertions about environmental and company outcomes. Overall, twenty-five cases in this landscape make one of these four types of claims, while thirty-seven make two of these types of claims and nine cases make three types. The two cases in the bottom right corner of figure 6.3 are FishWise

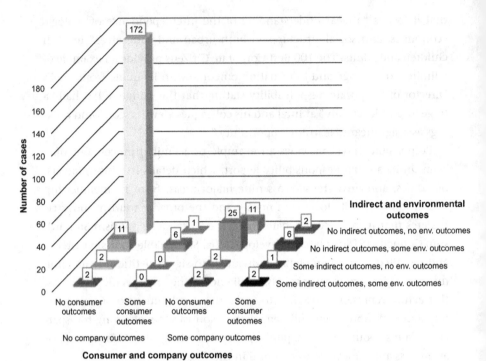

**Figure 6.3**

The landscape of outcome transparency. *Note:* The 172 cases in the back left-hand corner make no claims about having produced any consumer outcomes, company outcomes, environmental outcomes, or indirect outcomes. The two cases in the front right-hand corner claim to have produced all four of these types of outcomes.

and Responsible Travel, and they make all four types of claims about their effectiveness.

## The Information Realism Perspective

The interviews summarized in this chapter contained strains of both information pessimism and optimism. Some participants viewed eco-labels and sustainability ratings as having positive effects on corporate behavior, serving as a "motivational tool," a source of "innovation stimulation," and a mechanism for learning about the opportunities associated with sustainability efforts. Others described their beneficial effects on consumers, increasing their familiarity with environmental issues and increasing sales of environmentally labeled products. Some interviewees highlighted the increased procurement of these products by government agencies. They also emphasized the greater efficiency of information-based initiatives compared to conventional regulations, which has enabled government agencies to be more collaborative and adaptive as they pursue these approaches. Other participants expressed confidence that environmental certifications and rankings complemented rather than undermined existing regulatory approaches.

However, expressions of pessimism were not uncommon in the interviews. Some interviewees voiced concern about the mixed and limited results of these information-based efforts in terms of both consumer and corporate behavior. Others highlighted the problems of information overload and consumer confusion, the unappreciated costs of collaborative standard development processes, the dominant role industry plays in those processes, the distracting effects of these programs on regulatory agencies, and their potential unintended consequences. Several participants were skeptical of a complementary effect between regulation and information-based strategies, and were more concerned about the latter serving as a poor substitute for the former.

The data on outcome transparency presented in this chapter would likely be heralded by these information pessimists as further evidence of the inadequacy of information-based governance. Over two-thirds of the cases make no attempt to discuss their outcomes, and nearly two-thirds of those that do limit themselves to general and unsubstantiated claims about their effectiveness. No third-party verification is provided for the vast majority of the more specific claims that are made; we are supposed to trust these self-evaluating organizations even though they have a clear conflict of interest. Better to not trust any of them, and invest in real governance strategies that will create real results. Or so the information pessimist might argue.

Information optimists, however, would probably view the data with a more forgiving eye. Measuring the effects and effectiveness of any governance strategy is challenging and fraught with difficulties, and we should not expect anything different with information-based approaches. On the contrary, given the complexity of their effects and the multiplicity of mechanisms by which they accomplish their goals, we should assume that evaluating their effectiveness will be even more difficult, and should make appropriate allowances. These information optimists would assert that many of the outcomes that the interviewees identified, such as educating the public, influencing corporate behavior, and catalyzing public policy, are inherently broad, diffuse, and multicausal, making them particularly resistant to measurement. But that does not mean they are not worth pursuing, not least because of their breadth and generality. Just because something cannot be confirmed by science does not necessarily mean it is not effective.

These optimists would also point out that the modern phenomenon of information-based governance is relatively new and that these programs are relatively young, and need more time to both generate results that can be measured and create methods to measure them. With this perspective, it is therefore truly impressive that so many cases have at least made the attempt to document their outcomes, even if many are still aspirational and not yet quantified. The specific claims that some of the cases do make are often nontrivial and quite impressive, whether it is the number of hectares certified by the Forest Stewardship Council or the Sustainable Forestry Initiative, the number of buildings certified by LEED or Green Globes, or the number of acres certified by USDA Organic or Rainforest Alliance.

An information realist would acknowledge all of these points, and would agree that evaluating the effects and effectiveness of these initiatives is indeed challenging. She would also assert that regulatory approaches are often not held to these same standards of performance. However, that is no excuse for information strategies to back away from such standards; instead, they must embrace them fully and pursue them tenaciously. The stakes are too high—both in terms of their own viability and the seriousness of the environmental challenges they are purportedly trying to solve—to do otherwise. The goals of these initiatives must be made explicit and progress toward these goals must be implemented rigorously and reported transparently. Such intentionality and openness will allow stakeholders, from the man standing in the grocery store to the government official sitting at his desk, to more effectively evaluate competing options and indeed, the entire idea of using information to change the world. Such information realism may be a painful tonic to drink for these programs, but it is a necessary one.

Given the complexity of these programs and the diversity of perspectives on their most important effects, it is clear from the preceding discussion that we are unlikely to find one single goal or metric of performance. Such a lack of a single definition of effectiveness may be frustrating and dissatisfying, but this unfortunately is the result of the irreducible complexity of these initiatives. The information realist acknowledges this complexity, and rather than rail against it as a pessimist would or wave it away as an optimist might, she embraces the challenge of developing a variety of mechanisms and metrics that capture that complexity. The field of participatory multi-attribute decision analysis (MADA) has arisen precisely to help policymakers and stakeholders tackle this challenge.[28] Increasingly utilized in environmental policymaking processes, MADA can incorporate a variety of criteria, including input, process, output, and outcome variables, into complex but intelligible evaluations of performance. These evaluations can include direct environmental outcomes as well as consumer, company, and policy outputs and more indirect process, awareness, and knowledge outcomes.

Fortunately, the potential positive effects of these programs are numerous, and every initiative does not have to achieve all of them, at least not initially. They can focus on goals that are most appropriate and pressing for their given context and sector, taking into account the existing regulatory, information, and stakeholder landscapes. They can focus on particular audiences to influence and actions for them to take, rather than trying to change all of them at once. And they can learn from examples of efforts to set goals, measure performance, and report progress by existing initiatives. The next section will highlight a few of these promising practices.

### Promising and Problematic Practices

One of the first places to start is clearly and comprehensively reporting to the public any efforts to evaluate an initiative's effectiveness. A regular and well-written report that documents these efforts is a good idea, although not often found in the EEPAC Dataset. One rare and notable example is Responsible Travel's annual sustainability report that I mentioned earlier. A twenty-two-page document with an introductory letter from the cofounder and managing director, the report is well designed and well organized, and reports specific numbers where possible and provides qualitative anecdotes where relevant. It includes photos and sections on each of the major areas in which the organization has set goals. Indeed, a related, promising practice is explicitly setting specific goals for the program. Climate Savers Computing Initiative provides an excellent example of articulating a specific and ambitious objective for itself: "By 2010, we seek to reduce global $CO_2$ emissions from the operation of computers by 54 million tons per year, equivalent to the annual output of 11 million cars or 10–20 coal-fired power plants. With your help, this effort will lead to a 50 percent reduction in power consumption by computers by 2010, and committed participants could collectively save $5.5 billion in energy costs."

While such a detailed and comprehensive approach is laudable, it also helps to have a brief summary of the key outcomes of the organization. Rainforest Alliance, Bird Friendly Coffee, WaterSense, Best Workplaces for Commuters, Responsible Shopper, and several other programs all provide detailed summary data on the environmental, company, and/or consumer outcomes of their programs. Some programs, including EPEAT and ENERGY STAR, provide calculators on their website that allow visitors and institutions to calculate the impact of the certification based on their own inputs. For example, the Best Workplaces for Commuters website has a calculator that estimates the financial, employee, environmental, traffic, parking, and tax benefits of enrolling in the program. After visitors input data about their own workplace, a benefits page is produced displaying the reduced urban air pollutant emissions (lbs/year), increased worker productivity ($/year), building cost savings ($/year), and data on twenty other outcomes associated with the certification.[29] Similarly, DOE's Alternative Fuels and Advanced Vehicles website provides a Vehicle Cost Calculator that allows users to compare autos on their annual fuel use, electricity use, operating costs, costs per mile, and annual emissions of greenhouse gases (and provides an interactive graph of this data over time).[30]

While lacking many specific numbers, the Earth Island Institute's Dolphin Safe Tuna Program provides a concise list of the "general accomplishments" of its International Monitoring Program, which include preventing deceptive labeling of canned tuna and securing commitments from tuna processors to supply tuna without chasing and netting dolphins. While numerical data that establishes the breadth of a program's effects is preferable, qualitative information can still be useful to stakeholders, and is certainly better than nothing—as long as it is accurate.

The issue of accuracy brings us to another excellent practice, which is third-party verification of any of a program's important outcomes. As mentioned earlier, the Forest Stewardship Council provides a short summary of independent assessments of its certified forests, many of which were conducted by academic researchers. Michael Kraft, Mark Stephan, and Troy Abel provide one of the most comprehensive and rigorous independent evaluations of an information-based governance initiative in their book on the Toxics Release Inventory (TRI).[31] Their work demonstrates the importance of analyzing the distribution of the impacts of these programs; while overall they find that the development of the TRI has led to a substantial decrease in toxic emissions in the United States, these reductions vary significantly by industry, state, community, and individual facilities. Following the logic of chapters 3 and 4, such external verification that pays attention to both micro- and macro-level impact distributions is no guarantee of validity, but it can nonetheless increase the trustworthiness, legitimacy, and credibility of an organization's claims about its own efficacy.

As I mentioned before, it is important to start somewhere, and such third-party verification may feel out of reach for many initiatives, at least initially. But it is something to aspire to and stakeholders should expect it from the most established initiatives. Likewise, we might not expect a new program to make use of all of the different effectiveness pathways discussed in this chapter. But the most impressive effectiveness claims will nevertheless be those that encompass a wide range of mechanisms and take advantage of these multiple pathways to bringing about change. Thus the fact that FishWise and Responsible Travel at least mention environmental, consumer, corporate, and indirect outcomes on their website is noteworthy, even though the specificity and independence of their claims can be improved. The information realist is seeking initiatives that provide comprehensive, credible, and specific claims of effectiveness, but recognizing the general dearth of such claims, and the obstacles to making them, acknowledges any attempt by information initiatives to evaluate and report on their own efficacy.

Which brings us to problematic practices in this regard, of which there are two primary ones. The first is outright disinformation and fraud—claims of environmental performance when in fact it is lacking entirely. LG's manipulation of its refrigerators to pass ENERGY STAR tests and more recently, Volkswagen's manipulation of its cars to pass EPA emission tests are two examples of such practices.[32] The second problematic practice is simple silence—the fact that 70 percent of the programs do not mention their outcomes or effectiveness is deplorable. It does, however, represent an opportunity for programs to differentiate themselves from this silent majority by highlighting how they are indeed making a difference.

## The Effects and Effectiveness of Green Electronics

So in light of this discussion, what is Vernon to do? Fortunately, two of the certifications that are relevant to his decision, EPEAT and ENERGY STAR, happen to be leaders in the area of outcome transparency. As mentioned previously, both programs document the cumulative reduced impacts of their certified products on their websites. Both provide detailed environmental benefit reports, and both provide links to detailed calculators that estimate the benefits of purchasing their certified products (similar to the ones described earlier). Vernon had been stumped by figuring out how to compare and choose between these two options, and it is indeed challenging given that they are both leaders in performance reporting. Both programs provide aggregated results data across all of their certified products, and only limited information about specific product categories, such as the computer monitors that Vernon is analyzing. ENERGY STAR published figure 6.4 in 2012 showing the difference between ENERGY STAR and non-ENERGY STAR certified computers and monitors since 1992.[33] Figure 6.5 is an infographic from EPEAT's website summarizing the electricity and resource material reductions associated with EPEAT-certified computers and monitors.[34] As these figures show, they are both impressive, but not very comparable.

The comparison is complicated by the fact that ENERGY STAR certification is a requirement for any level of EPEAT certification.[35] Thus from Vernon's perspective the certifications are equivalent in terms of energy savings at the product level. However, the initiative-wide results for the two certifications may not necessarily be equivalent. While EPEAT reports that 55 million EPEAT-registered computers and displays were sold in the U.S. in 2012, ENERGY STAR data indicates that approximately 65 million computers and displays were shipped with ENERGY STAR certification that same year.[36] While EPEAT does not disaggregate their data further on their website, ENERGY STAR reports that its certified products in 2012 included 22 million LCD displays (representing 83 percent market penetration).[37] As

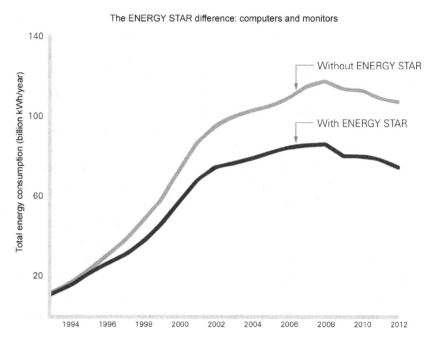

**Figure 6.4**
ENERGY STAR 2012 benefits infographic. *Note:* Adopted from ENERGY STAR, "Product Retrospective: Computers and Monitors."

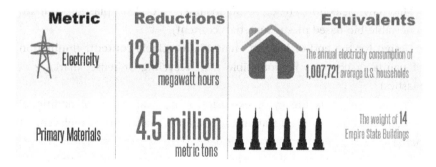

**Figure 6.5**
EPEAT 2013 benefits infographic. *Note:* Adopted from EPEAT, "Environmental Benefits." *Source:* Green Electronics Council.

of January 2016, EPEAT lists 853 displays registered on its website (201 Silver and 653 Gold), while ENERGY STAR lists 1,565 certified displays (both include professional and signage displays).[38] So even though they are certifying energy savings at the same level, as a program, ENERGY STAR is certifying more products and more of these products are being shipped to consumers and organizational purchasers.

However, this is not the end of the story. EPEAT has a significantly broader geographic scope than ENERGY STAR. While it has developed partnerships with a handful of countries and the European Union to promote products that it has certified, ENERGY STAR is primarily focused on the U.S. market.[39] In contrast, EPEAT has created product registries in 43 different countries, and in 2012, unit shipments of EPEAT-registered products in countries outside the United States surpassed unit shipments of EPEAT-registered products within the United States for the first time (bringing the worldwide total to 114 million).[40] EPEAT also has both required and optional criteria (that enable products to achieve Silver and Gold certification) that cover additional areas that go beyond energy conservation, including:

• Reduction/elimination of environmentally sensitive materials (e.g., elimination of intentionally added cadmium, mercury, lead, hexavalent chromium, and certain flame retardants and plasticizers, and PVC)
• Materials selection (e.g., postconsumer recycled plastic content, renewable/bio-based plastic materials content)
• Design for the end of life (e.g., reusable/recyclable content, elimination of paints or coatings incompatible with recycling or reuse, marking of plastics)
• Product longevity/life cycle extension (e.g., availability of additional three-year warranty, upgradeability, modularity, availability of replacement parts)
• End of life management (product take-back service, recycling vendor auditing)
• Corporate performance (environmental management system, corporate sustainability reporting, environmental policy consistent with the International Standards Organization's ISO-14001 management standard)
• Packaging (e.g., reduction/elimination of intentionally added toxics, recyclable packaging, take-back program for packaging)[41]

This list brings us back full circle to chapter 2 and our discussion of values and the value of eco-labels and sustainability ratings. At the end of that chapter, we found that all other things being equal, comprehensiveness is an important criterion by which to evaluate information-based governance strategies. EPEAT has a broader geographic scope than ENERGY STAR, and includes criteria that extend beyond the blended good of energy savings that is the sole focus of ENERGY STAR. This type of good creates both private benefits (in the form of reduced energy costs) and public benefits (in the form of reduced pollution from energy production), while EPEAT is focused on a broader range of more specifically environmental and public-health-related public goods.

EPEAT has quantified these benefits both annually and cumulatively since the inception of the program. For example, it claims that EPEAT-registered products purchased in 2013 will "reduce use of primary materials by 4.5 million metric tons, equivalent to the weight of 14 Empire State Buildings" over their lifetimes.[42] If Vernon's boss counts these nonenergy-related benefits as "making a difference" and is interested in the relative impact per product (as opposed to the cumulative impact of the entire certification program), it appears that their agency should focus on purchasing EPEAT (and preferably EPEAT Gold) displays, particularly since the certification is not associated with a price premium.[43] This is perhaps why the second Bush administration made it a requirement for all federal agencies to purchase EPEAT-registered products, and may be the best option for Vernon's state agency as well.

ENERGY STAR, however, might be able to change this calculus by clearly demonstrating that it excels in important areas that are also important to Vernon and his boss. For example, perhaps they value the complementary impact that certifications can have on public policies discussed earlier in this chapter. While it does not yet make this case explicitly, ENERGY STAR could argue that it is helping raise the regulatory floor of energy efficiency standards for entire product categories. For many electronic appliances, these mandatory standards have been increasing over the past several decades,[44] and ENERGY STAR may have served a critical role in demonstrating that it was indeed possible to achieve these standards. This effect on public policy may be the most enduring and important outcome of information-based governance strategies, as it ensures that whole sectors and product categories are more environmentally friendly, rather than only those products and companies that are certified or rated. Although a younger program, EPEAT may be able to make a similar case with regard to some of its criteria as well.

Likewise, if TCO begins both documenting and publicizing specific outcomes from its certification program, it might be able to make a strong case to Vernon and his agency. With such documentation, TCO may be particularly compelling if Vernon's boss has a strong interest in the ergonomic qualities of their computer monitors. Likewise, he may be attracted to it if he wants to certify that his agency's monitors are manufactured in socially responsible ways that respect the rights of workers and prioritize their safety. These are criteria that neither ENERGY STAR nor EPEAT currently cover.

Unlike TCO, these two programs also share a history of controversy. Both have come under criticism in terms of the trustworthiness and validity of their certifications, for example. In 2010, the Government Accountability Office issued a report asserting that ENERGY STAR is "for the most part a self-certification program vulnerable to fraud and abuse" and documenting how fifteen bogus products, including a gas-powered alarm clock, were able to become ENERGY STAR certified.[45] The program responded by instituting a requirement that all ENERGY STAR products must be certified by an independent third-party "certification body."[46] This standard surpasses EPEAT's verification process, which randomly selects a subset of products from its registry to investigate and confirm that they meet the program's standards. These unannounced investigations are conducted regularly, and any discovered nonconformance is made public to embarrass the companies involved.[47] Nevertheless, it remains a system based on self-declarations and lacks the independence and comprehensiveness of ENERGY STAR's new third-party testing requirement.

In 2012, questions were indeed raised about the rigorousness with which EPEAT's standards were being applied when Apple's Retina MacBook Pro was allowed to keep its Gold certification, even though the certification relied on the product being considered upgradeable with common tools.[48] As Kyle Wiens argues, a product with "proprietary screws, glued-in hazardous batteries, non-upgradeable memory and storage, and several large, difficult-to-remove circuit boards would fail all three tests" associated with this criteria.[49] EPEAT asked its Product Verification Committee to address this question, and it clarified that upgradability can be satisfied if the product

contains "an externally-accessible port," such as a USB port.[50] While perhaps technically correct, critics blasted this decision as "eviscerating" the
original purpose of the certification and amounting to a "greenwashing" of
Apple's products.[51] This controversy raised the possibility that rather than
continuously raising the bar for electronics companies, EPEAT's standards
may be lowering it instead.

The point is that EPEAT's breadth of criteria and reporting of outcomes,
while laudable, does not inoculate it from criticism on the other important dimensions of information-based governance that we have discussed
in this book. Far from giving them a free pass in these areas, strong claims
of effectiveness and value invite the spotlight from consumers, regulators,
competitors, and the media. Particularly in the areas of institutional trustworthiness and methodological validity, it is important to pay attention to
all components of the information value chain. Weaknesses in that chain
raise questions about the validity of any claims made about the outcomes
of the program, and create opportunities for other initiatives to surpass
them.

The next and final chapter will discuss how different stakeholder groups
can take a holistic approach to designing and evaluating information-based
environmental governance strategies and overcome the challenges and
trade-offs associated with this form of governance.

# 7 Green Realism: Limits, Linkages, and Outcomes

## Green Decisions 2.0

Now that we have completed our exploration of the information value chain and discussed its possible outcomes, we can return to our five friends from chapter 1 and see where they are with the various green decisions facing them. After thinking through the different types of value discussed in chapter 2, Mark has decided to purchase the organic milk over the conventional milk, as he believes it does indeed reflect his values. The combination of both public environmental benefits and personal health benefits has convinced him that it is indeed worth the extra cost. As for Carrie, after considering the various dimensions of trustworthiness discussed in chapter 3, she has emailed the reporter that she cannot fully trust either the Forest Service Council (FSC) or the Sustainable Forestry Initiative (SFI) due to the conflicts of interest implicit in their business models. Overall, however, she has less distrust in FSC because its director general shares her background as an environmental advocate. She recognizes that if she were a business owner, she might trust SFI more given the more business-oriented experience of its CEO. Lynn, however, after considering the criteria of validity highlighted in chapter 4 and discussing them with her colleagues, has more unequivocally recommended the Living Building Challenge as the certification program that has the best combination of innovativeness, comprehensiveness, and methodological validity for her institution.

Likewise, Anu and her team have applied the lessons learned from chapter 5 to create a highly salient and easy-to-use information delivery system that integrates product and company certifications (Cradle to Cradle, Bio-Preferred, Roundtable on Sustainable Palm Oil, and B Corporation) with an accessible website that interactively highlights her company's sustainability goals, progress, and challenges. And after weighing the different types of effects and definitions of effectiveness discussed in chapter 6, Vernon has met with his boss and recommended that the agency only purchase EPEAT-registered computer monitors, with the caveat that they should revisit the decision in a year to review any newly available social and environmental outcome data from TCO.

In order to make these decisions, each of our five friends has applied theoretical concepts and empirical knowledge relevant to particular components of the information value chain. While their analyses have been relatively comprehensive and in-depth with respect to these components, they nevertheless have also been incomplete. Mark may want to take into account the trustworthiness of the organization behind the organic label (the U.S. Department of Agriculture), and not just its relevance to his values. As chapter 3 suggests, Carrie's evaluation of forest certifications may in large part ultimately depend not on her assessment of their organizational legitimacy but on her evaluation of their certification methods. Lynn and Vernon's decisions also extended beyond methodological and effectiveness criteria back to questions of value and values, while Anu must not only consider the usability of her new initiative but also its perceived credibility, importance, and validity.

This common need for holistic assessments of these information value chains is nowhere more necessary than in the context of socially responsible investing (SRI). Just as consumers like Mark, activists like Carrie, academics like Lynn, executives like Anu, and policymakers like Vernon struggle to evaluate sustainability information, so too do investors who are committed to making investments that reflect their values. Indeed, many of the corporate sustainability ratings discussed in this book and included in the EEPAC Dataset, such as the Calvert Social Index and Corporate Knight's Global 100, are designed with such investors in mind. These important corporate stakeholders have several mechanisms by which they can exert power in the marketplace, from excluding poor performers from their portfolios (i.e., negative screens) to exclusively investing in the most sustainable firms (i.e., positive screens) to sponsoring shareholder resolutions on key social or environmental issues relevant to various companies.[1]

Financial assets that are managed with environmental, social, and governance factors explicitly incorporated into their investment analyses have increased dramatically in the past twenty years, from $12 billion in 1995 to $6.2 trillion in 2014 (a 500-fold increase), and now represent one out of every six dollars under professional management in the United States.[2] In 2014, a total of 480 institutional investors and 308 money managers were using environmental, social, and governance criteria in their investment decisions.[3] But what are those criteria, and how valid and trustworthy are they? How relevant are they to the values of concerned investors, and do they both measure and generate meaningful environmental outcomes? How can these investors be sure they are using the right information to invest in the most sustainable firms?

The growth of SRI thus clearly illustrates the challenges and limitations of information-based governance strategies more generally. The recognition of these challenges and limitations is at the heart of information realism and its close cousin, green realism. Green realism is the application of information realism in the field of sustainability, and requires attention to the issues and insights discussed in this book. The complexity of these value chains and their constituent parts makes it difficult to create successful information-based governance initiatives, particularly in the context of society's current environmental challenges, and demonstrates the importance of implementing the promising practices explored in each of the previous chapters. These practices increase the probability that these programs will have at least some positive outcomes.

The inherent limitations of these initiatives also point to the necessity of forging linkages with other forms of governance that leverage the value of the information they provide. This chapter summarizes the nature of both these limitations and linkages, and provides recommendations for stakeholders based in information realism on how they might minimize the effects of these limitations and maximize the benefits of potential linkages. It suggests several pathways that Mark, Carrie, Lynn, Anu, Vernon and others like them can use to push information-based governance to extend its influence beyond their individual decision-making contexts (as important as they are). Such pathways can enable this form of governance to affect decision making at a much broader scale and catalyze much more extensive social, political, and technological change. We can call such a shift in impact "Green Decisions 2.0." The chapter also offers some final thoughts about who ultimately should be responsible for ensuring the further maturation of information-based governance, and why such a responsibility is important, not only for the future of sustainability efforts but also for citizen engagement and democratic governance more generally.

## A Glass Half Full, a Glass Half Empty

Many information-based environmental governance initiatives have already created significant and important value for society and the environment. Keeping in mind the diverse effects of information examined in chapter 6, let us focus for the moment on the environmental outcomes of these programs, which was the most commonly mentioned component of effectiveness in my interviews discussed in that chapter. Between 1992 and 2011, for example, certified organic farmland (including both pasture and cropland) in the United States increased from 935,450 acres to 5,383,119 acres, an increase of over 500 percent.[4] Arguably the most rigorous assessment of the environmental benefits of organic farming, Allan Blackman and Maria Naranjo's study found that compared to similar noncertified farms, certified organic farms were 61–71 percent less likely to apply herbicides, 43–45 percent less likely to use chemical fertilizers, and 14–18 percent less likely to apply pesticides.[5] Likewise, they found certified farms were 59–63 percent more likely to use organic fertilizers and 15–29 percent more likely to utilize soil conservation practices. Given the positive environmental benefits of these practices and the negative environmental costs of the toxic chemical applications listed previously, the fivefold increase in organic acreage in the United States since 1992 has created undeniable and important benefits for society and the environment. The other programs discussed in previous chapters—from LEED to EPEAT to FSC and SFI—can cite similar statistics about their also very real outcomes.

However, the magnitude of these outcomes pales in comparison to the overarching size of the relevant market and sector. Agricultural land in the United States, for example, covers nearly a billion acres.[6] That impressive 500 percent increase? It accounts for approximately one-half of 1 percent of that total acreage. What about green buildings? ENERGY STAR has done only slightly better, having certified 4 percent of the 87 billion square feet of commercial space in the United Sates (as of January 2015), while LEED slightly worse—just three-tenths of 1 percent of that total square footage (as of July 2015).[7] How about forests? Collectively, forest certification programs have certified approximately 10 percent of global forests—FSC accounts for approximately one-third of those forests and the Programme for the Endorsement of Forest Certification (of which SFI is a member) cover the other two-thirds.[8] These certified forests cover 20 percent of production and multiple-use forests around the world. Certainly, this is more impressive than the agriculture and building data, but still 80–90 percent of global forests are not being certified (this number is similar for U.S. forests).[9]

And of course the many methodological caveats and validity concerns discussed in chapter 4 must be kept in mind when considering these figures. Research on other voluntary information-based programs also suggests their environmental benefits are limited or nonexistent.[10] For example, George Mason University scholars Nicole Darnall and Stephen Sides concluded from a meta-analysis of nine studies that industry participants in four self-monitored programs—EPA's 33/50 Program, DOE's Climate Challenge Program, the chemical industry's Responsible Care Program, and the ski industry's Sustainable Slopes Program—had significantly lower environmental performance rates than nonparticipating firms.[11] The environmental performance of participants in ISO 14001, which requires independent certification, was statistically indistinguishable from nonparticipants.[12] Using a propensity score matching approach, Duke University professor William Pizer and his Resources for the Future colleagues Richard Morgenstern and Jhih-Shyang Shi find that the effects of EPA's Climate Wise Program and the DOE's Voluntary Reporting of Greenhouse Gas Program are limited—"no more than 10 percent and probably less than 5 percent."[13]

The phenomenon of information-based governance thus presents a perplexing paradox for us—on the one hand, their rapid growth and in many cases real environmental outcomes are encouraging. But on the other hand, their absolute levels of coverage relative to the scope of their sectors is underwhelming and their poor performance in at least some cases is concerning. We can reframe this paradox using the classic metaphor of a glass of water that is simultaneously half full and half empty. The rapid growth and tangible benefits are the glass half-full interpretation, and is not incorrect. The relative and absolute context of those benefits is the glass half-empty perspective, and is also not wrong.

This is the first fundamental insight of information realism—we must recognize and embrace this paradox, rather than only see one side of it.

Information realists see the reality of the benefits being created by many—although not all—information-based governance strategies, but at the same time acknowledge their limited progress and inherent limitations. Green grades—eco-labels, sustainability ratings, and so on—have indeed grown up and matured significantly over the past thirty years, and that success should be recognized. At the same time, however, they have a lot more growing up to do. Indeed, most of the cases in my sample are not nearly as developed and impactful as the ones I have focused on in this and previous chapters.

This point leads to the second key supposition of information realism—that the landscape of information-based governance strategies is not uniform or homogenous but defined by significant differences across a wide range of relevant characteristics. The figures depicting this landscape in chapters 2 through 5 highlight this diversity, showing how the 245 cases in the EEPAC Dataset differ in their coverage of issues, signals of credibility, grants of legitimacy, transparency of methods and outcomes, discussion of validity and reliability, and use of information forms and pathways. While it is difficult to prove conclusively that particular constellations of characteristics are more effective than others, each of the previous chapters has identified both promising and problematic practices that social science theory and empirical research suggest will lead to stronger outcomes for these programs. While the information realist acknowledges implementing all of these promising practices simultaneously is challenging, she will nevertheless seek out and reward those programs that cover a broad range of environmental criteria, establish their organizational trustworthiness and outcome effectiveness, demonstrate their methodological validity and reliability, and deliver information that is prominent, intelligible, and feasible to use.

The information realist, however, also recognizes that even the most well-designed information-based governance strategy will face intrinsic limits on its impact. This third supposition of information realism is based on five key factors. The first is the inherent trade-offs that exist between many of these characteristics. These trade-offs came up, for example, in chapter 4 in our discussion of methodological validity and reliability—a highly reliable metric may have low validity because it does not capture the complexity of sustainability performance. They also were apparent in chapter 5 in debates about the right amount of information to provide audiences—too much can overload and dishearten them, while too little can mislead and frustrate them.

Such trade-offs exist at a higher level as well. In the interviews discussed in chapter 6, I asked participants to rate the importance of the four general attributes of organizational reputation, methodological validity, information intelligibility, and issue importance to determining the effectiveness of information-based environmental governance strategies. While the vast majority found them all to be at least somewhat important drivers of effectiveness, their rankings of their relative importance varied across the board. Such disagreement suggests underlying tensions and trade-offs among these different attributes. For example, a trade-off often exists between maintaining the scientific rigor of an evaluation and making the final results simple enough for the public to understand and appreciate. Similarly, limiting the involvement of companies to enhance the trustworthiness and independence of an initiative may reduce both the relevance of its information and the validity of its data. An independent program may not address issues that are important to corporate actors or may not make use of information only available to these actors. Initiatives that emphasize formal rules, systematic auditing, and scientific validity can hinder the development of trust and productive relationships among stakeholders at the local level.[14]

Likewise, on the one hand, a certification based on a relatively unambitious standard may be attractive to mainstream companies and consumers but be viewed as greenwashing by activists and consumers who are deeply committed to sustainability. On the other hand, a highly rigorous and trustworthy label that no company endorses or puts on its products may have limited effectiveness. As work by David Cash and his colleagues at the Harvard Kennedy School of Government shows, such trade-offs are implicit in knowledge production systems:

We find that the most successful efforts to connect knowledge to action are those that are not only effective at engendering favorable perceptions of salience, credibility, and legitimacy, but are also effective at balancing tradeoffs among these three attributes such that none of the three attributes falls below thresholds that will trigger the rejection of information or the resistance to recommended action. We have yet to identify a successful formula for this balancing act, but self-conscious efforts to balance salience, credibility and legitimacy within and across the multiple boundaries of a knowledge production system are more likely to influence action than efforts that ignore these problems.[15]

While information optimists would likely insist that some programs have successfully found this balance and it is relatively straightforward to achieve, information realists are less sanguine about this possibility. They realize that while a few programs might come close to finding this "successful formula," due to resource constraints most do not and likely will not. This conclusion relates to four other factors that drive the information realist's understanding of the inherent limitations of these programs. These include their voluntary nature, their development costs for organizations (which may often be lower than other forms of governance but are never zero), their transaction costs for audiences (the time necessary to learn about, understand, and apply them to particular decisions), and the resistance of many important stakeholders to adopting them. Each of these factors likely contribute to the low uptake of information-based governance strategies relative to their relevant markets.

And regarding the information optimists' likely assertion that these programs are relatively new and just need more time, perhaps this will be true in some isolated cases. But information realism posits that for the most part these programs are not ever going to "take over the world." Most consumers are not going to spend five minutes in the supermarket like Mark looking up information about organic milk, most professors are not going to exhaustively research green building options like Lynn did, most corporate executives are not going to develop a holistic information delivery system for their companies as Anu did, and most government officials are not going to follow Vernon's lead and carefully weigh the pros and cons of electronics certifications. Most people would be overwhelmed by all of this information and all of these decisions, and would likely conclude that the costs of such engaged decision-making outweigh the benefits. They would therefore prefer to focus on other less frustrating and more rewarding things in their lives. This is the argument that Omri Ben-Shahar and Carl Schneider make in their book about mandated disclosures, and in many ways it is equally applicable to voluntary ones as well.[16]

This is the glass half-empty side of information realism—being realistic about the extent and potential of the direct successes and effects of information-based governance strategies. So where does this leave us? How should we pass over the razor's edge of this information paradox? Shall we give up and go back to more well-trodden trails of governance as the information pessimists might advise, or shall we plod on and willfully ignore these obstacles, as the information optimists would likely suggest? As the next section explains, both options will likely lead to a dismal future for both information-based environmental governance initiatives and environmental governance more generally. A third option grounded in information realism offers a more positive future, one that recognizes but is not paralyzed by the limitations of these strategies. Instead, it builds on and leverages their ongoing but still incomplete development as a legitimate and effective form of governance.                               .

## A Tale of Three Futures

### The Progress of the Pessimist: A Mile Wide, but an Inch Deep

Chapter 1 describes the green debate between information pessimists and optimists, and some likely outcomes if one or the other is perceived as winning that debate. If the pessimists gain the upper hand in funding and public opinion, more emphasis will be placed on other environmental governance strategies. Some will push for new and stronger environmental regulations like a carbon tax, as climate activists such as Al Gore have advocated for.[17] Others such as Bill Gates will assert that new technologies are essential to solving society's current environmental challenges, and will advocate for increased funding dedicated to large-scale research and development initiatives.[18] Following the lead of Pope Francis, still others will proclaim that these challenges are moral and ethical issues and ultimately require citizens to muster the personal and political will to solve them in both our public and private lives.[19]

While each of these approaches is valuable and can play an important role in environmental governance, it is important to recognize they too have their limitations. Both regulation-based and technology-based governance strategies require enormous political will and organization to succeed, as they need to convince voters, policymakers, and funders that they are necessary and worthwhile. Presidents can sign executive orders, government agencies can write new rules under existing mandates, and wealthy investors can fund clean energy developments, and at least in the short term some important progress can be made. But over the longer term, they risk being reversed, rewritten, and ignored if they do not have broad political support, and progress on issues such as climate change and biodiversity loss could be set back years, if not decades. Some scholars of environmental politics, such as MIT's Judith Layzer, have argued that such retrenchment has already occurred and is still occurring at multiple levels of government.[20] Likewise, morality-based strategies risk being dismissed as uninformed and unrealistic if they are not based on credible information and not linked to tangible actions that can be taken. As I will discuss later in this chapter, when strategically deployed, information-based governance initiatives can provide an important foundation for these other governance approaches and help mitigate their own shortcomings.

**The Outcomes of the Optimist: Solitary, Ignored, False, Brutish and Short**
The approach of the information optimists would likely fare no better. They would likely fail to address the limitations of information-based strategies and continue with the status quo. Product eco-labels and sustainability ratings would continue to proliferate with limited connections among them, and their quality and effectiveness would continue to be highly varied, underanalyzed, and unpredictable. While some improvements in their design might be made over time, they would still likely be ignored by most people and only have a marginal effect on solving society's greatest environmental problems.

That's the more positive outlook; more ominously, the deeper fears of the information pessimists may come to fruition. These programs may give people a false sense of accomplishment and undermine broader efforts to confront these problems more directly and effectively. Alternatively, the bad apples among them may give the entire enterprise of information-based governance a bad reputation. Such an impression might darken the public's attitude toward the effectiveness of any and all efforts to confront environmental challenges, and encourage a sense of disillusionment and fatalism.

**The Pathways of the Realist: Overcoming Limitations and Recognizing Linkages**
What then is the alternative to these dismal scenarios? Information realism offers several different pathways to a more encouraging future. Each begins with a deep awareness of the challenges and limitations of information-based governance that this book has elucidated. But they are also grounded in a realization that all forms of governance face their own such challenges

and limitations—no one approach has a monopoly on feasibility. Given the magnitude of the challenges facing society, it is critical that we work to make all of these approaches to governance—including but not limited to information-based strategies—more effective than they currently are.

**Aggregation and Consolidation**  The first possible pathway to doing so is to combine and aggregate as many different initiatives as possible. This proposal focuses on the problem of stakeholder overload and confusion, and diagnoses proliferation as the most proximate cause. To some extent, this pathway is already being followed in several different contexts. For example, organizations that rate the social responsibility of companies primarily for investors went through a major consolidation between 2007 and 2010. In 2007, the proxy advisory firm Institutional Shareholder Services (ISS) acquired environment, society, and governance (ESG) research firm Investor Responsibility Research Center (IRRC).[21] In 2009, RiskMetrics in turn acquired ISS, as well as the ESG research firms KLD and Innovest, while Thomson Reuters acquired ESG data provider Asset4. That same year, the ESG research firms SustainAnalytics and Jantzi Research also completed a merger.[22] Then in 2010, MSCI purchased RiskMetrics for $1.5 billion.[23] Some consolidation has occurred in the consumer space as well, with UL Environment acquiring EcoLogo in 2009, GreenGuard in 2011, and Good-Guide in 2012.[24]

Such aggregation can simplify the range of choices for relevant audiences, and increase the prominence of the remaining options. It does not always reduce the range of choices, however; UL Environment still offers GreenGuard and EcoLogo certified products on its website, for example. In this case, the benefit of the consolidation arises from the added usability of a single site to visit (e.g., UL's "Sustainable Product Guide") and the added credibility of the umbrella brand.[25] Underwriters Laboratory (UL) has been testing and certifying products since 1894 and has 67,000 corporate customers in more than 100 countries.[26] Companies that might be hesitant about smaller certification firms may be more likely to trust and work with a larger and more established certifier like UL.

Other mechanisms short of full acquisition can also help raise the prominence, credibility, and usability of particular groups of initiatives. The ISEAL Alliance, for example, is a group of sustainability standards and accreditation bodies that have met the organization's Codes of Good Practice and related requirements.[27] Stakeholders can choose to use the standard systems that are members of ISEAL if they believe these requirements sufficiently reflect their own standards of validity and credibility. Yale University professor Ben Cashore proposed a system in 2008 that would go one step further and create a "Better World" consumer label that would be co-branded with the individual labels meeting its standards (e.g., "Better World-FSC," "Better World-Fair Trade"). The labels included in this system would not change but be united by a common effort to raise awareness about the Better World initiative. Cashore suggests that such a single label would be more recognizable and easier to market than each individual label.[28]

These different forms of aggregation are examples of what I call, with a hat tip to the *Lord of the Rings* trilogy, "one ring to rule them all" strategies.[29] They are all intriguing information realism strategies that attempt to directly address the limitations of information-based governance. They can reduce information overload, simplify decision making, and establish clear standards and mechanisms of accountability for labels and ratings that have been brought under a single source of authority. These approaches may work and be helpful to stakeholders in some contexts, but several factors may limit their effectiveness. First of all, mechanisms must be developed to ensure that each of the individual certifications maintains the standards of the umbrella organization. Such mechanisms will likely be challenging to implement both politically (i.e., resistance will be strong from some stakeholders) and operationally (i.e., coordination and enforcement will be logistically complex).

Regarding the active consolidation approach, there is also nothing limiting later fragmentation and proliferation after they have been consolidated. This is indeed what has happened in the SRI field, for example; in 2009, Bloomberg launched its own ESG data services,[30] and in 2013, TrueValue Labs was founded to deliver "powerful tools to measure ESG and sustainability performance."[31] A 2012 survey by Integrity Research Associates found that thirty-five different organizations were still providing ESG research to institutional investors.[32] And even if the consolidation holds, problems generally associated with monopolies and cartels may arise. Without outside competition, the quality of the standards may deteriorate (or at least not improve), and the possibility of corruption and fraud may increase. As the saying goes, absolute power corrupts absolutely. Nevertheless, mechanisms can be introduced to avoid or at least mitigate these potential issues, such as requiring the certification of tangible "on the ground" sustainability outcomes, credible third-party verification, reliable tracking systems, and robust dispute resolution procedures.[33]

**Choice Editing**    An alternative pathway that also builds on the insights of information realism is to restrict the range of choices not at the supply side of the information but at its demand side. This approach is generally known as decision or *choice editing,* and involves institutions restricting the range of product choices available to their customers or employees. This idea comes from the field of behavioral economics, and is being implemented by a range of both private sector and public sector organizations. Whole Foods, Target, and Walmart, for example, have committed to only selling sustainably sourced seafood,[34] while the federal government has instructed its agencies to purchase EPEAT-registered electronic products for 95 percent of its needs.[35] As the UK Sustainable Consumption Roundtable explains, "Choice editing for sustainability is about shifting the field of choice for mainstream consumers: cutting out unnecessarily damaging products and getting real sustainable choices on the shelves."[36] Tim Lang, a professor of food policy at City University London, questions the logic of putting the responsibility of figuring out what constitutes "sustainability" on consumers. This is too much to ask of a busy public; Lang suggests instead that it is the job of governments and retailers to filter out the bad options—to edit our choices so we do not have to.[37]

This strategy also focuses on the information overload and frustration problem discussed in chapter 5, and can be implemented in a variety of ways. Michael Maniates, a social sciences professor at Yale-NUS College in Singapore, identifies four main types of choice editing: (1) completely eliminating offending choices (e.g., the phase-out of products using ozone-depleting CFCs [chlorofluorocarbons]); (2) gradually trimming away the worst options (e.g., Japan's increasing energy efficiency standards); (3) strategically making worst choices less attractive or accessible (e.g., the placement of fatty foods on hard-to-see shelves); and (4) creatively using defaults, cues, and other "choice architecture" to orient consumers toward sustainable choices (e.g., the automatic subscription of utility customers to renewable energy sources).[38] These options range from more overt bans that require government legislation to more subtle nudges that consumers can reject if they are so inclined.[39]

On the surface, these strategies appear to be giving up on the effectiveness of information-based governance; they are designed to reduce the number of sustainability decisions people have to make. But unlike information pessimists, proponents of choice editing are only dubious about requiring mainstream consumers to play a primary role in driving the success of this form of governance. In the choice editing paradigm, sustainability information is still necessary, but its primary audience is not consumers like Mark but institutional decision makers like Vernon and Anu. These corporate and governmental decision makers are responsible for designing the policies that determine what products will be available to consumers and employees in the first place. To the extent that they have the capacity to influence available choices within their organizations, individuals like Lynn and Carrie working at civil society institutions can also engage in such choice editing (for example, only offering organic or fair trade coffee in cafeterias and meetings).

This approach has the potential to vastly extend the reach of information-based governance strategies, even more so than the aggregation approach discussed earlier. It is arguably the most radically realist approach as it is based in a deep skepticism about placing responsibility for progress toward a more sustainable world on the shoulders of individual consumers. This skepticism is grounded in a belief that despite the rhetoric around free markets, consumers actually have very limited power in the modern capitalist economy. As the UK Sustainable Consumption Roundtable concludes, "There is not enough evidence that green consumers on their own are able to change mainstream product markets."[40] Proponents of this strategy, however, are confident that institutional choice editors, with proper training, have the capacity to recognize the best sources of information and use them to select the most sustainable products and companies available (and filter out the worst).

However well founded and well designed, the choice editing tactics that are informed by this skepticism face an important challenge that is similar to the strategies advocated for by information pessimists already discussed. In the short term they may extend the reach and increase the effectiveness of information initiatives, but in the long run they risk a backlash from citizens and politicians who perceive them as manipulative and unfounded. A case in point is the outcry about a component of the 2007 Energy Independence Act and Security Act that set energy efficiency standards for light bulbs (a classic case of choice editing). In 2010, several Republican congressional representatives introduced a bill to repeal this act of "eco-fascism," as several conservative commentators have called it.[41] It did not pass, but at least some consumers were motivated by the controversy to buy up as many incandescent bulbs as they could before they were phased out.[42] While many libertarian-minded citizens will oppose such choice editing in principle, resistance will be particularly strong among the general public if it does not understand the motivations for limiting their options or does not trust the choice editors—whether they are retailers, bureaucrats, or politicians.

**An Information Marketplace**   A third pathway to improving information-based governance strategies that is also based in information realism takes a different tack. Instead of moving away from consumer involvement in this governance mechanism, it embraces their collective power in the marketplace. From this perspective, the problem has been that consumers have not been equipped to make informed decisions and deploy their power effectively. What is needed is not choice architecture that limits or hides choices, but choice architecture that makes them more transparent, intelligible, and feasible. The logic of this approach is that sustainability information about products and companies is itself a product. And just as consumers evaluate a new television, consumers need information about this information to evaluate it. With the exception of the verifiably unsafe or fraudulent, consumers do not rely on the government to decide what televisions they should buy. Increasingly, with the growth of online shopping options that provide thousands of choices at their fingertips, they are also resistant to retailers making those decisions for them. Proponents of this approach assert that consumers want to be able to evaluate and use relevant information to edit their own choices and select products themselves based on their own filters and preferences.

Efforts to enable consumers and other stakeholders to do so are growing. One of the earliest initiatives along these lines is Consumer Reports online Eco-Label Center, which provides expert opinions and ratings regarding the meaning and value of a wide range of product certifications.[43] Another example is Big Room's EcoLabel Index, which provides data on 463 eco-labels from 199 countries and twenty-five industry sectors. Big Room, Inc. succinctly articulates the philosophy of this approach, stating that "markets need trustworthy and accessible information in order to grow. Since 2007, Ecolabel Index has been the provider of that information for the eco-label market. This unique platform collects and structures data on eco-labels globally, increasing transparency and helping buyers and sellers use them more effectively."[44]

Similar efforts include Standards Map, a website developed by the International Trade Center that enables users to learn about and compare more than 170 sustainability standards, and provides a range of interactive means to visualize the underlying data.[45] Another is Ratings Hub, an online database of more than 200 corporate sustainability ratings, rankings, and indexes created by the Global Initiative for Sustainability Ratings (GISR). GISR is also developing an accreditation process by which programs in the database could verify that they meet a series of twelve process and content principles, from transparency and continuous improvement to materiality and comprehensiveness.[46] A more grassroots approach to evaluating sustainability claims has been pioneered by EnviroMedia Social Marketing and the University of Oregon School of Journalism and Communication. Their Greenwashing Index enables ordinary consumers to post ratings and reviews of green claims about products, and then aggregates those ratings into a Greenwashing Index Ratings Scale.

These all represent interesting efforts to improve the intelligibility of the sustainability information being provided to them. But they do little to make it more feasible to use and more compatible with and applicable to their normal routines and decision-making processes. A key challenge is making this information about sustainability claims available when people actually encounter them and when they are deciding between different products and companies making these claims. GoodGuide, for example, attempts to make this connection by allowing users to filter products by certifications they hold and providing expandable information about those certifications. Another challenge is reducing the amount of time users need to assess the available information about their choices. One strategy to do so is personalization mechanisms that remember users' preferences and offer recommendations based on earlier decisions. Just as businesses from luxury hotels to local barber shops can use technology to remember their customers' preferences,[47] so can retailers and sustainability raters keep track of what criteria and certifications they most value and trust.

The Ethical Consumer, for example, allows users to set and save their own criteria weights, which personalize the ratings of companies available on the website.

The ultimate goal of this approach is to create an accessible and transparent "information marketplace" that allows users to find sustainability information—and by extension products and companies—that matches their values. Just as choice editing seems to be motivated by information pessimism, this information marketplace idea may appear to be driven by information optimism. What grounds this approach in information realism is its recognition of the heterogeneity of this marketplace and the possibility that some consolidation and "weeding out" of poor claims may be necessary. The information optimist generally believes that more certifications and ratings are needed and beneficial—given the failure of other governance mechanisms and the gaps in issue coverage discussed in chapter 2. He is also committed to reaching as many people and audiences as possible—every certification can and should become universally recognized and used.

Information marketplace proponents, in contrast, are more modest in their goals, and acknowledge that some information may have niche markets. They also recognize the likelihood of some actors spreading misinformation and disinformation through these programs, and view well-designed information marketplaces as a mechanism for rooting out fraudulent and inaccurate claims. Their goal is not for all initiatives to reach every person on earth, but for the best ones to be able to influence their key audiences. Just as policy entrepreneurs often reach out not to the general public but to particular "issue publics" to push forward specific policies, information entrepreneurs can and should orient themselves toward particular "information publics."[48] This orientation pushes them to focus not only on the direct effects of a well-functioning information marketplace, but also on its indirect influence on stakeholders who can extend its impact.

**Public Policy Linkages**   This point leads to a final pathway based in information realism, which emphasizes the importance of linkages between information-based governance and public policy. This pathway highlights the potential for stakeholders who have been sensitized to particular environmental issues through their exposure to them through sustainability certifications and ratings to become engaged with the political process and advocate for stronger environmental laws. This approach is motivated by a keen awareness of the implicit limitations of information-based strategies outlined earlier in this chapter. UC Berkeley's David Vogel reinforces the importance of linking voluntary corporate social responsibility efforts with mandatory ones. He asserts that "corporate social responsibility needs to be redefined to include the responsibilities of business to strengthen civil society and the capacity of governments to require that all firms act more responsibly."[49] Similarly, in their articulation of ten key design principles, Fung, Graham, and Weil from the Harvard Kennedy School suggest effective disclosure programs must not only "provide information that is easy for ordinary citizens to use" but also "leverage other regulatory systems."[50] George Washington University Professor of Strategic Management and Public Policy Jorge Rivera and his coauthors reinforce this point generally in their book on voluntary programs and specifically in their work on tourism certification in Costa Rica.[51] And Michael Kraft and his colleagues conclude that even mandatory information disclosure programs like the TRI "can be successful only when coupled with ongoing and predictable regulatory efforts that companies and facilities take seriously."[52]

Such linkages can range from public policies directly related to information initiative themselves to those dealing with environmental challenges more generally. For example, stakeholders might support greater funding and policy support for the FTC efforts to bring suits against environmental performance claims that it perceives as fraudulent.[53] In recent years, the FTC has been more active in this area, but given the diversity of claims still existing in the marketplace (and documented in this book), the field of information-based governance would likely be served well if it took a more muscular and proactive approach to discouraging unsubstantiated green product and company claims. Alternatively, stakeholders concerned about the climate impacts of the products they are purchasing might lobby Congress to pass comprehensive regulations limiting greenhouse gas emissions. Effective information-based strategies can create policy windows for a wide range of governance proposals and help get them on the agendas of policymakers.[54]

To some extent, this approach overlaps with the choice-editing pathway discussed earlier. But where choice editing that involves regulations is focused exclusively on creating standards that exclude certain products from the marketplace, this broader focus on public policy linkages includes the full range of possible regulatory options—from energy efficiency and safety product standards to cap and trade regulations, incentives for reduced pesticide applications, and penalties for logging in high-value conservation forests. The logic is that learning about ENERGY STAR and USDA Organic and other information-based governance initiatives motivates stakeholders to recognize the need for public policies to complement the work that these programs are doing to deal with major environmental challenges. Building on the concept of transformational leadership, policy entrepreneurs can use these programs to ratchet up and transform expectations about what aspirational companies and well-designed regulations are capable of achieving.[55]

The required criteria of EPEAT, for example, could become legal requirements of a new electronics sustainability law. The voluntary standards of the EPA's WaterSense program could be the basis of new mandatory standards for all new toilets, showerheads, and other products that use water. The criteria developed by the Global Reporting Initiative, Climate Counts, and other corporate sustainability ratings could be incorporated into new disclosure requirements being considered by the Securities and Exchange Commission.[56] Certification from the Forest Stewardship Council or Sustainable Forestry Initiative could be required for forests managed by federal and state agencies.[57] Requirements of building certifications such as LEED, Green Globes, or ENERGY STAR could be integrated into local building codes, as has been done in Washington, DC.[58] And the requirements of ENERGY STAR and USDA Organic could be used by the EPA, DOE, and U.S. Department of Agriculture to update their standards for energy efficiency and agricultural practices.

Such linkages would be important outcomes of information-based governance initiatives. However, several significant obstacles must be overcome if these outcomes are to become more than aspirational. The first challenge is the possibility of an opposite effect occurring; rather than becoming more engaged, stakeholders retreat further away from the public sphere. Their time and interest in environmental politics and governance may be "crowded out" by their engagement with environmental information and consumption. This is a primary concern of the information pessimists, and is well articulated by Andrew Szasz, a professor at the University of California, Santa Cruz. Szasz introduces the concept of the inverted quarantine, which refers to the sense of safety that consumers may feel when they engage in individual behaviors that protect them from threats to their personal health and safety. Everything within this quarantine—one's own body or house, for example—is safe, while everything outside of it is still dangerous (and everyone outside of it is also still in danger). Such perceived quarantines can be created by drinking bottled water, buying organic food, or using environmentally-friendly house cleaners.[59]

Szasz asserts that such perceived quarantines are not only misleading— bottled water is not necessarily safer than tap water—but deeply problematic because they disconnect citizens from the public sphere and demobilize them from advocating for broader public policies. If a consumer has protected his own health and home, why should he waste his energy on such time-consuming and frustrating activism? While research is mixed on whether such crowding out actually occurs,[60] it is clearly a possibility that should be taken into account. But even if the opposite crowding-in effect prevails, a second obstacle arises—newly motivated stakeholders now require a new set of information about how best to engage with the political process. What should they do? Who should they contact? How should they mobilize? Effective citizen action is no less complex than effective consumer action, and requires an equal degree of strategy and planning. While beyond the scope of this book, the concepts and frameworks introduced in the previous chapters may be applicable to that context as well— information for citizens must also be salient, trustworthy, valid, and usable in order to effectively inform consumers and other stakeholders on how to become successful policy advocates.

## Recommendations

The broader ramifications of these insights are discussed at the end of this chapter, but first I want to summarize the implications of this book's analysis for specific stakeholder groups, beginning with the designers of eco-labels and green ratings.

### Implications for Designers

While providing a comprehensive design manual is beyond the scope of this book, I can highlight a few general points for designers of information-based governance strategies. First, the concept of the information value chain presented in chapter 1 provides a valuable roadmap to the design decisions that must be made in building information-based governance strategies. Second, the effects pathway diagram discussed in chapter 6 can help designers think through their theory of social change and how their programs are going to contribute to the creation of public goods. Articulating your definition of effectiveness, identifying your key audiences, clarifying your goals, and measuring your progress toward them are critical steps in this process. The fact that the websites of 20 percent of the cases in the EEPAC Dataset are now defunct demonstrates the fluid nature of this space and the real risks of failure that information-based initiatives face. By utilizing the frameworks presented and paying careful attention to the challenges, opportunities, and trade-offs discussed in this book, designers can manage those risks more effectively and increase their chances of developing a successful program.

The discussion of the landscapes of existing initiatives and the promising and problematic practices that they employ in chapters 2 through 6 can also reveal some of the more specific design choices facing designers, as well as the gaps in coverage and performance that exist in this space. Chapter 2 emphasizes the importance of connecting theories of value and values to the information being created, and using as clear and comprehensive a set of criteria as possible. Chapter 3 highlights the importance of sending signals of credibility to and recruiting grants of legitimacy from your primary audiences, whether they are the general public, corporate executives, activists, or experts. Chapter 4 outlines the various factors that influence perceptions of the quality of the underlying data used in these initiatives, and explains the importance of methodological replicability, reliability, and validity.

Chapter 5 introduces a range of mechanisms that designers can use to make that data more salient, intelligible, and easy to use. It suggests that it is important that these initiatives make information about themselves accessible on their websites—they should not bury important methodological descriptions deep in a webpage hierarchy or PDF file that few people will access. In terms of the design of the information itself, do not fall into a myopic focus on simplicity. Data presentation should be driven by the needs of your audience and the nature of the data, and often tiered, hierarchical interfaces that provide both summary and detailed information can effectively serve multiple functions and audiences simultaneously. In terms of the broader outcomes of information-based initiatives, chapter

6 discusses the many possible effects of these programs and highlights the importance of considering both the intended and unintended consequences of the information they are providing.

## Implications for Companies

Many of the preceding recommendations apply equally to companies that are implementing, using, or designing environmental certifications or ratings. Having clear goals and a directed strategy, for example, is also important for corporate actors. More specifically, thinking through what types of public and private goods they are trying to create is critical, as is whether these goals are being accomplished through the mechanisms of internal management, social responsibility, politics, or governance. Information-based management is more focused on using information to directly create private goods, information-based politics on using information to create those private goods by leveraging public resources, and information-based governance and corporate responsibility efforts on using information to create public goods. It is important to create a strategy that properly balances these uses of information, recognizes their strengths and weaknesses, and directs them appropriately.

Companies often view environmental and social concerns as a threat to their brand, reputation, and bottom line, and the cases of Exxon, Nike, BP, and Volkswagen justify this viewpoint. But they can also be seen as an opportunity for cost savings, brand enhancement, and other benefits that both employees and board members can embrace. As documented in this book and elsewhere, consumer interest in environmental certifications such as USDA Organic and ENERGY STAR is growing rapidly. According to a 2015 Nielsen survey of 30,000 people across sixty countries, 66 percent of consumers are "willing to pay more for products and services that come from companies that are committed to positive social and environmental impact," which is sixteen points more than in 2013).[61] An even higher proportion of millennials and respondents aged fifteen to twenty—nearly 75 percent in 2015—expressed similar support for corporate sustainability commitments.

Companies can reinforce these trends and benefit from them by disseminating trustworthy, valid, salient, and usable forms of sustainability information. Consumers can spread this information themselves through their social networks, and those networks can in turn and over time become powerful supporters of the most credible companies and sources of information. The longstanding and growing popularity of the kosher label, which has developed a vast network of support and accountability both within and beyond the Jewish diaspora, demonstrates the value of such civil society connections to businesses and their efforts to build successful information-based strategies.[62]

## Implications for Activists

Many nonprofit organizations have also been demanding third-party independent verification of claims made by corporations, but chapters 4 and 6 indicate that they have been less successful in making such independence a common and popular trait of environmental certifications and ratings. Those interested in promoting independent data verification and generation should find ways to reduce their opportunity and financial costs, perhaps through joint marketing campaigns, pooling of resources, or sharing of monitoring mechanisms. This raises a question about the value of collaboration. The fact that fewer than a quarter of the initiatives surveyed involve more than one organization indicate that there may be more opportunities for working together. Several nonprofit interviewees mentioned the issue of resources as a potentially limiting factor on effectiveness, and so collaborative initiatives may enable greater efficiencies and cost sharing. Depending on the specific context, such joint efforts may not always be strategically appropriate or possible, and sometimes a single organization may be able to get more done more quickly, but collaborative opportunities should nevertheless at least be considered in the design of these initiatives.

Activists (and other stakeholders as well) can also use the data presented in chapter 2 to analyze gaps in the coverage of specific issues that may warrant more attention. There is a heavy focus on manufacturing and food sectors, and criteria relevant to pollution, climate change, and economic costs, and so new initiatives might focus on less-covered sectors and issues, such as water use, wildlife, and product quality. Also, these initiatives have been more focused on evaluating individual products than entire companies, and this in some ways makes sense because people are used to making decisions about products. However, shifting attention toward the overall performance of companies—and rewarding those that are green throughout their operations and supply chains, not just for a few product lines—may be a more effective strategy in the long run and can avoid a myopic focus on individual products.

## Implications for Policymakers

Many of the government-run initiatives, such as USDA Organic and ENERGY STAR, are leaders in their areas and are employing some of the most promising practices we have discussed. Continuing the government's utilization of information-based governance mechanisms therefore may be a smart strategy, although understanding that involvement in the broader context of its legislative mandates, the information needs of the public, and other forms of governance is critical to fully evaluating the government's role in this space. In particular, tracking both the catalytic and depressive effects of information on regulations, technology innovation, and other government initiatives should be a priority in the ongoing assessment of these programs. Such tracking can help resolve the question of whether these programs serve as ceilings or floors of performance over time. They should also evaluate their effects on the less commonly-used effect pathways described in chapter 6 and explore how these other pathways might be utilized more effectively.

More broadly, the different information realism pathways discussed here all suggest a potential role for the government to help improve information-based governance strategies. For example, policymakers should be considering proposals on how to make the marketplace of eco-labels more transparent, more accountable, and more functional. If designed correctly, such a functional marketplace can enable competition to occur that orients them toward the broader purpose of good governance—mobilizing collective action to create public goods and deliver environmental outcomes. In particular, such a marketplace would stimulate competition not only among programs providing descriptive information from firms and other organizations about product and corporate environmental performance, but also among programs providing evaluative information that enable audiences to interpret the meaning and relative value of different levels of performance. Encouraging the documentation of both the aggregate impact and the distribution of impacts of these programs, as Michael Kraft and his colleagues did in their analysis of the TRI, is important as well.[63] Such an orientation would help everyone—government actors and otherwise—have more tools and capacity to deal with the increase in information and transparency that such competition would necessarily cause.

## Implications for Researchers

There are two main implications for scholars from this research. The first is that even though many of these initiatives have been studied extensively in the past, a large number of them have not. Those that have been studied have been analyzed using a relatively narrow lens and usually in isolation. Most past studies have also focused on large, well-known programs, but it is the less popular and even failing ones that may yield as many lessons about what works and what does not work as those perceived to be successful. Secondly, many of their specific characteristics have not been studied in detail, and especially not in a comparative sense. A deeper analysis of any one of the 100+ characteristics that are coded for in the EEPAC Dataset would yield interesting results.

More research on the specific signals of credibility and the types of effectiveness that different stakeholder groups prefer and prioritize would also be valuable. And scholars should continue to examine the complex dynamics and effects of information-based governance initiatives—how exactly are their results used and what are their specific outcomes? What are their short- and long-term effects on consumers and other stakeholders? Improving our understanding of the relationship between these initiatives and other forms of governance is another important area of inquiry, as is comparing their effectiveness. Information-based governance has still not been studied as rigorously and systematically as it should be, and this book hopefully provides useful frameworks and data that can inform future work in this area.

## Implications for Consumers

For consumers, the main implication is that not all eco-labels and ratings are created equal, and it is indeed possible to distinguish among them. Understanding the processes and attributes of their information supply chains can help consumers differentiate among different claims and understand what they are getting from them. It can be worthwhile to think about the goals and motivations behind these initiatives, as information-based strategies and the feedback loops associated with them can both positively and negatively affect other forms of governance, such as regulations. If consumers like or do not like certain features of particular programs, they should let them know—by choosing products with their preferred labels, posting their preferences on social media, or writing directly to companies and evaluation organizations.

In particular, if consumers want these initiatives to improve in the future, they should encourage greater transparency about their methods, sources of funding, and environmental outcomes, as that will make it easier to evaluate them and keep them accountable. With such greater transparency individual shoppers will be able to use the ones they like and ignore the others. This in turn will enable consumers to signal their preferences to manufacturers and the designers of these programs. Also, individuals can encourage more systematic efforts to make the marketplace of eco-labels more transparent and functional as a whole, so that they compete directly against each other and consumers can more easily make choices among them that are based on their own preferences, rather than on the limited information they choose to provide.

## Conclusions

The preceding sections provide a number of tactical suggestions for different stakeholders to pursue, and hopefully will be useful for people like Mark, Carrie, Lynn, Anu, and Vernon as they work to mitigate the limitations and maximize the benefits of information-based strategies in their specific contexts. The chapter also discusses four general pathways that information realists can use to improve the effectiveness of these strategies more generally—through consolidation efforts, choice-editing policies, information marketplaces, and linkages to public policy. Each of these approaches have their own strengths and limitations, and likely some combination of them will be necessary to enable these information initiatives to further mature as a form of effective governance.

A final question remains, however, and that is who is ultimately responsible for pursuing these efforts? And for holding these information-based programs accountable, minimizing their misinformation, and ensuring they improve over time? Such accountability is necessary regardless of one's view of this form of governance. For the information pessimists who fundamentally doubt their efficacy and see them as evidence of policymakers avoiding the responsibility of making tough policy choices themselves (as

Ben-Shahar and Schneider do), it is important to evaluate their effectiveness and identify whether they are serving their original purposes.[64] For the information optimists who are convinced they are indeed making a significant difference, to garner further support for them it is important to provide evidence to support this claim. But again, who among us is responsible for doing so? As Plato famously asked, *"Quis custodiet ipsos custodes? Who watches the watchers?"*[65]

While this is a question that I cannot fully address here, I can present a few thoughts for further consideration. First of all, there have been many calls to improve, simplify, regulate, and govern this world of information, and some organizations have already begun acting on these calls for action. These recent developments represent useful categories of actors who might play this role of "watcher of the watchers." The first category is *market facilitators*, which include organizations such as the International Trade Commission, Big Room, and GoodGuide that are trying to create systematic marketplaces of information. The second is *democratic organizers*, which include organizations such as EnviroMedia and Citizens Market that have tried to build grassroots movements to evaluate corporate claims of sustainability. The *people's representatives* are the third category of potential watchers, and include federal agencies such as the FTC and the Consumer Product Safety Commission that have been created by elected representatives to combat fraud and protect consumers.[66] These elected or appointed representatives of the people are arguably already tasked with regulating information-based governance strategies for the greater good.

Alternatively, *information experts*, building on their scientific and domain-specific knowledge, can independently evaluate the validity and falsifiability of different claims made by these initiatives. Consumer Reports' Eco-Label Center and Greener Choices website are perhaps the best examples of this approach.[67] A fifth category, *club coordinators*, is based on the idea of clubs, which only confer membership to those who meet certain standards and requirements. The Global Ecolabeling Network (GEN) and ISEAL perhaps most closely match this model of quality control.[68] Just as eco-labels and voluntary programs can serve as clubs of companies,[69] these meta-organizations can serve as clubs for the eco-labels and voluntary programs themselves. And finally, there are *industry representatives*, which act through industry associations or individual corporations to monitor and regulate the claims that are made about products in their supply chains or companies in their industries. Examples include the chemical industry's Responsible Care Program and Walmart and Procter and Gamble's sustainability initiatives.[70]

Each of these options have their strengths and weaknesses, which parallel the trade-offs discussed earlier that arise in the construction of the initiatives they would be evaluating. Tensions exist between their credibility and trustworthiness, usability and salience, and democratic openness and focused expertise, and a similar analysis of these trade-offs is necessary. But a deeper question may also be important, and that is one of legitimacy. How would any of these models earn grants of legitimacy from the general public? It will likely require some combination of all of the attributes discussed in this book—trustworthiness, credibility, and so on—but it may also need something more. Remembering Weber's typology of legitimacy, it may require some combination of charismatic, traditional, and legal authority that builds on these core attributes to inspire a broad cross-section of citizens to embrace a particular approach.[71]

Such authority may be helpful or even necessary to provide the activation energy for one of these models to become dominant in the short term, but its legitimacy must be maintained over time. Which of these models then will be the most sustainable, durable, and adaptive? The future of information-based governance must be thought about in the long term—where do we want to see this space going in the next five, ten, or fifty years? Obviously huge uncertainties arise as time horizons increase, but having at least some vision of the future is important. Otherwise, society risks muddling through without making any substantive progress over time on improving these information initiatives or solving existing environmental challenges.

Regardless which of these specific regimes becomes dominant, the gaps and limitations documented in this book indicate that stronger mechanisms of accountability are necessary for the world of information-based environmental governance. Perhaps an endless loop of accountability among competing organizations, the government, and citizens—a proverbial separation of powers—is the best we can hope for. Or perhaps it will be some combination of mechanisms that prevails—an increase in required transparency by government and industry associations, continuing voluntary efforts by cutting-edge companies, continuing short-term pressure by activists demanding change at the forefront of issues, and continuing development of standards by certification organizations over the longer term. In any case, continued research will be needed to track and evaluate the progress of these efforts, and assess them against both theoretical and comparative counterfactuals. This research should be both independent and iterative, and can help us answer the question posed of who watches the watchers. And how well are they doing it?

To paraphrase Theodore Roosevelt, ultimately the answer in a democratic society is us—you and I.[72] Earlier I declared that most people will not be as engaged as our five friends Vernon, Anu, Lynn, Carrie, and Mark, who we have seen take deep dives into the complexities of sustainability claims throughout this book. But their efforts are nevertheless important and admirable—they are our unsung information heroes. Whether you are a government official, corporate executive, academic researcher, civil society advocate, or ordinary citizen, you can engage in the process of keeping information claims accountable, credible, and accurate. Not every claim on every product, of course, but paying attention to just one claim on one product at a time can make a difference. Pick something you care about and are interested in, as Mark did, and do a little research. Make a decision about how you can engage the issue as both a citizen and a consumer, tell your friends about it, and move onto something else. Soon enough you will have exerted at least a little bit of power over these programs, and pushed them to become better.

I make this appeal because ultimately information-based governance expresses what most deeply defines a healthy democracy—a faith in citizen wisdom. That wisdom includes a commitment to both participating in civic life and delegating authority to our representatives. We as citizens engage in politics, vote in elections, and volunteer in our communities despite the fact that such engagement often feels like it makes little difference at the individual level. But we recognize that it is the collective difference that all of our contributions make that matters in a democracy. And then we also elect leaders to make contributions and decisions that we cannot easily make ourselves.

Just as in our broader civic lives, we must find that balance in the context of information-based governance. For example, we may decide to delegate some authority to retailers or government agencies to edit our choices for us, but we must still remain engaged with these initiatives at some level, both for our own benefit and the benefit of society. Such engagement keeps us informed about their progress, keeps us knowledgeable about the impacts of our purchases, and keeps those organizations accountable. This may seem like a burden, but it is also the privilege and benefit of being autonomous citizens. This is one of the key insights of civic republicanism—whether we engage in the public sphere in pursuit of a stronger democracy or a more sustainable society, such engagement can enrich both our own lives and the life of the republic.[73]

And so the story of green grades is not just about these initiatives growing up and having to continue to grow up, but also of our own maturation and development as citizens and consumers. Many of us have supported and encouraged the growth of eco-labels and sustainability ratings through our choices in the supermarket, the workplace, or the voting booth, and these contributions are laudable. But we must continue to take responsibility for their evolution and the quality of the information that we encounter on a daily basis, and not outsource that responsibility entirely to governments or NGOs or scientists or businesses. And for those who work in these institutions, it is important to view their work not only from their parochial organizational perspectives, but also as delegates of the public, as the information we are exposed to is one of our most important and powerful public goods.

The analytical framework and empirical results presented in this book can help us with this task. They also have important implications for the environmental movement more generally. When effectively designed, information-based governance strategies can inform and educate stakeholders about the important sustainability challenges facing society in creative and uniquely personal ways. Visualizing the effects of a product you have personally bought on an endangered species or a community inundated by pollution or a child coping with exposure to toxic waste can motivate us in ways that no grand calls to action ever will. Conversely, realizing that a product you thought was sustainably produced has been "greenwashed" can disillusion us and cause a level of apathy that is also difficult to replicate. Getting information-based governance right is therefore critical for those interested in making progress in the broader environmental domain.

It is also important in a host of other domains as well, from movie content to bond ratings to health care. Understanding the value chains of information in these contexts can help designers improve their information initiatives, users evaluate the information they are providing, and regulators ensure that they are meeting baseline levels of performance and constantly improving over time. Information realism provides a principled and evidence-based perspective on these initiatives that recognizes their past contributions, potential impacts, and inherent limitations. It is a useful alternative to the information pessimism and information optimism that have dominated discussions of information-based governance strategies, as it offers multiple pathways for leveraging their past successes into future achievements. Whether these pathways are pursued by governments, businesses, nonprofits, or individuals, information realism suggests they can be pragmatically applied to not only environmental concerns but also social and labor issues, medical and health settings, and economic and financial contexts.

So returning to the overarching question of the book, can information "save the Earth?" The information realist would say certainly not any type of information, and certainly not by itself. But relevant, trustworthy, valid, and usable information produced by information-based governance strategies that are based on well-designed information value chains can indeed make a difference. However, they are not a cure-all and must be linked to other forms of governance to avoid their pitfalls and achieve their full potential. Like any form of collective action, the effectiveness of information-based governance is not guaranteed, and can benefit from all of our efforts to contribute to their success and ensure they inform our ongoing quest for a more sustainable world and a more democratic society.

# Appendix I: Background on the EEPAC Dataset

The primary objective in constructing the Environmental Evaluations of Products and Companies (EEPAC) Dataset was to rigorously collect data on the organizational, methodological, and procedural characteristics of a broad range of not only product eco-labels but also sustainability ratings, rankings, awards, boycotts, databases, and reviews related to both products and companies. These are all mechanisms that different stakeholders can use to evaluate the environmental performance of products and companies, and focusing only on one form does not capture the full range of information available to them. Rather than collecting data that is either too general or too detailed, data was collected at the level of detail that is most likely to signal credibility to the public. Instead of relying on nonrandom responses to questionnaires, data was collected directly from the websites of a broad range of 245 initiatives. The text from these websites was systematically downloaded, coded, and analyzed to create the EEPAC Dataset. This appendix describes the methods used to create this dataset, including the sampling process, data collection protocols, data quality assurance procedures, and data analysis process.

In order to enable an in-depth analysis of the environmental information about products and companies that is being provided to American consumers and other stakeholder groups by a broad range of initiatives, including not only more well-known but also less well-known programs, the sampling frame was limited to initiatives that are relevant to the U.S. market. The United States has the world's largest economy and the largest number of relevant eco-labels, making it an important case to examine.[1] While a comparison of programs across different types of economies might yield interesting insights and different results, this dataset was designed instead to enable thorough and relatively comprehensive analyses of initiatives within one particular economy where they are particularly prevalent. As Raynolds, Long, and Murray also note, national laws, market characteristics, and social movement pressures shape the need for private regulations, and further justify an empirical focus on initiatives available within a specific country.[2]

The study's sample was selected through a multistep process that first involved aggregating several online databases of environmental evaluations and lists of relevant programs, including Ecolabelling.org, Ecolabels. org, AllGreenRatings.com, the Global Ecolabelling Network, and ISEAL. The initial sample also included initiatives identified in news reports, academic articles, blogs, and similar sources of information between 2006 and 2008. Also included were initiatives identified through a series of systematic keyword searches on Google for "eco-labels," "green ratings," and other keywords across a set of 10 product categories (e.g., electronics, toys, etc.). This process resulted in a list of 471 initiatives, identified through the end of 2008 (thus programs introduced after 2008 are not included). They include programs developed by government agencies, nonprofit organizations, academic institutions, for-profit enterprises, and media outlets.

In order to ensure that the EEPAC Dataset cases are an accurate and unbiased sample of information-based environmental initiatives, programs were excluded that did not meet the study's sampling frame, which limits the included cases to *"information-based environmental governance initiatives that generate publicly-available environmental evaluations of products or companies that make products that are generally available in the United States marketplace."* In this context, a governance initiative is an intentional, planned effort to exert power over others to encourage collective action and create public goods. This definition excludes both internally oriented information-based environmental management programs that are not made public as well as anonymous, hearsay, and generic claims such as "natural" or "recyclable" that do not have a single, traceable source. The definition also excludes corporate sustainability reports, which are generally descriptive and not evaluative forms of information.[3]

Excluding duplicates, initiatives that did not meet this sampling frame, and initiatives that overlap, replicate, or are part of a broader program, reduced the final sample size to 245 cases. Data about this sample were then collected through a rigorous and comprehensive content analysis process. Content analysis is a "research technique for making replicable and valid inferences from texts (or other meaningful matter) to the contexts of their use,"[4] and has been used extensively to analyze corporate annual reports and sustainability disclosures.[5] While utilizing a similar method, this study instead focuses on analyzing the content provided in environmental evaluations of products and companies. It uses a research design incorporating both mechanistic and interpretative elements in order to both identify the presence or absence of particular themes and interpret the meaning of those themes.[6] This process involved identifying a set of characteristics relevant to these themes, defining these characteristics using a set of variables or "codes," and using these codes to analyze the text from the websites of the sample cases. While some initiatives communicate with their audiences through other media as well, the Internet is the primary means by which all of these programs provide comprehensive information about themselves to the public and is therefore the most suitable source of data for this study.

To develop the initial set of codes, an analytic inductive process building on insights from the literature was used to code the website text from an initial set of forty cases.[7] Text segments were coded by theme and not limited to individual words, sentences, or paragraphs, allowing the coding process to capture signals at multiple levels of textual resolution.[8] Similar to the two-step approach used by Beck, Campbell, and Shrives, four other coders used this same set of codes to code the text of the same forty cases, and discrepancies in the results revealed that transparency, independence, and expertise have more dimensions than previous research has identified.[9] This process allowed for a more nuanced interpretation of the meaning associated with these themes and a more granular definition of their associated codes. The multiple dimensions of these characteristics, which will be described in more detail later in this appendix, were incorporated into the coding system used to analyze the full sample of cases.

This coding system was documented in a detailed codebook, which includes 223 binary codes indicating the presence of particular characteristics relating to the methods, content, and organizational backgrounds of the sample cases. Two coders used this codebook and the qualitative coding software MaxQDA to manually code 2,535 webpages and PDF documents downloaded from the websites of the 245 cases in the sample. This process was completed between April 2009 and September 2010, and produced a dataset of 9,829 coded text segments. Because the websites of these initiatives had such extensive amounts of text that coders had to manually download and code, programs introduced after 2008 and website text updated after September 2010 are not included in the dataset.

The analysis process involved compiling the data produced from this coding analysis, testing the inter-rater reliability of this data, and checking it for errors and inconsistencies. In order to ensure the replicability and inter-rater reliability of the data, both coders coded a random sample of twenty-five cases, or approximately 10 percent of the overall sample, and compared the results for discordances. The average level of agreement for the coded data is 91 percent, with an average Kappa score of 0.28, indicating a fair level of agreement.[10] An analysis of the z scores for the Kappa calculations reveals that the average probability that the observed level of agreement is due to chance is 14 percent across all of the codes for which this statistic could be calculated (z scores cannot be computed for codes that have a Kappa value of 0).

A more granular analysis of these statistics finds that this probability is less than 1 percent for 42 percent of the codes, less than 5 percent for 6 percent of the codes, less than 10 percent for 5 percent of the codes, and could not be calculated for 29 percent of them. The remaining 17 percent (or thirty-one codes) have probabilities ranging from 0.11 to 0.69, and are listed in table 8.1. These thirty-one codes include fifteen organization-related codes, five content-related codes, ten methods-related codes, and one interface-related code. They include all three expertise codes and six of the ten codes related to use or endorsement of a case. There are no apparent similarities among these less reliable codes. Given the fact that these thirty-one codes are less likely to be accurate and replicable than the other 194 codes in the dataset, I note their lower reliability when they are mentioned in the text.

Further analysis of the data reveals that these discrepancies are likely due to the prevalence of negative codes (i.e., codes indicating a case does not have a particular trait) relative to positive codes (i.e., codes indicating a case has a particular trait). If there are either a lot more or less positive agreements than negative agreements, then the likelihood of chance agreements is also high and the Kappa coefficient will be lower. The Prevalence Index for all thirty-one of these lower reliability codes is above 0.5, suggesting that the low frequency of these codes occurring may be a strong contributor to their lower Kappa values. The 29 percent of codes for which "due to

Table 8.1

Additional reliability statistics for 31 codes (out of 223) with "due to chance" probabilities > 0.1

| Code category | Code group | Code name | Kappa | Prob > Z |
|---|---|---|---|---|
| Evaluation focus | Facility | Facility | -0.0546 | 0.6712 |
| Sector | Finance and insurance | Banks | -0.0246 | 0.5825 |
| | Manufacturing | Appliances | -0.0309 | 0.6469 |
| | | Pharmaceuticals | -0.0246 | 0.5825 |
| Private benefits | Economic benefits | Limited economic benefit | 0.0000 | 0.5000 |
| Up to date | Criteria updating | Limited criteria update | -0.0586 | 0.6622 |
| | Data age | Data age (Some post-January 2008) | -0.0563 | 0.6183 |
| Independence | Independent data Verification | All data verification by evaluation organization | -0.0417 | 0.6183 |
| Peer review | Data peer review | Description of data peer review with relevant expertise | -0.0417 | 0.5825 |
| | | Mention of data peer review | -0.0294 | 0.5825 |
| Expertise | Expertise | Academic expertise | -0.0417 | 0.6170 |
| | | Relevant academic expertise | -0.0684 | 0.6602 |
| | | Relevant expertise | 0.0543 | 0.2882 |
| Transparency | Goal transparency | Limited organizational goal transparency | 0.2091 | 0.1319 |
| | Method transparency | Strong method transparency | 0.1259 | 0.2224 |
| Initiative Type | Boycott | Boycott | -0.0369 | 0.6146 |
| Organizational structure | Organizational structure | Multiple, dependent organizations | 0.0000 | 0.5000 |
| | | Multiple, independent organizations | -0.0700 | 0.6593 |
| | | Organizational coalition | -0.0700 | 0.6593 |

**Table 8.1** (continued)

| Code category | Code group | Code name | Kappa | Prob > Z |
|---|---|---|---|---|
| Academic | Academic association | Academic association (explicit and generic) | 0.0000 | 0.5000 |
| | Academic use | Academic use (implicit and specific) | -0.0309 | 0.6183 |
| Retailer | Retailer use | Retailer use (explicit and generic) | -0.0471 | 0.6119 |
| | | Retailer use (implicit and specific) | 0.0000 | 0.5000 |
| Government | Government association | Government association (implicit and specific) | -0.0204 | 0.5825 |
| | Government data | Use of government data (explicit and generic) | 0.0566 | 0.3479 |
| | | Use of government data (implicit and specific) | -0.0852 | 0.6893 |
| | Government use | Government use (explicit and generic) | 0.0000 | 0.5000 |
| | Past government Involvement | Past government involvement (implicit and specific) | -0.0294 | 0.5825 |
| Rated organization | Rated organization use | Rated organization use (implicit and specific) | 0.0000 | 0.5000 |
| Nonprofit | Nonprofit involvement | Nonprofit involvement (explicit and generic) | 0.2424 | 0.1127 |
| | Nonprofit use | Nonprofit use (explicit and generic) | -0.0471 | 0.6119 |

chance" probabilities could not be calculated also have high Prevalence Index values, with an average of 0.95 and no value lower than 0.8, which would explain their relatively high Kappa values. While it is possible that the agreement found for these codes is indeed due to chance, the likelihood appears to be small given these high prevalence levels and the fact that the observed agreement for all of these codes is above 92 percent and the average number of discordances is only 2.0.

As noted in chapter 5, my research assistant and I revisited the websites of all 245 cases in 2016 and coded them for a subset of information delivery-related attributes, from links to social media to the presence of search engines and mobile apps. Data for this smaller set of variables was recorded in a spreadsheet and 13 percent of the cases were coded by both of us. Comparisons of our results revealed an average 89 percent level of agreement and an average Kappa score of 0.63 across the twenty-one codes for which we collected data (Kappa scores could not be calculated for two of these codes). This analysis suggests the inter-rater reliability of these data is very high and raised no concerns about any particular codes.

In summary, the inter-rater reliability of the EEPAC Dataset is generally very high. The lower levels of reliability associated with thirty-one of the codes is likely due to the prevalence of those codes in the dataset and may not be due to systematic coding error. Nevertheless, it may be possible to refine and improve the coding process in order to increase its accuracy and reliability across multiple coders. It may also be true, however, that given how rigorous and systematic this coding process was, many of these attributes may be inherently difficult to classify and make legible. If two coders who were trained to identify these attributes could not always agree on what they were finding, it is unlikely that two ordinary consumers would be able to. This raises the possibility that individuals may be visiting the same websites but coming to radically different conclusions about them. This possibility demonstrates the need for these programs to focus more on their information transparency and usability, an insight which I discuss throughout the text.

The main point, however, is that while the reliability of some of the coded data may be limited, the methods used to gather that data are more rigorous than most, if not all, of those used by similar efforts. In particular, this data is more reliable and less biased than data gathered through surveys voluntarily filled out by eco-labeling organizations, since responses to those surveys are not made public, are not responded to by all organizations, and do not necessarily provide information that matches up with the information (and the structure of that information) that the organizations provide publicly.

Beyond the possibility that the reliability of some of the coding data may be improved, other opportunities exist to build on and make use of the EEPAC Dataset. Comparing this data on what initiatives are claiming about themselves on their websites to what they actually have done would be revealing, as would looking at changes in these attributes over time. Cluster and factor analyses of the existing data can be conducted. More attention can be paid to particular types of programs, such as green shopping sites. Different forms of information, such as ratings and certifications, can be more intensively compared. The effects of different methods of data simplification and aggregation can be studied. The dynamics of these programs can be compared across different product categories and economic sectors and across national and regional boundaries. The effects of trade, transnational networks, and harmonization efforts can be assessed, as can the effects of different types of collaborations and alliances among stakeholder groups. Looking at how internally oriented information-based management programs may differ from externally oriented information-based politics and governance approaches might be fruitful as well. And tracking the relationships among information-based strategies and other forms of governance, such as regulation, would be highly informative about the overall effects of these initiatives.

# Appendix II: Interview Sampling Methods

For the interviews discussed in chapter 6, I selected a stratified sample of consumers and representatives from nonprofit organizations, companies, government agencies, academic institutions, and evaluation organizations. In total, I interviewed sixty-eight individuals for approximately one hour each. The interviews were conducted either in person or over the phone, depending on the location and availability of the participant. Approximately half were conducted in person and half were conducted by phone. The overall response rate for the organizational interviews was 53 percent. Interviews with organizational representatives focused on understanding the participant's perspectives on and knowledge of different types of information-based environmental initiatives, and involved a set of both structured Likert-based questions as well as more open-ended semi-structured questions. Interviews with consumers used a computer-assisted personal interviewing (CAPI) method to explore their knowledge of these initiatives and their perceptions of the popularity and effectiveness of a specific set of eco-labels and green rating programs.

In selecting the company representatives, I limited the sample to staff working at companies in the consumer electronics sector. Given the large number of companies that exist in the United States, this sampling frame enabled me to focus on perceptions of eco-labels within one sector and the nature of effectiveness within that sector. The consumer electronics industry has a long history and wide range of eco-labels and green ratings, and provides a rich case study from which we can learn. In order to ensure a representative sample of electronics companies, I contacted companies with large, medium, and small percentages of market share across nine different product categories—televisions, cell phones, printers, personal computers, cameras, audio-visual equipment, home theater equipment, gaming consoles, music players—as well as computer manufacturing more generally. Market share was determined by consulting reports from the Mintel Group on each of these different product categories.[1] The company with the largest market share for each of these categories was contacted, as was at least one company with a small- or medium-sized market share for each category. I also contacted several retailers of electronics equipment and other companies involved in the consumer electronics supply chain (e.g., Google, Intel). In general, my objective was to contact people in these companies who were knowledgeable about both their own internal environmental programs as well as external eco-label and green rating initiatives that are relevant to the consumer electronics sector. In some cases, people I initially contacted referred me to colleagues who had more expertise in these two areas.

I was able to conduct interviews with representatives from nine of the twenty-seven companies I contacted, for a 33 percent response rate, which is comparable to or higher than that of other attitudinal and industry surveys of businesses and executives.[2] Interviewees include employees at the #1 seller (at the time) of music products (Apple), the #1 seller of personal computers (Dell), the #1 seller of audio-visual equipment (Sony), the #2 computer manufacturer (IBM), the #2 seller of mobile phones (Nokia), the #1 consumer electronics retailer (BestBuy), the #4 online retailer (Office Depot), and the #6 seller of televisions (Polaroid).[3] The full list of companies contacted and interviewed is provided in table 9.1.

I also interviewed individuals working at organizations who are implementing eco-label or green rating initiatives related to the electronics sector. I identified these organizations by searching the EEPAC Dataset, which includes twelve programs directly related to consumer electronics. I attempted to contact individuals with leadership roles in these programs and who are most likely to be aware of their histories and operations. I was able to interview individuals at nine of these organizations, for a 75 percent response rate. Those interviewed were typically either at the Vice President or Director level in larger organizations, or at the Executive Director level at smaller organizations. They included individuals involved with the implementation of 80Plus, ENERGY STAR, the SVTC Computer Report Card, Climate Savers Computing Initiative, EPEAT, Greener Electronics Guide (by Greenpeace), TV Recycling Report Card, TCO Certified,

**Table 9.1**

Company sample

| Company | Location | Contacted | Interviewed |
|---|---|---|---|
| Apple | Cupertino, CA | Yes | Yes |
| Best Buy | Richfield, MN | Yes | Yes |
| Dell | Austin, TX | Yes | Yes |
| IBM | Tampa, FL | Yes | Yes |
| Nokia | Finland | Yes | Yes |
| Office Depot | Boca Raton, FL | Yes | Yes |
| Polaroid | Somerset, NJ | Yes | Yes |
| Sony | San Diego, CA | Yes | Yes |
| Bose | - | Yes | No |
| Canon | - | Yes | No |
| Dell | - | Yes | No |
| Eastman Kodak | - | Yes | No |
| Epson | - | Yes | No |
| Google | - | Yes | No |
| Harman | - | Yes | No |
| Hewlett-Packard | - | Yes | No |
| Intel | - | Yes | No |
| Lexmark | - | Yes | No |
| Microsoft | - | Yes | No |
| Motorola | - | Yes | No |
| Nikon | - | Yes | No |
| Panasonic | - | Yes | No |
| Philips | - | Yes | No |
| Pioneer | - | Yes | No |
| Samsung | - | Yes | No |
| Toshiba | - | Yes | No |
| Vizio | - | Yes | No |

and GREEN-SPECS. A list of the organizations contacted and interviewed
is provided in table 9.2.

For the other stakeholder groups, I did not limit my sampling to the
electronics sector, primarily because not as many individuals in these
groups are exclusively focused on electronics. In selecting government
representatives to contact, I first identified a range of federal agencies and
congressional agencies that do work relevant to eco-labels and environmental governance. These included the U.S. Environmental Protection
Agency (EPA), U.S. Department of Energy (DOE), The White House Council
on Environmental Quality (CEQ), U.S. Government Accountability Office

**Table 9.2**
Rating organization sample

| Eco-label/ Rating program | Implementing organization | Location | Contacted | Interviewed |
|---|---|---|---|---|
| 80Plus | Ecos Consulting | Portland, OR | Yes | Yes |
| Climate Savers Computing Initiative | Climate Savers Computing Initiative | San Jose, CA | Yes | Yes |
| Computer Report Card | Silicon Valley Toxics Coalition | San Jose, CA | Yes | Yes |
| ENERGY STAR | EPA | Washington, DC | Yes | Yes |
| EPEAT | Green Electronics Council | Portland, OR | Yes | Yes |
| Greener Electronics Guide | Greenpeace | Oakland, CA | Yes | Yes |
| GREEN-SPECS | Greenelectronics. com | Seattle, WA | Yes | Yes |
| TCO Certified | TCO Development | Chicago, IL | Yes | Yes |
| TV Recycling Scorecard | Computer TakeBack Coalition | San Jose, CA | Yes | Yes |
| Eco-Highlights Label | Hewlett-Packard | - | Yes | No |
| Green IT | Fujitsu | - | Yes | No |
| The Eco Declaration (ECMA 370) | ECMA International | - | Yes | No |

(GAO), U.S. Federal Trade Commission (FTC), and U.S. Senator Dianne Feinstein's Office, which had recently indicated interest in creating a national eco-label. I did not contact any state or local officials, although they would be interesting to include in future research. I sought to contact a balance of regulators, analysts, program managers, and higher-level administrators to solicit a diverse range of opinions. I aimed to speak with people who are both directly involved in managing eco-labels implemented by the government as well as with people who are more generally involved in environmental regulation or analysis. I also wanted to include a balance of participants with and without experience with the electronics sector.

In total, I contacted thirty-eight people across the three branches of government, three agencies in the executive branch, and seven main offices within those agencies. I received responses and was able to conduct interviews with sixteen of these individuals (for a 42 percent response rate). These included individuals representing one congressional agency (the GAO) and three executive agencies (the EPA, FTC, and DOE). They included six individuals who have been involved in implementing specific government-supported eco-labels or recognition programs (e.g., EPEAT, Indoor Air Quality, Responsible Appliance Disposal, Environmentally Preferable Purchasing, Climate Leaders), five individuals focused on more general program planning and strategic analysis, three individuals with broader administrative responsibilities, and two individuals responsible for enforcing specific regulations and laws. At least five of the participants have extensive experience with environmental issues in the electronics sector. Four interviewees have office director-level status, three have division director-status, three have program chief or coordinator status, and seven work on specific programs. The list of the agencies and offices contacted and interviewed is provided in table 9.3. I contacted a larger number of staff in EPA's Office of Air and Radiation because it has a larger number of eco-label and rating programs, although the number of people interviewed within that office is comparable to those from other offices.

I also selected a sample of representatives from environmental nonprofit organizations to explore their opinions about the effects and effectiveness of certifications and ratings in the environmental arena. Since these individuals as a group were meant to represent the diversity of attitudes in the NGO community toward these programs, I sought to include a balance of

**Table 9.3**
Government agency sample

| Agency | Office | Contacted | Interviewed |
|---|---|---|---|
| DOE | Office of Energy Efficiency and Renewable Energy | 3 | 2 |
| EPA | Office of Administration and Resource Management | 1 | 1 |
| | Office of Air and Radiation | 11 | 3 |
| | Office of Chemical Safety and Pollution Prevention | 6 | 4 |
| | Office of Enforcement and Compliance Assurance | 5 | 1 |
| | Office of Solid Waste and Emergency Response | 5 | 3 |
| FTC | Bureau of Consumer Protection | 1 | 1 |
| GAO | Natural Resources and Environment | 2 | 1 |
| U.S. Senate | Senator Dianne Feinstein (D-CA) | 2 | 0 |
| White House | Council on Environmental Quality | 2 | 0 |
| Total | | 38 | 16 |

representatives from both well-known and less well-known organizations focusing on a range of different environmental issues, including toxics, biodiversity, climate change, general environmental concerns, and consumer concerns. I also aimed to include both advocacy organizations and organizations more focused on environmental research and analysis. Similar to my criteria for my government sampling frame, I sought a balance of people with and without direct experience creating or analyzing eco-labels or ratings. And given my focus on the electronics sector, I wanted to recruit both individuals who have worked extensively on environmental issues in that sector as well as those with more general experience relevant to other sectors.

I therefore first compiled lists of the most reputable, richest (in terms of amount of donations), and largest (in terms of membership) nonprofit organizations from the American Institute of Philanthropy (AIP),[4] the Public Broadcasting Service,[5] and U.S. News and World Report.[6] I also identified nonprofits working on electronics environmental and consumer issues from different alliance websites (e.g., the Computer TakeBack Coalition), and I identified nonprofits involved in eco-label and rating programs using my database of eco-label and rating programs (discussed in chapter 3). I then created a master list of these nonprofits and categorized the listed nonprofits on the list by their general area of focus.

To select the final sample of organizations, I first identified three organizations from each area of focus, one of which had an electronics focus, one with eco-label/rating experience but no specific focus on electronics, and one with no electronics or eco-label/rating experience. Six additional organizations with a general focus and three with a focus on toxics were selected (with the same distribution of types), given their relevance to the electronics sector. In selecting this sample, I included a balance of large and small nonprofits—"large" being measured by whether they are one of the most reputable, richest, and/or largest organizations listed previously. Where there was more than one option per type of organization, organizations were selected first by excluding any that have a regional/local/non-U.S. or non-environment focus, and then selecting randomly from those remaining.

This process resulted in a sample of twenty-five organizations to contact, twelve of which are "small" and thirteen of which are "large" (i.e., on at least one of the most reputable, richest, or largest lists). Nine are associated with a nonelectronics specific eco-label or rating program, eight are associated with an electronics eco-label or rating, and eight are not associated with any eco-labels or ratings. The original sample includes the richest organization, five of the top twelve most respected (as rated by AIP), and six of the top twenty largest (by membership size). The sample also includes eight organizations with a general focus, five with a health and toxics focus, four with a research focus, three with a biodiversity focus, three with a climate focus, and two with a consumer focus.

I was able to interview individuals at ten of these nonprofit organizations. These included two organizations with a climate focus (Climate Counts and the Climate Conservancy), two with a focus on environmental health (Center for Environmental Health and Center for Health, Environment, and Justice), one with a focus on biodiversity (Rainforest Alliance), two with a more general environmental focus (Union of Concerned Scientists and EarthJustice), two with a focus on research (World Resources Institute and the Keystone Center), and one with a consumer advocacy focus (Consumer Federation of America). Four of the ten are on the most reputable, richest, and/or largest lists of environmental organizations, four have done work related specifically to the electronics sector, four have done work related to eco-labels more generally, and two have not done any specific work related to either electronics or eco-labels. Table 9.4 shows the list of organizations contacted and interviewed.

I used a similar stratification method for selecting my sample of academic experts. I categorized these experts in terms of their type of expertise and whether they have conducted specific research on electronics or have been involved in the design of any eco-labels. I then randomly contacted a representative subset of these individuals, and was able to conduct interviews with twelve of them (for a 75 percent response rate), six of whom have conducted research on electronics and five of whom have been involved in the design of a specific eco-label or green rating. Four have backgrounds in engineering, three have backgrounds in economics,

**Table 9.4**
Nonprofit organization sample

| Organization | Location | Contacted | Interviewed |
|---|---|---|---|
| Center for Environmental Health | Oakland, CA | Yes | Yes |
| Center for Health, Environment and Justice | Falls Church, VA | Yes | Yes |
| Climate Conservancy | Palo Alto, CA | Yes | Yes |
| Climate Counts | Manchester, NH | Yes | Yes |
| Consumer Federation of America | Washington, DC | Yes | Yes |
| EarthJustice | New York, NY | Yes | Yes |
| Keystone Center | Keystone, CO | Yes | Yes |
| Rainforest Alliance | New York, NY | Yes | Yes |
| Union of Concerned Scientists | Berkeley, CA | Yes | Yes |
| World Resources Institute | Washington, DC | Yes | Yes |
| Alliance for Climate Protection | - | Yes | No |
| Clean Production Action | - | Yes | No |
| Clean Water Action | - | Yes | No |
| Consumers Union | - | Yes | No |
| Earth Island Institute | - | Yes | No |
| Environmental Defense | - | Yes | No |
| Natural Resources Defense Council | - | Yes | No |
| Resources for the Future | - | Yes | No |
| World Wildlife Fund | - | Yes | No |

three have backgrounds in political science, public policy, or planning, and two have backgrounds in marketing or management. They come from academic institutions that include Arizona State University, Lawrence Berkeley National Laboratory, San Jose State University, Michigan Technological University, Ohio State University, University of Maine, Baruch College, Harvard University, Georgia Institute of Technology, University of California, Berkeley, Yale University, and Duke University. The fields and academic institutions of the individuals contacted and interviewed are listed in table 9.5.

I also selected a sample of consumers to interview using a stratified random sampling method. I first identified interested subjects using the UC Berkeley Psychology Department's Research Subject Volunteer Program (RSVP) list of prescreened and prequalified volunteer subjects, which includes more than 1,700 people from around the Bay Area (65 percent

**Table 9.5**
Academic expert sample

| Institution | Field | Contacted | Interviewed |
| --- | --- | --- | --- |
| Lawrence Berkeley National Laboratory | Engineering | Yes | Yes |
| San Jose State University | Planning | Yes | Yes |
| Arizona State University | Engineering | Yes | Yes |
| Michigan Technological University | Economics | Yes | Yes |
| Ohio State University | Economics | Yes | Yes |
| University of Maine | Economics | Yes | Yes |
| Baruch College/CUNY Zichlin School of Business | Marketing | Yes | Yes |
| Harvard Kennedy School | Policy | Yes | Yes |
| Georgia Institute of Technology | Engineering | Yes | Yes |
| University of California, Berkeley | Engineering | Yes | Yes |
| Duke University | Management | Yes | Yes |
| Yale University | Political Science | Yes | Yes |
| University of Arkansas | Engineering | Yes | No |
| Carnegie Mellon University | Engineering | Yes | No |
| Harvard Business School | Business | Yes | No |
| Arizona State University | Engineering | Yes | No |

**Table 9.6**
Consumer sample

| Participant | Gender | Age | Education | Green |
|---|---|---|---|---|
| 1 | Female | <40 | College degree | Less "green" |
| 2 | Female | <40 | Graduate degree | More "green" |
| 3 | Female | <40 | In college | Less "green" |
| 4 | Female | 40 or older | College degree | More "green" |
| 5 | Female | 40 or older | College degree | More "green" |
| 6 | Female | 40 or older | High school degree | Less "green" |
| 7 | Male | <40 | Graduate degree | Less "green" |
| 8 | Male | <40 | In college | More "green" |
| 9 | Male | 40 or older | College degree | More "green" |
| 10 | Male | 40 or older | Graduate degree | More "green" |
| 11 | Male | 40 or older | High school degree | Less "green" |
| 12 | Male | 40 or older | High school degree | More "green" |

are not affiliated with UC Berkeley).[7] These potential subjects were asked to fill out a pre-interview survey that identified their age, gender, educational level, race/ethnicity, and levels of "green" or environmental interest. Responses from this screening process were used to select a random sample of twelve consumers, stratified by gender, age, educational level, and environmental activism. The final sample, shown in table 9.6, included six men and six women, seven aged 40 or older individuals and five under 40, seven relatively "green" and five relatively "not green," and three high school-educated, two in college, four college-educated, and three graduate school-educated.

# Notes

## 1 The Green Debate

1. For more on the certification and partnership work of these two organizations, see Environmental Defense Fund, "Partnerships: The Key to Scalable Solutions"; Rainforest Alliance, "Certification and Assurance Services."

2. For more on the political and legislative strategies of these two organizations, see Natural Resources Defense Council, "Policy Library"; Sierra Club, "Politics and Elections."

3. Parts of this section have been adapted from Bullock, "Green Grades," and Bullock, "Information-Based Governance Theory."

4. This definition builds on discussions of the term by several scholars of governance, including Bevir, *Key Concepts in Governance*, 3; Kjær, *Governance*, 10; Young, *On Environmental Governance*, 3.

5. Smith, "Power."

6. Weber, *The Theory of Social and Economic Organization*, 152.

7. Grigsby, *Analyzing Politics*.

8. Other scholars have classified the different forms of governance in a range of different ways. Bevir (*Key Concepts in Governance*) for example, discusses authority-based, network-based, market-based, and participative governance types. Rosenau ("Governance in a New Global Order") classifies governance in terms of its processes (either unidirectional or multidirectional) and structures (either formal, informal, or mixed) and identifies six primary governance types—top-down governance, network governance, bottom-up governance, side-by-side governance, market governance, and mobius-web governance. Treib, Bähr, and Falkner ("Modes of Governance") suggest four types of governance—coercion, voluntarism, targeting, and framework regulation, while Jordan, Wurzel, and Zito ("The Rise of 'New' Policy Instruments in Comparative Perspective") differentiate between forms of governance that use, respectively, regulation, voluntary agreements, market-based

instruments, eco-labels, and environmental management systems. Hysing ("From Government to Governance?") delineates five different "governing instruments and styles" that include command and control (legal sanctions), incentive-based instruments (taxes and grants), delegated public functions (outsourcing, decentralization, privatization), information instruments (consultations, counseling, education), and voluntary instruments (agreements and labeling).

These typologies provide intriguing insights into the dynamics of governance strategies, but do not provide a clear articulation of their fundamental differences. Also, only Hysing's work explicitly mentions information, but does not explore how it operates, other than describing it as a "soft" instrument. The governance typology in this book builds on this earlier work by more directly addressing the role of information and defining the range of governance approaches more clearly. To accomplish this task, it uses two of the most basic forms of power—physical force and economic incentives—to differentiate between four basic types of governance.

9. For a straightforward explanation of public vs. private goods, see EconPort, "EconPort—Handbook—Public Goods—Classification Table."

10. Houldsworth and Jirasinghe, *Managing and Measuring Employee Performance*; Gunasekaran, Patel, and McGaughey, "A Framework for Supply Chain Performance Measurement"; Kaplan and Norton, *The Balanced Scorecard*.

11. Lasswell, *Politics: Who Gets What, When and How*; "Politics, N," *OED Online*.

12. It should be noted that "information-based governance" is distinct from Arthur Mol's (*Environmental Reform in the Information Age*) concept of "informational governance." Mol's work makes a valuable contribution toward integrating the two fields of information and governance, and provides a helpful synthesis of much of their relevant theoretical and empirical work. Mol ("Environmental Governance in the Information Age") uses the idea of informational governance to describe the "idea that information (and informational processes, technologies, institutions, and resources linked to it) is fundamentally restructuring processes, institutions, and practices of environmental governance, in a way which is essentially different from that of conventional modes of environmental governance." Mol argues that the modern flows of information have enabled new sets of actors and networks to be more engaged in both authoritative and network-driven governance processes.

While this concept provides an insightful lens on the use of information in environmental governance, it does not clearly differentiate between two distinct phenomena—the general use of information in all governance strategies versus the specific use of information as the primary instrument of power in a subset of governance strategies. Unlike informational governance, the concept of information-based governance introduced in this book focuses on the specific instances of governance that not only use information, but also use it as their primary mechanism for driving collective action. While all governance strategies use information to some degree, "information-based governance" begins with and depends on the provision of information to effect change. While informational governance provides an overarching description of "information in governance," information-based governance is a more specific and operational definition of "information as governance." Thus Mol's informational governance is a different type of classification that transcends and pervades the governance types described in the preceding typology. According to Mol (*Environmental Reform in the Information Age*), information increasingly is an essential part of all forms of governance, and is changing its basic nature.

13. Fung, Graham, and Weil (*Full Disclosure*), for example, examine eight information disclosure policies that are mandated by government regulations, but do not consider voluntary disclosure programs. Cashore, Auld, and Newsom (*Governing through Markets*) focus on nonstate actors in their comprehensive analysis of forest certification programs, but do not discuss government certifications and ratings programs. Conroy (*Branded!*) surveys a broad range of certification programs and characterizes them as products of different social movements. Dauvergne and Lister (*Eco-Business*) discuss these programs more from the perspective of business and corporate strategy. De Vinney, Auger, and Eckhardt (*The Myth of the Ethical Consumer*) use original survey research to focus on consumer perspectives on eco-labels, but do not explore their broader institutional dynamics.

14. Darnton, "An Early Information Society"; Yeo, "Reading Encyclopedias"; Raymond, *News, Newspapers, and Society in Early Modern Britain*; Chartier, *The Order of Books*.

15. Brown and Duguid, *The Social Life of Information*; Nunberg, "Farewell to the Information Age"; Gleick, *The Information*.

16. Agar, *The Government Machine*; Cullen, *The Statistical Movement in Early Victorian Britain*; Frankel, *States of Inquiry*; Goldman, *Science, Reform, and Politics in Victorian Britain*; Nunberg, "Farewell to the Information Age"; Bimber, *Information and American Democracy*.

17. Richardson, "Brand Names before the Industrial Revolution."

18. Ibid.

19. Ibid.

20. De Munck, "The Agency of Branding and the Location of Value."

21. Campbell-Kelly, "Information Technology and Organizational Change in the British Census, 1801–1911"; Cullen, *The Statistical Movement in Early Victorian Britain*.

22. Higgs, *The Information State in England*, 2–3.

23. Campbell-Kelly, "Information Technology and Organizational Change in the British Census, 1801–1911," 36; Levitan, *A Cultural History of the British Census*, 18–19.

24. Levitan, *A Cultural History of the British Census*, 5–6; Frankel, *States of Inquiry*.

25. Levitan, *A Cultural History of the British Census*, 5–8; Higgs, *The Information State in England*, 67, 89.

26. Tyler, *Look for the Union Label*; Bird and Robinson, "The Effectiveness of the Union Label and 'Buy Union' Campaigns"; Spedden, *The Trade Union Label*; Duguid, "Trade Marks, Innovation, & the Union Label."

27. Bird and Robinson, "The Effectiveness of the Union Label and 'Buy Union' Campaigns."

28. Ibid.

29. For more information about these rating initiatives, see U.S. News and World Report, "Best Colleges"; Morningstar, "Equity Research Methodology"; Consumer Reports, "About Us"; Nielsen, "TV Ratings"; Charity Navigator, "Overview"; Fair Isaac Corporation, "What Is a Credit Score?"

30. Lamoureux, "Advertising"; CMOsurvey.org, "CMO Survey Report," 23.

31. Barnhart, *The Guide to National Professional Certification Programs*.

32. Consumers Union, "Mission"; Charity Navigator, "How Do We Rate Charities?"; Freedom House, "About Freedom in the World"; Transparency International, "What We Do—Research."

33. Friedman, "Consumer Boycotts"; Baby Milk Action, "The Nestlé Boycott."

34. Morse, "The Birth of the College Rankings"; National Research Council, "A Data-Based Assessment of Research-Doctorate Programs in the United States," n.d.

35. Fung, Graham, and Weil, *Full Disclosure*, 77–87.

36. Ibid., 82, 88.

37. U.S. Consumer Product Safety Commission, "CPSC Recall Announcements and Product Safety Alerts."

38. U.S. Environmental Protection Agency, "ENERGY STAR"; U.S. Environmental Protection Agency, "Green Power Partnership National Top 100."

39. Big Room, "Ecolabel Index."

40. For more information about these environmental information initiatives, see ENERGY STAR, "Facts and Stats"; Forest Stewardship Council, "Our History"; U.S. Green Building Council, "LEED."

41. Global Ecolabelling Network, "What Is Ecolabelling?"

42. For more information about the work of these organizations, see Greenpeace, "Guide to Greener Electronics"; Bendell and Kleanthous, "Deeper Luxury"; Union of Concerned Scientists, "Automaker Rankings 2014"; Kropp, "SRI Field Continues to Shift with RiskMetrics' Acquisition of KLD"; Fortune, "10 Green Giants"; Newsweek, "2016 Green Rankings."

43. Hsu et al., "Environmental Performance Index"; Earth Day Network, "Ecological Footprint Quiz"; Global Footprint Network, "Footprint Basics."

44. Portney, *Taking Sustainable Cities Seriously*, 40, 71–80.

45. Friedman, *Consumer Boycotts*; Bartley, "Certifying Forests and Factories"; Gulbrandsen, "The Effectiveness of Non-State Governance Schemes"; Greenpeace, "Guide to Greener Electronics."

46. Cashore, Auld, and Newsom, *Governing through Markets*; King and Lenox, "Industry Self-Regulation without Sanctions."

47. Khanna and Damon, "EPA's Voluntary 33/50 Program."

48. U.S. Environmental Protection Agency, "List of Voluntary Partnership Programs."

49. Chasek, Downie, and Brown, *Global Environmental Politics*, 108–110.

50. Prescott, "The Rio Summit—Success or Failure?"; Chasek, Downie, and Brown, *Global Environmental Politics*, 35–37.

51. Bartley, "Certifying Forests and Factories"; Chasek, Downie, and Brown, *Global Environmental Politics*, 224.

52. Chasek, Downie, and Brown, *Global Environmental Politics*, 156; Prescott, "The Rio Summit—Success or Failure?"

53. Bartley, "Certifying Forests and Factories."

54. Layzer, *The Environmental Case*, 4th ed., 36–58; Rosenbaum, *Environmental Politics and Policy*, 203–234.

55. Layzer, *The Environmental Case*, 4th ed., 50–52.

56. Vig, "Presidential Powers and Environmental Policy," 90; Vaughn, *Environmental Politics*, 225.

57. U.S. Environmental Protection Agency, "EPA National Air Quality and Emissions Trends Report, 1987," 3; U.S. Environmental Protection Agency, "EPA National Air Pollutant Emission Estimates 1970–1979," 2–4.

58. U.S. Environmental Protection Agency, "National Water Quality Inventory Report 1988 Report to Congress," xii.

59. Vaughn, *Environmental Politics*, 206, 225–226.

60. Fiorino, *The New Environmental Regulation*, 47–57.

61. Hamilton, *Regulation through Revelation*.

62. Obach, *Organic Struggle*, 75–76; Fiorino, *The New Environmental Regulation*, 50.

63. Casteel, "Boren Targets Clinton Btu Tax with Own Plan"; Erlandson, "The Btu Tax Experience."

64. Vig, "Presidential Powers and Environmental Policy," 94–95; Layzer, *The Environmental Case*, 3rd ed., 397–404.

65. Lyon and Maxwell, "Environmental Public Voluntary Programs Reconsidered"; Stanwick and Stanwick, "Exploring Voluntary Environmental Partnerships"; Erlandson, "The Btu Tax Experience."

66. Vig, "Presidential Powers and Environmental Policy"; Steelman and Rivera, "Voluntary Environmental Programs in the United States"; Rosenbaum, *Environmental Politics and Policy*, 13.

67. Layzer, *The Environmental Case*, 4th ed., 410–412; Davenport, "A Climate Deal, 6 Fateful Years in the Making"; Gillis and Davenport, "Leaders Roll Up Sleeves on Climate, but Experts Say Plans Don't Pack a Wallop"; Schwartz, "Another Inconvenient Truth."

68. "Ideology, N."

69. Google, "Ngram Trends for 'Information Society' and 'Information Age.'"

70. Schmidt and Cohen, *The New Digital Age*, 15.

71. Gilder, *Knowledge and Power*, 15; MacFarquhar, "The Gilder Effect."

72. Sunstein, *Infotopia*.

73. Rodriguez Garcia, "Scientia Potestas Est–Knowledge Is Power." On March 51, 2014, I found 4,430,000 Google results for "knowledge is power," as opposed to 721,000 Google results for "information is power."

74. Gutting, "Michel Foucault."

75. Merriam-Webster, "Information."

76. Stenmark, "The Relationship between Information and Knowledge"; Choo, *The Knowing Organization*.

77. Frank, *Microeconomics and Behavior*, 375.

78. Hayek, "The Use of Knowledge in Society."

79. Spence, "Job Market Signaling"; Stiglitz, "The Theory of 'Screening,' Education, and the Distribution of Income."

80. Fung, Graham, and Weil, *Full Disclosure*, 31.

81. Ibid., 31–32; Akerlof, "The Market for 'Lemons'"; Spence, "Job Market Signaling"; Stiglitz, "The Theory of 'Screening,' Education, and the Distribution of Income."

82. Mann and Wüstemann, "Public Governance of Information Asymmetries."

83. Conroy, *Branded!*; Klein, *No Logo*.

84. Florini, *The Right to Know*, 8.

85. Graham, *Democracy by Disclosure*, 138.

86. Ibid.

87. Fung, Graham, and Weil, *Full Disclosure*, 32–33; Fung, "Infotopia."

88. For a comprehensive discussion of different interpretations of objectivity, see Porter, *Trust in Numbers*, 1–8.

89. Schudson, "The Objectivity Norm in American Journalism."

90. Ibid.

91. Haraway, "Situated Knowledges."

92. Dauvergne and Lister, *Eco-Business*, 2.

93. Ibid., 2–3.

94. Ibid., 111.

95. Bowen, *After Greenwashing*, 44–48, 239–241.

96. O'Rourke, *Shopping for Good*.

97. Ibid., 37.

98. Delmas and Burbano, "The Drivers of Greenwashing."

99. O'Rourke, *Shopping for Good*, 38.

100. These arguments are clearly articulated by Scott Nova, Juliet Schor, Lisa Ann Richey, Stefano Ponte, Andrew Szasz, Richard Locke, and Margaret Levi in O'Rourke, *Shopping for Good*; and by Vogel, *The Market for Virtue*.

101. Boström and Klintman, *Eco-Standards, Product Labelling and Green Consumerism*, 71, 79.

102. Ibid., 79–82.

103. Ibid., 73–75.

104. U.S. Environmental Protection Agency, "National Environmental Performance Track."

105. Office of the EPA Inspector General, "Performance Track Could Improve Program Design and Management to Ensure Value," iii.

106. Jackson, "Next Steps for the National Environmental Performance Track Program and the Future of Environmental Leadership Programs."

107. Herrera, "EPA Stuns Industry with Plans to Kill Climate Leaders Program."

108. Conroy, Branded!, 287–297; O'Rourke, Shopping for Good, 7–8, 92–98.

109. Jackson, "Next Steps for the National Environmental Performance Track Program and the Future of Environmental Leadership Programs."

110. Conroy, Branded!, 25; O'Rourke, Shopping for Good, 13–16, 96–97.

111. Wiser and Pickle, "Green Marketing, Renewables, and Free Riders."

112. Kotchen, "Green Markets and Private Provision of Public Goods."

113. Gereffi, "The Global Economy."

114. These attributes have been discussed extensively in the broader governance, management, and information literatures. This framework also builds on earlier work in the policy evaluation and sustainability science fields. Cash et al. ("Salience, Credibility, Legitimacy and Boundaries"), for example, have developed a useful framework for analyzing environmental policy initiatives that focuses on their credibility, legitimacy, and salience. While not directly correspondent, it reinforces the importance of these attributes in characterizing and evaluating different types of governance efforts.

## 2  Valuing Green

1. Prices from the website of Harris Teeter ("Harris Teeter"), a regional grocery store headquartered in Charlotte, N.C.

2. Henry, "What Does Organic Really Mean, and Is It Worth My Money?"

3. Consumer Reports, "When to Buy Organic Food."

4. Ibid.

5. Greene, Scowcroft, and Tawse, "Top 10 Reasons to Support Organic in the 21st Century."

6. Hudson, "Should You Buy Organic?"; Jacobson, "Is Organic Milk Worth the Price?"

7. Zelman, "Organic Food—Is 'Natural' Worth the Extra Cost?"

8. Ibid.; Henry, "What Does Organic Really Mean, and Is It Worth My Money?"

9. Chong and Druckman, "Framing Theory"; Fung, Graham, and Weil, *Full Disclosure*; Romaniuk and Sharp, "Conceptualizing and Measuring Brand Salience"; Schwartz, "An Overview of the Schwartz Theory of Basic Values"; van Dam and van Trijp, "Relevant or Determinant"; Ajzen, "The Theory of Planned Behavior."

10. Harper, "Value (N.)."

11. Whitaker, *History and Criticism of the Labor Theory of Value in English Political Economy*, 10–11, provides an in-depth discussion of the labor theory of value. Quote is from Ricardo, *On the Principles of Political Economy, and Taxation*, 5.

12. One of the first articulations of this conception of value was made by Menger (*Principles of Economics*, 114–175.). First published in German in 1871, the first English translation was published in 1950 by The Free Press.

13. Many areas of philosophy, from moral philosophy to aesthetics, explore the meaning of "value," and several, including axiology and value theory, explicitly focus on this topic. The Stanford Encyclopedia of Philosophy provides an accessible overview of this rich area of inquiry; see Schroeder, "Value Theory."

14. Korsgaard, "Two Distinctions in Goodness," 170.

15. Ibid.

16. For an accessible overview and empirical analysis of these different forms of values (which were first articulated in the 1960s), see Walsh, Loomis, and Gillman, "Valuing Option, Existence, and Bequest Demands for Wilderness."

17. While beyond the scope of this chapter, scholars have debated whether these different forms of value are mutually exclusive, or whether they "supervene" on each other. For more on this topic, see Zimmerman, "Intrinsic vs. Extrinsic Value."

18. Schwartz and Bilsky, "Toward a Universal Psychological Structure of Human Values."

19. Krystallis, Vassallo, and Chryssohoidis, "The Usefulness of Schwartz's 'Values Theory' in Understanding Consumer Behaviour towards Differentiated Products"; Schwartz et al., "Extending the Cross-Cultural Validity of the Theory of Basic Human Values with a Different Method of Measurement"; Schwartz, "Universals in the Content and Structure of Values"; Beierlein et al., "Testing the Discriminant Validity of Schwartz' Portrait Value Questionnaire Items"; Datler, Jagodzinski, and Schmidt, "Two Theories on the Test Bench."

20. Schwartz, "An Overview of the Schwartz Theory of Basic Values."

21. Stern, Dietz, and Guagnano, "A Brief Inventory of Values."

22. Ibid.

23. Stern et al., "A Value-Belief-Norm Theory of Support for Social Movements"; Oreg and Katz-Gerro, "Predicting Proenvironmental Behavior Cross-Nationally."

24. Diaz and Rosenberg, "Spreading Dead Zones and Consequences for Marine Ecosystems"; Rabotyagov et al., "The Economics of Dead Zones," 1.

25. O'Connor and Whitall, "Linking Hypoxia to Shrimp Catch in the Northern Gulf of Mexico."

26. de Groot and Steg, "Value Orientations to Explain Beliefs Related to Environmental Significant Behavior"; de Groot and Steg, "Mean or Green"; de Groot and Steg, "Value Orientations and Environmental Beliefs in Five Countries"; Honkanen and Verplanken, "Understanding Attitudes towards Genetically Modified Food."

27. Evans et al., "Self-Interest and Pro-Environmental Behaviour"; De Vos, "Certification and Eco-Labeling."

28. Ottman, "Focus on Consumer Self-Interest to Win Today's Green Customer."

29. Schneider, "11 Steps to Mainstream Your Green Products."

30. Blamey, "The Activation of Environmental Norms"; Larson, *Persuasion*; Schwartz, "Normative Influences on Altruism."

31. Verplanken and Holland, "Motivated Decision Making."

32. Hahnel et al., "What Is Green Worth to You?"

33. A study by Krystallis, Vassallo, and Chryssohoidis ("The Usefulness of Schwartz's 'Values Theory' in Understanding Consumer Behaviour towards Differentiated Products") provides a useful example of this phenomenon. Their analysis of more than eight thousand survey responses for fourteen different values questions (from eight EU countries) concludes that the sample can be classified as "collectivistic" or "individualistic," partially because some value domains may be "empirically missing" and "not practically important." They also find that clusters of respondents with higher percentages of regular or occasional organic buyers hold collectivistic values more strongly. However, they acknowledge that the difference between the clusters with the highest and lowest percentage of organic buyers is only 4.8 percent. Clearly their two-factor categorization is not adequately explaining the diversity of factors driving organic food purchases in Europe. Furthermore, their assumption that some value domains are not practically important does not acknowledge that those domains may be activated under the right circumstances and may play an important role in driving green purchases.

34. Bauer, Heinrich, and Schäfer, "The Effects of Organic Labels on Global, Local, and Private Brands."

35. Olsen, Thach, and Hemphill, "The Impact of Environmental Protection and Hedonistic Values on Organic Wine Purchases in the US."

36. Chinnici, D'Amico, and Pecorino, "A Multivariate Statistical Analysis on the Consumers of Organic Products."

37. Ma and Lee, "Understanding Consumption Behaviours for Fair Trade Non-Food Products."

38. Shaw et al., "An Exploration of Values in Ethical Consumer Decision Making."

39. Autio et al., "Consuming Nostalgia?," 567.

40. Chinnici, D'Amico, and Pecorino, "A Multivariate Statistical Analysis on the Consumers of Organic Products," 197.

41. Chen, "Consumer Attitudes and Purchase Intentions in Relation to Organic Foods in Taiwan"; Dean, Raats, and Shepherd, "Moral Concerns and Consumer Choice of Fresh and Processed Organic Foods."

42. Bauer, Heinrich, and Schäfer, "The Effects of Organic Labels on Global, Local, and Private Brands"; Lockie et al., "Eating 'Green'"; Loureiro, McCluskey, and Mittelhammer, "Assessing Consumer Preferences for Organic, Eco-Labeled, and Regular Apples"; Hjelmar, "Consumers' Purchase of Organic Food Products"; Pino, Peluso, and Guido, "Determinants of Regular and Occasional Consumers' Intentions to Buy Organic Food"; Kriwy and Mecking, "Health and Environmental Consciousness, Costs of Behaviour and the Purchase of Organic Food."

43. Hjelmar, "Consumers' Purchase of Organic Food Products"; Gamlund, "Who Has Moral Status in the Environment?"

44. Krystallis, Vassallo, and Chryssohoidis, "The Usefulness of Schwartz's 'Values Theory' in Understanding Consumer Behaviour towards Differentiated Products"; Karp, "Values and Their Effect on Pro-Environmental Behavior"; Aertsens et al., "Personal Determinants of Organic Food Consumption"; Ma and Lee, "Understanding Consumption Behaviours for Fair Trade Non-Food Products."

45. Schwartz, "An Overview of the Schwartz Theory of Basic Values."

46. Krystallis et al., "Societal and Individualistic Drivers as Predictors of Organic Purchasing Revealed through a Portrait Value Questionnaire (PVQ)-Based Inventory"; Beierlein et al., "Testing the Discriminant Validity of Schwartz' Portrait Value Questionnaire Items"; Schwartz and Boehnke, "Evaluating the Structure of Human Values with Confirmatory Factor Analysis."

47. Shaw et al., "An Exploration of Values in Ethical Consumer Decision Making."

48. Dietz, Fitzgerald, and Shwom, "Environmental Values"; Stern, Dietz, and Guagnano, "A Brief Inventory of Values"; de Groot and Steg, "Mean or Green"; Krystallis

et al., "Societal and Individualistic Drivers as Predictors of Organic Purchasing Revealed through a Portrait Value Questionnaire (PVQ)-Based Inventory."

49. Devinney and his colleagues make this point and expand on it in their work. DeVinney, Auger, and Eckhardt, *The Myth of the Ethical Consumer*.

50. Petty and Cacioppo, "The Elaboration Likelihood Model of Persuasion"; Fishbein and Ajzen, *Predicting and Changing Behavior*.

51. TerraChoice, "The Sins of Greenwashing."

52. While the social issues covered by these cases are important, the focus of this research (and this book) is on the environmental dimensions of information-based governance strategies.

53. TerraChoice, "The Sins of Greenwashing."

54. Verplanken and Holland, "Motivated Decision Making"; Blamey, "The Activation of Environmental Norms Extending Schwartz's Model."

55. Gallup, "Environment."

56. Wong, "Mother Nature's Medicine Cabinet"; France, "Drought Hurts Agriculture, Costs West Over $40 Billion."

57. Shaw et al., "An Exploration of Values in Ethical Consumer Decision Making."

58. Oxfam, "Battle of the Brands: The Annual Scorecard Update"; Coster, "Ranking the World's Most Sustainable Companies."

59. U.S. Census Bureau, "NAICS—North American Industry Classification System."

60. The thirty-eight product categories were identified and named during the coding process from the perspective of the consumer and the products covered by each case. Once the coding was complete, I then manually mapped these product categories to the two-digit NAICS sector that most directly covered each category. The NAICS sectors are broader and more systematically constructed, and represent a more general industry and sector level perspective on the coverage of these cases.

61. It should be noted that all of these classification percentages are approximate because categories covered by different cases do not always correspond exactly, categories can overlap significantly, and it is not always clear what categories and sectors are included in an initiative's scope. Also, the product categories were developed inductively throughout the coding process, and while codes were applied retroactively to previously coded text after the final case was coded, the iterative nature of this process makes this sector-based data potentially less accurate and reliable than the rest of the coded data presented below.

62. Grunert, Hieke, and Wills, "Sustainability Labels on Food Products."

63. DeVinney, Auger, and Eckhardt, *The Myth of the Ethical Consumer*.

64. King, "Materiality Assessments"; KPMG, "The Essentials of Materiality Assessment."

65. Allison-Hope, "How to Navigate the Maze of Materiality Definitions."

66. "Common Cause Communication: A Toolkit."

67. TerraChoice, "The Sins of Greenwashing."

68. Consumer Reports, "From Crop to Table Report."

69. Nielsen, "Health and Wellness through the Eyes of the Diverse Consumer."

70. Ibid.

71. Dettmann and Dimitri, "Who's Buying Organic Vegetables?"

72. Best, "The Surprising Truth about Who's Really Buying Organic."

73. Crawford, "Who Buys Organic and Where They Buy It Is Evolving."

74. Consumer Reports National Research Center, "Organic Food Labels Survey 2014 Nationally-Representative Phone Survey"; Gallup, "Forty-Five Percent of Americans Seek out Organic Foods."

75. Gallup, "Forty-Five Percent of Americans Seek out Organic Foods."

76. Organic Trade Association, "U.S. Organic Sales Post New Record of $43.3 Billion in 2015"; Carlson and Jaenicke, "Changes in Retail Organic Price Premiums from 2004 to 2010."

77. Brownstone, "Americans Will Pay More for Organic, but They Also Have No Idea What 'Organic' Means."

## 3 Trusting Green

1. ForestEthics, "SFI: Certified Greenwash"; ForestEthics, "Follow the Greenwash Money."

2. ForestEthics, "NGO Opposition Letter to SFI."

3. Forest Stewardship Council, "Our History"; Forest Stewardship Council, "Governance."

4. Hance, "The FSC Is the 'Enron of Forestry' Says Rainforest Activist"; Greenpeace, "Holding the Line with FSC"; Grover, "Friends of the Earth Backs Away from Forest Stewardship Council."

5. FSC-Watch, "About"; FSC-Watch, "The 10 Worst Things about the Forest Stewardship Council."

6. FSC-Watch, "The 10 Worst Things about the Forest Stewardship Council."

7. Forest Stewardship Council, "Financial Reporting."

8. Much of the information in this chapter is adapted from two articles I published in *Political Research Quarterly* and *Business and Politics* (Bullock, "Independent Labels?"; Bullock, "Signaling the Credibility of Private Actors as Public Agents").

9. Fukuyama, "Social Capital and the Global Economy."

10. Algan and Cahuc, "Trust and Growth"; Fukuyama, *Trust.*

11. Sandlund, "Trust Is a Must."

12. Sztompka, *Trust*; Levi and Stoker, "Political Trust and Trustworthiness"; Kramer, "Trust and Distrust in Organizations."

13. Levi and Stoker, "Political Trust and Trustworthiness"; Marsh and Dibben, "The Role of Trust in Information Science and Technology."

14. Nannestad, "What Have We Learned about Generalized Trust, If Anything?"

15. Ibid.

16. Ibid.

17. Levi and Stoker, "Political Trust and Trustworthiness."

18. Hovland, Janis, and Kelley, *Communication and Persuasion*; Berlo, Lemert, and Mertz, "Dimensions for Evaluating the Acceptability of Message Sources."

19. Marsh and Dibben, "Trust, Untrust, Distrust and Mistrust—An Exploration of the Dark(er) Side."

20. "Misinformation vs Disinformation—What's the Difference?"

21. Levi and Stoker, "Political Trust and Trustworthiness."

22. Ibid.

23. Ibid.

24. Edelman, "Trust in Institutions Drops to Level of Great Recession."

25. Bicchieri, Xiao, and Muldoon, "Trustworthiness Is a Social Norm, but Trusting Is Not."

26. Black, "Constructing and Contesting Legitimacy and Accountability in Polycentric Regulatory Regimes," 150.

27. Spence, "Job Market Signaling."

28. Merriam Webster, "Credibility."

29. Cash et al., "Salience, Credibility, Legitimacy and Boundaries"; Fogg and Tseng, "The Elements of Computer Credibility."

30. Alagona, "Credibility"; Shapin, "Cordelia's Love"; Boström, "Establishing Credibility."

31. Suchman, "Managing Legitimacy," 574.

32. Uphoff, "Distinguishing Power, Authority & Legitimacy," 310.

33. Deegan, "The Legitimising Effect of Social and Environmental Disclosures"; Roberts, "Determinants of Corporate Social Responsibility Disclosure."

34. Suchman, "Managing Legitimacy."

35. Mena and Palazzo, "Input and Output Legitimacy of Multi-Stakeholder Initiatives"; Suchman, "Managing Legitimacy."

36. Mena and Palazzo, "Input and Output Legitimacy of Multi-Stakeholder Initiatives," 528.

37. Blackman and Naranjo, "Does Eco-Certification Have Environmental Benefits?"; Rondinelli and Vastag, "Panacea, Common Sense, or Just a Label?"

38. Mena and Palazzo, "Input and Output Legitimacy of Multi-Stakeholder Initiatives."

39. Scharpf, "Economic Integration, Democracy and the Welfare State," 19 (emphasis in original).

40. Mena and Palazzo, "Input and Output Legitimacy of Multi-Stakeholder Initiatives."

41. Suchman, "Managing Legitimacy."

42. Darby and Karni, "Free Competition and the Optimal Amount of Fraud"; Nadaï, "Conditions for the Development of a Product Ecolabel"; Nelson, "Information and Consumer Behavior"; Roe and Sheldon, "Credence Good Labeling."

43. Walker, "Consumers Don't Trust Green Product Claims, Survey Says."

44. Peattie and Crane, "Green Marketing."

45. Delmas and Burbano, "The Drivers of Greenwashing."

46. Hensley, "Costs Temper Demand for Organic Foods"; Steinmetz, "Why Many Consumers Think 'Organic' Labels Are Hogwash."

47. Thøgersen, Haugaard, and Olesen, "Consumer Responses to Ecolabels."

48. Shapiro, "Richer, More Equal Countries Are More Trusting, Study Finds."

49. Peattie, "Green Consumption," 179.

50. Most studies do not distinguish between mistrust and distrust, and so I use them somewhat interchangeably in the text here. However, as noted earlier, it is important to recognize that lack of trust can stem from both phenomena.

51. Sustainable Brands, "Poll Finds US Shoppers Willing to Spend 31% More Per Week on Responsibly Produced Food."

52. Mountford, "Why Businesses Suffer from a Trust Gap"; Sadowski, Whitaker, and Buckingham, "Rate the Raters Phase Two."

53. Teisl, Rubin, and Noblet, "Non-Dirty Dancing?"

54. Banerjee and Solomon, "Eco-Labeling for Energy Efficiency and Sustainability," 109.

55. Sønderskov and Daugbjerg, "The State and Consumer Confidence in Eco-Labeling," 513.

56. McKnight, Choudhury, and Kacmar, "The Impact of Initial Consumer Trust on Intentions to Transact with a Web Site."

57. Dando and Swift, "Transparency and Assurance"; Boström, "Establishing Credibility"; Mueller, Dos Santos, and Seuring, "The Contribution of Environmental and Social Standards Towards Ensuring Legitimacy in Supply Chain Governance."

58. Bullock, "Green Grades."

59. Sadowski, Whitaker, and Buckingham, "Rate the Raters Phase Two: Taking Inventory of the Ratings Universe."

60. Bullock, "Green Grades."

61. Banerjee and Solomon, "Eco-Labeling for Energy Efficiency and Sustainability"; Boström, "Establishing Credibility"; Costa et al., "Quality Promotion through Eco-Labeling"; Dando and Swift, "Transparency and Assurance"; Jose and Lee, "Environmental Reporting of Global Corporations"; Maury, "A Circle of Influence"; Nilsson, Tunçer, and Thidell, "The Use of Eco-Labeling like Initiatives on Food Products to Promote Quality Assurance"; Smith, Palazzo, and Bhattacharya, "Marketing's Consequences"; Starobin and Weinthal, "The Search for Credible Information in Social and Environmental Global Governance"; U.S. Federal Trade Commission, "Proposed Revisions to the Green Guides."

62. Smith, Palazzo, and Bhattacharya, "Marketing's Consequences," 625.

63. Jose and Lee, "Environmental Reporting of Global Corporations," 318.

64. Berlo, Lemert, and Mertz, "Dimensions for Evaluating the Acceptability of Message Sources"; Hovland, Janis, and Kelley, *Communication and Persuasion*.

65. Berlo, Lemert, and Mertz, "Dimensions for Evaluating the Acceptability of Message Sources."

66. Edelman, "2013 Edelman Trust Barometer Finds a Crisis in Leadership"; Sadowski, Whitaker, and Buckingham, "Rate the Raters Phase Two: Taking Inventory of the Ratings Universe."

67. Greenpeace, "Hazardous Substances Reduced but Not Eliminated from Laptops."

68. "All products" refers to all product types—a sample of products from each product type is selected for testing by the laboratory.

69. One-sided Fisher's exact test = 0.575. Cases implemented by firms include any case in which the organization conducting the evaluation is evaluating its own products or performance. This includes manufacturers as well as retailers who evaluate their own branded products and manufacturer's products of other companies.

70. It should be noted that the Kappa values from the inter-rater reliability analysis (discussed in appendix I) are relatively low for the data peer-review variables. This is likely due to the fact that very few cases have such peer review.

71. Dando and Swift, "Transparency and Assurance"; Gifford, "Effective Shareholder Engagement."

72. Uphoff, "Distinguishing Power, Authority & Legitimacy."

73. Dando and Swift, "Transparency and Assurance," 196.

74. Mattli and Büthe, "Accountability in Accounting?"

75. Ericsson and Smith, *Toward a General Theory of Expertise*; Chi, Glaser, and Farr, *The Nature of Expertise*; Jasanoff, "(No?) Accounting for Expertise."

76. Berlo, Lemert, and Mertz, "Dimensions for Evaluating the Acceptability of Message Sources"; Hovland, Janis, and Kelley, *Communication and Persuasion*.

77. Suchman, "Managing Legitimacy."

78. Sadowski, Whitaker, and Buckingham, "Rate the Raters Phase Two."

79. Fisher's one sided exact test = 0.015. It should be noted that the inter-rater reliability for two of these expertise-related attributes was among the lowest in the broader set of data collected. The probability that the agreement between the raters was due to chance is between 60 percent and 66 percent for the academic and relevant academic expertise codes, and their Kappa scores were both less than 0 (-0.04 and -0.07, respectively). As appendix I highlights, this may be more due to the low prevalence of these characteristics (the calculated Prevalence Index was 0.88 and 0.84, respectively, for these two criteria), and less due to the reliability of the coding process. However, expertise may indeed be difficult for both coders and more general audiences to recognize and agree on, especially given the many different forms it can take.

80. Ilchman and Uphoff, *The Political Economy of Change*.

81. Their framework also includes physical force and legitimacy, but these two forms of legitimacy are beyond the scope of this chapter. While information-based governance strategies by definition do not directly use physical force to accomplish their aims, they may depend on the government's latent threat of physical force through the enforcement of copyright, trademark, and other relevant laws. For more on this dynamic, see Bullock, "Independent Labels?," 56. As for legitimacy, while the transfer of power resources reflects an implicit recognition of their legitimacy by the resource providers, a more direct and broad-based measure of the perceived legitimacy of these initiatives would require surveys and interviews with different stakeholder groups that are beyond the scope of this chapter.

82. Uphoff, "Distinguishing Power, Authority & Legitimacy."

83. Collins English Dictionary, "Public Sector."

84. Investopedia, "Private Sector."

85. Anheier, *Civil Society*.

86. The private sector codes in the EEPAC Dataset are limited to suppliers and retailers, and do not include other types of for-profit firms (e.g., media organizations such as Newsweek, investment firms such as Calvert, or certifiers such as TCO Development) that are associated with these cases. The focus of the coding process was on private sector organizations that are producing or selling products that might be evaluated by these cases (or likely to be evaluated by cases focusing on firm-level sustainability performance).

87. Bullock, "Independent Labels?"

88. Koppell, "Pathologies of Accountability"; Romzek and Ingraham, "Cross Pressures of Accountability"; Romzek and Dubnick, "Accountability."

89. Black, "Constructing and Contesting Legitimacy and Accountability in Polycentric Regulatory Regimes," 149.

90. Baur and Palazzo, "The Moral Legitimacy of NGOs as Partners of Corporations."

91. Some have described a "third party" as an external group separate from manufacturers, industry associations, and governmental bodies (Gereffi, Garcia-Johnson, and Sasser, "The NGO-Industrial Complex"), while others have described such a party as "external" or "outside" but still paid for by the company (Prakash and Potoski, *The Voluntary Environmentalists*, 22; Starobin and Weinthal, "The Search for Credible Information in Social and Environmental Global Governance").

92. U.S. Green Building Council, "Distinguishing Your Expertise."

93. Boström, "Establishing Credibility"; Dando and Swift, "Transparency and Assurance"; Mueller, Dos Santos, and Seuring, "The Contribution of Environmental and

Social Standards Towards Ensuring Legitimacy in Supply Chain Governance"; U.S. Environmental Protection Agency, "EPA Environmentally Preferable Purchasing Program Pilot to Assess Standards and Ecolabels for EPA's Recommendations to Federal Agencies: Final PILOT Assessment Guidelines."

94. Romzek and Ingraham, "Cross Pressures of Accountability."

95. Mattli and Büthe, "Accountability in Accounting?"

96. Pew Research Center, "Beyond Distrust."

97. Boström and Klintman, *Eco-Standards, Product Labeling and Green Consumerism*, 142.

98. Forest Stewardship Council, "Board of Directors"; Sustainable Forestry Initiative, "SFI Board Members."

99. Forest Stewardship Council, "Debrief of FSC General Assembly 2014."

100. A 2014 article in the *Economist* provides a useful overview of these issues relating to financial accounting ("The Dozy Watchdogs").

101. Fair Labor Association, "Charter Document," 24.

102. SCS Global Services, "Certification Manual."

103. These assessments are based on information on their websites in 2009 and 2010.

104. Natural Resources Defense Council, "NRDC: A Shopper's Guide to Home Tissue Products."

## 4  Measuring Green

1. U.S. Green Building Council, "LEED: Leadership in Energy & Environmental Design."

2. Ibid.

3. Ibid.

4. Ibid.

5. E3, "U.S. Green Building Council."

6. Alter, "In Defence of LEED"; Schnaars and Morgan, "In U.S. Building Industry, Is It Too Easy to Be Green?"; Bainfield, "As Good and Important as It Is, LEED Can Be So Embarrassing."

7. Swearingen, "LEED-Certified Buildings Are Often Less Energy-Efficient Than Uncertified Ones"; Schnaars and Morgan, "In U.S. Building Industry, Is It Too Easy

to Be Green?"; Bainfield, "As Good and Important as It Is, LEED Can Be So Embarrassing."

8. Schnaars and Morgan, "In U.S. Building Industry, Is It Too Easy to Be Green?"; Scofield, "Efficacy of LEED-Certification in Reducing Energy Consumption and Greenhouse Gas Emission for Large New York City Office Buildings."

9. Gifford, "The Four Sins of LEEDwashing."

10. GBI, "Green Building Initiative."

11. Whelan, "Competition for LEED."

12. Weeks, "Anti-Greenwashing Group Launches with Letter to GBI"; Averill, "Up the Green Mountain."

13. U.S. General Services Administration, "Green Building Certification System Review Co-Chairs Memorandum."

14. Swearingen, "LEED-Certified Buildings Are Often Less Energy-Efficient Than Uncertified Ones"; Wilner, "LEED Buildings 'Less Energy Efficient'—Says Who?"

15. International Living Future Institute, "Living Building Challenge 3.0."

16. U.S. Green Building Council, "LEED."

17. Grimm and Grimm, "Little Snow-White."

18. Investopedia, "Sunshine Laws"; Williams, "The Securities and Exchange Commission and Corporate Social Transparency."

19. Fung, "Infotopia"; Stasavage, "Transparency, Democratic Accountability, and the Economic Consequences of Monetary Institutions"; Florini, *The Right to Know.*

20. An exception is the work of Auld and Gulbrandsen ("Transparency in Nonstate Certification," 2010), which distinguishes between input and output transparency.

21. Remler and Ryzin, *Research Methods in Practice*, 118–124.

22. I use both representation and metric here because theoretically such operationalizations do not necessarily have to be numeric or even textual, but can be abstract and artistic as well. For the purposes of this chapter, however, I am focused primarily on the validity of sustainability metrics, and so will generally use that term.

23. Drost, "Validity and Reliability in Social Science Research."

24. Auld and Gulbrandsen, "Transparency in Nonstate Certification"; Dando and Swift, "Transparency and Assurance"; Fung, Graham, and Weil, *Full Disclosure*; Hess, "Social Reporting and New Governance Regulation"; Mena and Palazzo, "Input and Output Legitimacy of Multi-Stakeholder Initiatives"; Nilsson, Tunçer, and Thidell, "The Use of Eco-Labeling like Initiatives on Food Products to Promote Quality Assurance—Is There Enough Credibility?"

25. Boström, "Establishing Credibility"; Nilsson, Tunçer, and Thidell, "The Use of Eco-Labeling like Initiatives on Food Products to Promote Quality Assurance—Is There Enough Credibility?"

26. Hess, "Social Reporting and New Governance Regulation," 471.

27. Fung, "Infotopia."

28. Auld and Gulbrandsen, "Transparency in Nonstate Certification."

29. The one-sided Fisher's exact tests for these two results were 0.003 and 0.087, respectively.

30. This excludes references to the reliability or consistency of other programs or sources of information.

31. Full disclosure: I am a cofounder of GoodGuide and served as its first director of content development.

32. U.S. Environmental Protection Agency, "Life Cycle Assessment: Principles and Practice."

33. The Oxford English Dictionary provides two definitions of "reliability"—the first is focused on trustworthiness and confidence and maps to the more colloquial usage, and the second reflects a more statistical understanding of it as "the degree to which repeated measurements of the same subject under identical conditions yield consistent results," which is the focus of this chapter ("Reliability, N.").

34. Van Amstel et al., "The Reliability of Product-Specific Eco-Labels as an Agrobiodiversity Management Instrument."

35. Sadowski, Whitaker, and Ayars, "Rate the Raters Phase 3."

36. Nature, "Availability of Data, Material and Methods."

37. Jacoby, "The AJPS Replication Policy."

38. Achenbach, "The New Scientific Revolution."

39. Baker, "First Results from Psychology's Largest Reproducibility Test."

40. Broockman and Kalla, "We Discovered One of Social Science's Biggest Frauds."

41. Global Reporting Initiative, "G4 Sustainability Reporting Guidelines."

42. Szwajkowski and Figlewicz, "Evaluating Corporate Performance."

43. Walls, Phan, and Berrone, "Measuring Environmental Strategy."

44. National Research Council, "A Data-Based Assessment of Research-Doctorate Programs in the United States."

45. National Research Council, "A Data-Based Assessment of Research-Doctorate Programs in the United States: Excel Data Table."

46. Chartier et al., "Sensitivity and Stability of Ranking Vectors."

47. Dando and Swift, "Transparency and Assurance."

48. Bullock and Wilder, "The Comprehensiveness of Competing Higher Education Sustainability Assessments."

49. Smith et al., "The Validity of Food Miles as an Indicator of Sustainable Development."

50. Bennear, "What Do We Really Know?"

51. Chartier, Langville, and Simov, "March Madness to Movies."

52. Ibid. This is just one example of the insights available from the growing literature on the science and mathematics of ratings and rankings. While the technical details of this literature are beyond the scope of this chapter, an excellent starting point for further exploration is Langville and Meyer, *Who's #1?*

53. Rahman and Post, "Measurement Issues in Environmental Corporate Social Responsibility (ECSR)."

54. Chatterji, Levine, and Toffel, "How Well Do Social Ratings Actually Measure Corporate Social Responsibility?"

55. Blackman and Naranjo, "Does Eco-Certification Have Environmental Benefits?"

56. Remler and Ryzin, *Research Methods in Practice*, 132.

57. See U.S. Green Building Council, "Fallsway Housing and Service Center," for an example of the published LEED scorecard; ENERGY STAR, "ENERGY STAR Certified Buildings & Plants."

58. Green Building Initiative, "Certified Buildings Directory"; International Living Future Institute, "Registered Living Building Challenge Projects."

59. International Living Future Institute, "A Living Education."

60. U.S. Green Building Council, "LEED Technical Advisory Groups Charge."

61. ENERGY STAR, "Learn about Benchmarking"; ENERGY STAR, "Eligibility Criteria for the 1–100 ENERGY STAR Score."

62. Green Building Initiative, "Overview"; Green Building Initiative, "Green Globes Program Features."

63. U.S. Green Building Council, "Guide to LEED Certification."

64. International Living Future Institute, "The Certification Process."

65. ENERGY STAR, "ENERGY STAR Certification for Your Building."

66. U.S. Green Building Council, "LEED v4, the Newest Version of LEED Green Building Program Launches at USGBC's Annual Greenbuild Conference"; ENERGY STAR, "Version 3 Overview," 0; International Living Future Institute, "Living Building Challenge 3.0."

67. Law, "LEED vs. Green Globes"; Hand, "LEED vs. Green Globes."

68. Roberts and Melton, "LEED vs. Green Globes."

69. Clark, "What Owners Need to Know about … Green-Building Rating Systems."

70. Wei, Ramalho, and Mandin, "Indoor Air Quality Requirements in Green Building Certifications."

71. Scofield, "Efficacy of LEED-Certification in Reducing Energy Consumption and Greenhouse Gas Emission for Large New York City Office Buildings."

72. Suh et al., "Environmental Performance of Green Building Code and Certification Systems."

73. Wang, Fowler, and Sullivan, "Green Building Certification System Review."

74. Ibid.

75. Tangherlini, "GSA Green Building Certification Systems Review Letter to Secretary of Energy."

76. U.S. General Services Administration, "Green Building Certification System Review Co-Chairs Memorandum."

77. Wang, Fowler, and Sullivan, "Green Building Certification System Review."

78. Ibid., xv; International Living Future Institute, "Living Building Challenge 3.0."

79. U.S. Green Building Council, "Building Life-Cycle Impact Reduction."

80. Green Building Initiative, "Green Globes for New Construction."

81. Owens et al., "LEED v4 Impact Category and Point Allocation Process."

82. Wang, Fowler, and Sullivan, "Green Building Certification System Review."

83. Clark, "What Owners Need to Know about … Green-Building Rating Systems"; Roberts and Melton, "LEED vs. Green Globes."

84. Schwartz, "Drexel University Releases Green Globes vs. LEED Cost Comparison Study."

85. Randazzo, "Sustainable Options: LEED™ and Green Globes™—a Comparison—Part One of a Three Part Series."

86. U.S. Green Building Council, "Terms and Conditions."

87. U.S. Green Building Council, "LEED Dynamic Plaque."

## 5  Delivering Green

1. Greenpeace, "Leaders and Laggards"; Scherer, "Investors Turn Up Heat over Carbon Emissions."

2. Myers and Alpert, "Semantic Confusion in Attitude Research"; van Dam and van Trijp, "Relevant or Determinant."

3. Romaniuk and Sharp, "Conceptualizing and Measuring Brand Salience"; Myers and Alpert, "Semantic Confusion in Attitude Research."

4. Chong and Druckman, "Framing Theory"; Nielsen, "10 Heuristics for User Interface Design."

5. Van Dam and van Trijp use construal theory to make this distinction between feasibility and desirability, while Chong and Druckman provide useful summaries of the concepts of applicability (from framing theory) and compatibility (from the literature on mandatory disclosure). The point about desirability being influenced by both reason and affect builds on insights from dual process models, which are clearly explained by Kahneman (van Dam and van Trijp, "Relevant or Determinant"; Chong and Druckman, "Framing Theory"; Kahneman, *Thinking, Fast and Slow*).

6. Oxford English Dictionary, "Prominence, N."

7. Rindova et al., "Being Good or Being Known."

8. International Standards Organization, "ISO 9241-11 Guidance on Usability. ISO 9241."

9. Lee and Kozar, "Understanding of Website Usability."

10. Nielsen, "10 Heuristics for User Interface Design."

11. Krug, *Don't Make Me Think, Revisited*, 22.

12. Van Duyne, Landay, and Hong, *The Design of Sites*.

13. Nielsen, "PDF: Unfit for Human Consumption."

14. Oxford English Dictionary, "Intelligibility, N."

15. Bless, Fiedler, and Strack, *Social Cognition*, 60.

16. Valiant, "A Theory of the Learnable"; Pinker, *Learnability and Cognition*; Leong and Randhawa, *Understanding Literacy and Cognition*; Sternberg and Zhang, *Perspectives on Thinking, Learning, and Cognitive Styles*.

17. Grzymala-Busse, *Managing Uncertainty in Expert Systems*, 50–51.

18. Fleming, "The VARK Modalities"; Newton, "The Learning Styles Myth Is Thriving in Higher Education."

19. Willingham, Hughes, and Dobolyi, "The Scientific Status of Learning Styles Theories"; Rohrer and Pashler, "Learning Styles."

20. Chong and Druckman, "Framing Public Opinion in Competitive Democracies"; Higgins and Kruglanski, *Social Psychology*.

21. Bless, Fiedler, and Strack, *Social Cognition*, 60; Chong and Druckman, "Framing Public Opinion in Competitive Democracies."

22. Krug, *Don't Make Me Think, Revisited*.

23. Bateman et al., "Useful Junk?"

24. This definition builds on discussions of interactivity in work by Anderson ("Getting the Mix Right Again.") and van Noort, Voorveld, and van Reijmersdal ("Interactivity in Brand Web Sites").

25. Delen, Liew, and Willson, "Effects of Interactivity and Instructional Scaffolding on Learning"; van Noort, Voorveld, and van Reijmersdal, "Interactivity in Brand Web Sites"; Sun and Hsu, "Effect of Interactivity on Learner Perceptions in Web-Based Instruction"; Croxton, "The Role of Interactivity in Student Satisfaction and Persistence in Online Learning."

26. Sun and Hsu, "Effect of Interactivity on Learner Perceptions in Web-Based Instruction."

27. van Noort, Voorveld, and van Reijmersdal, "Interactivity in Brand Web Sites."

28. Delen, Liew, and Willson, "Effects of Interactivity and Instructional Scaffolding on Learning."

29. Oxford English Dictionary, "Feasibility, N."

30. Chong and Druckman, "Framing Theory"; Higgins and Kruglanski, *Social Psychology*; Schuldt and Roh, "Of Accessibility and Applicability."

31. Todorov, "The Accessibility and Applicability of Knowledge."

32. Ibid.

33. Fung, Graham, and Weil, *Full Disclosure*, 56.

34. Ibid., 57.

35. Phillips and Bradshaw, "How Customers Actually Shop"; Stahlberg and Maila, *Shopper Marketing*.

36. Stahlberg and Maila, *Shopper Marketing*, 1.

37. Kaczynski, Wilhelm Stanis, and Hipp, "Point-of-Decision Prompts for Increasing Park-Based Physical Activity"; Soler et al., "Point-of-Decision Prompts to Increase Stair Use."

38. Nielsen, "10 Heuristics for User Interface Design."

39. Tognazzini, "First Principles of Interaction Design."

40. Krug, *Don't Make Me Think, Revisited*, 9.

41. Simon, *Economic Analysis and Public Policy, Volume 1*.

42. Popkin, *The Reasoning Voter*.

43. Fung and O'Rourke, "Reinventing Environmental Regulation from the Grassroots Up."

44. Grankvist, Dahlstrand, and Biels, "The Impact of Environmental Labelling on Consumer Preference."

45. Van Dam and De Jonge, "The Positive Side of Negative Labelling."

46. Kuklick, "The Analytic-Synthetic and the Descriptive-Evaluative Distinctions."

47. Ibid.

48. Bakos and Brynjolfsson, "Bundling and Competition on the Internet."

49. Ibid.

50. Pew Research Center, "Offline Adults."

51. Pew Research Center, "The Demographics of Social Media Users."

52. Ibid.

53. Ibid.

54. Alexa, "Top Sites in United States."

55. Pew Research Center, "U.S. Smartphone Use in 2015."

56. Ibid.

57. Hixon, "What Kind of Person Prefers An iPhone?"

58. Edwards, "These Maps Show That Android Is for People with Less Money."

59. Southwell, "Inducing Fear as a Public Communication Campaign Strategy"; Ruiter, Abraham, and Kok, "Scary Warnings and Rational Precautions."

60. Spence, "The Mobile Browser Is Dead, Long Live the App."

61. Cramer, "Should News Providers Focus on Mobile Browsers, Not Apps?"

62. Koetsier, "Wait, What? Mobile Browser Traffic Is 2X Bigger than App Traffic, and Growing Faster"; Sterling, "No, Apps Aren't Winning."

63. Whole Foods, "Whole Foods Market Empowers Shoppers to Make Sustainable Seafood Choices with Color-Coded Rating System."

64. WebpageFX, "Readability Test Tool;" Juicy Studio, "Readability Test."

65. Schwartz, "GoodGuide's Transparency Toolbar Tells You When You're about to Buy Something Toxic."

66. KnowMore.org, "KnowMore Extension."

67. CSRHub, "CSRHub Releases New Sustainable Ratings Widget WordPress Plugin."

68. B Corporation, "About B Lab."

69. Cradle to Cradle Products Innovation Institute, "Get Cradle to Cradle Certified—Platinum."

70. U.S. Department of Agriculture, "What Is BioPreferred?"

71. Roundtable on Sustainable Palm Oil, "Certification."

## 6  Being Green

1. ENERGY STAR, "Displays for Consumers"; ENERGY STAR, "Product Retrospective."

2. EPEAT, "Environmental Benefits."

3. Ibid.

4. TCO, "Criteria in TCO Certified."

5. Sinnott-Armstrong, "Consequentialism."

6. While some information disclosure may be government mandated, use of that information is still voluntary.

7. Freeman, *Strategic Management*, 46.

8. Vogel, *Market for Virtue*, 135, 172.

9. Sen, Gürhan-Canli, and Morwitz, "Withholding Consumption."

10. Vogel and Kagan, *Dynamics of Regulatory Change*.

11. Ibid.

12. Ottman, Stafford, and Hartman, "Avoiding Green Marketing Myopia."

13. Espach, "Private Regulation amid Public Disarray"; Haufler, "Self-Regulation and Business Norms"; Klein, *No Logo*; Lyon, "Green Firms Bearing Gifts."

14. Espach, "Private Regulation amid Public Disarray"; Haufler, "Self-Regulation and Business Norms"; Klein, *No Logo*; Lyon, "Green Firms Bearing Gifts."

15. Gunningham, Kagan, and Thornton, *Shades of Green.*

16. Klein, *No Logo*; Lipschutz and Fogel, "Regulation for the Rest of Us?"; Utting, "Regulating Business via Multistakeholder Initiatives."

17. Cashore, Auld, and Newsom, *Governing through Markets.*

18. Fung, Graham, and Weil, *Full Disclosure*, 54.

19. Gormley and Weimer, *Organizational Report Cards.*

20. Mena and Palazzo, "Input and Output Legitimacy of Multi-Stakeholder Initiatives."

21. Gormley and Weimer, *Organizational Report Cards*, 9, 10–11.

22. Sen, Gürhan-Canli, and Morwitz, "Withholding Consumption."

23. Mintel International Group Ltd, "Mintel Oxygen—Marketing Intelligence Reports."

24. ENERGY STAR, "ENERGY STAR Unit Shipment and Market Penetration Report Calendar Year 2014 Summary."

25. U.S. Environmental Protection Agency, "Executive Orders: ENERGY STAR."

26. Case, "President Bush Requires Federal Agencies to Buy EPEAT Registered Green Electronic Products."

27. Obama, "Executive Order—Planning for Federal Sustainability in the Next Decade."

28. Kiker et al., "Application of Multicriteria Decision Analysis in Environmental Decision Making"; Stirling, "Analysis, Participation and Power"; Huang, Keisler, and Linkov, "Multi-Criteria Decision Analysis in Environmental Sciences."

29. Best Workplaces for Commuters, "Business Benefits Calculator."

30. U.S. Department of Energy, "Vehicle Cost Calculator."

31. Kraft, Stephan, and Abel, *Coming Clean.*

32. Hakim, "VW Admits Cheating in the U.S., but Not in Europe"; Mouawad, "Beyond VW Scandal."

33. ENERGY STAR, "Product Retrospective."

34. EPEAT, "Environmental Benefits."

35. EPEAT, "Criteria."

36. EPEAT, "Environmental Benefits of 2012 EPEAT Purchasing"; ENERGY STAR, "ENERGY STAR Unit Shipment and Market Penetration Report Calendar Year 2012 Summary."

37. ENERGY STAR, "ENERGY STAR Unit Shipment and Market Penetration Report Calendar Year 2012 Summary."

38. EPEAT, "EPEAT Registry Quick Search"; ENERGY STAR, "ENERGY STAR Certified Displays."

39. ENERGY STAR, "International Partners."

40. EPEAT, "Global Reach;" EPEAT, "Environmental Benefits of 2012 EPEAT Purchasing."

41. EPEAT, "Criteria."

42. EPEAT, "Environmental Benefits."

43. DesAutels and Berthon, "The PC (Polluting Computer)."

44. U.S. Department of Energy, "History and Impacts of Appliance and Equipment Standards."

45. U.S. Government Accountability Office, "ENERGY STAR PROGRAM."

46. Wald and Kaufman, "U.S. Tightens Standards for EnergyStar Label"; ENERGY STAR, "Third-Party Certification."

47. EPEAT, "Verification."

48. Hruska, "Apple and the Environment Are Best Friends Again—But Mother Nature Blinked First."

49. Wiens, "The Retina MacBook Pro Was 'verified' EPEAT Gold, but It's Not Green."

50. EPEAT, "EPEAT Announces Findings in Ultrathin Notebook Investigations."

51. Hruska, "Apple Bullies EPEAT into Greenwashing Its Impossible-to-Repair MacBook Pro."

## 7  Green Realism

1. US SIF Foundation, "Report on US Sustainable, Responsible and Impact Investing Trends 2014."

2. Ibid.

3. Ibid.

4. USDA Economic Research Service, "Table 2—U.S. Certified Organic Farmland Acreage, Livestock Numbers, and Farm Operations, 1992–2011."

5. Blackman and Naranjo, "Does Eco-Certification Have Environmental Benefits?"

6. USDA Economic Research Service, "Table 3—Certified Organic and Total U.S. Acreage, Selected Crops and Livestock, 1995–2011."

7. ENERGY STAR, "Facts and Stats"; Massie, "The Top 10 Countries for LEED Projects"; U.S. Energy Information Agency, "A Look at the U.S. Commercial Building Stock."

8. Yale School of Forestry & Environmental Studies, "Forest Certification"; Biodiversity Indicators Project, "Area of Certified Forest."

9. Georgia-Pacific, "Forest Certification around the World."

10. DeLeon and Rivera, *Voluntary Environmental Programs*; Morgenstern and Pizer, *Reality Check*.

11. Darnall and Sides, "Assessing the Performance of Voluntary Environmental Programs."

12. Ibid.

13. Pizer, Morgenstern, and Shih, "The Performance of Industrial Sector Voluntary Climate Programs."

14. McDermott, "Trust, Legitimacy and Power in Forest Certification."

15. Cash et al., "Salience, Credibility, Legitimacy and Boundaries."

16. Ben-Shahar and Schneider, *More Than You Wanted to Know*.

17. Balluck, "Gore Warns of 'Climate Cliff,' Pushes Carbon Tax in 'Fiscal Cliff' Talks."

18. Volcovici and Jarry, "Microsoft's Gates to Start Multi-Billion-Dollar Clean Tech Initiative."

19. Tong, "Pope Francis Makes Climate Change a Moral Issue."

20. Layzer, *Open for Business*.

21. Kropp, "SRI Field Continues to Shift with RiskMetrics' Acquisition of KLD."

22. Ibid.; Kropp, "Jantzi–Sustainalytics Merger the Most Recent as the Consolidation of ESG Research Sector Continues"; Bosley, "Thomson Reuters Buys Swiss Data Provider ASSET4," 4.

23. "MSCI Buys RiskMetrics for $1.55 Billion."

24. Wheeland, "UL Acquires TerraChoice in Green Standard Consolidation"; Walker, "UL Environment Buys GoodGuide"; Floor Trends, "GreenGuard Part of UL's AQS Acquisition."

25. UL, "UL Environment Certified/Validated Products."

26. UL, "UL Company Fact Sheet."

27. ISEAL Alliance, "About Us."

28. Cashore, "Towards a Better World?"

29. Tolkien, *The Fellowship of the Ring*.

30. Casey, "Bloomberg's Sustainability Edge."

31. TruValue Labs, "Insight360"; CrunchBase, "TruValue Labs."

32. Integrity Research Associates, "Investment Banks Top Independents in ESG Research Survey."

33. Cashore, "Towards a Better World?"

34. Whole Foods, "Seafood"; "Target to Sell 100% Sustainable Fish by 2015"; Aubrey, "Sustainable Seafood Swims to a Big-Box Store Near You."

35. U.S. General Services Administration, "Environmentally Preferred Products."

36. Sustainable Consumption Roundtable, "Looking Back, Looking Forward."

37. Hickman, "Does the Consumer Really Know Best?"

38. Maniates, "Editing Out Unsustainable Behavior."

39. Thaler and Sunstein, *Nudge*.

40. Sustainable Consumption Roundtable, "Looking Back, Looking Forward."

41. Delingpole, *The Little Green Book of Eco-Fascism*; Nimmo, "New Eco-Fascist Light Bulbs to Cost $50 Each"; Sheppard, "The GOP."

42. Green, "Fearing the Phase-Out of Incandescent Bulbs."

43. Consumer Reports, "Food Labels." Consumer Reports has recently revamped its extensive Eco-Labels Center website, which first appeared in 2007, and renamed it "Food Labels."

44. Big Room, "Ecolabel Index."

45. International Trade Center, "Standards Map."

46. Global Initiative for Sustainability Ratings, "Accreditation."

47. O'Brien, "New App Lets Barbers Remember Your Haircut"; Belson, "Technology Lets High-End Hotels Anticipate Guests' Whims."

48. Shill, "Information 'Publics' and Equitable Access to Electronic Government Information."

49. Vogel, *The Market for Virtue*, 171.

50. Fung, Graham, and Weil, *Full Disclosure*, 177–180.

51. DeLeon and Rivera, *Voluntary Environmental Programs*; Rivera, "Institutional Pressures and Voluntary Environmental Behavior in Developing Countries."

52. Kraft, Stephan, and Abel, *Coming Clean*, 195.

53. U.S. Federal Trade Commission, "Enforcement"; Medcalf, "Marketing Green to Grab Green."

54. Kingdon and Thurber, *Agendas, Alternatives, and Public Policies*.

55. Bass and Riggio, *Transformational Leadership*.

56. Odom, "Investors Want Sustainability Disclosures in SEC Overhaul."

57. Pinchot Institute for Conservation, "National Forest Certification Study."

58. DC Department of Energy & Environment, "Green Building Act of 2006."

59. Szasz, *Shopping Our Way to Safety*.

60. Frey and Oberholzer-Gee, "The Cost of Price Incentives"; Berglund and Matti, "Citizen and Consumer"; Willis and Schor, "Does Changing a Light Bulb Lead to Changing the World?"

61. Nielsen, "Green Generation: Millennials Say Sustainability Is a Shopping Priority."

62. Starobin and Weinthal, "The Search for Credible Information in Social and Environmental Global Governance."

63. Kraft, Stephan, and Abel, *Coming Clean*.

64. Ben-Shahar and Schneider, *More Than You Wanted to Know*.

65. Plato, *The Republic*.

66. U.S. Consumer Product Safety Commission, "About CPSC"; U.S. Federal Trade Commission, "Enforcement."

67. Consumer Reports, "Greener Choices"; Consumer Reports, "Food Labels."

68. Global Ecolabeling Network, "GEN: Global Ecolabeling Network"; ISEAL Alliance, "About Us."

69. Prakash and Potoski, *The Voluntary Environmentalists*.

70. Herrera, "Walmart Sustainability Index Means Big Business"; GreenBiz, "P&G Launches the Latest Supplier Sustainability Scorecard"; International Council of Chemical Associations, "Responsible Care."

71. Weber, *From Max Weber*, 294–299.

72. Roosevelt said, "The government is us; we are the government, you and I" (Meyers, *Theodore Roosevelt, Patriot and Statesman*, 521).

73. Maynor, "Civic Republicanism."

## Appendix I

1. The Ecolabel Index (ecolabelindex.com) lists 201 eco-labels found in the United States, nearly twice the number found in the next two countries (Canada and Germany) combined.

2. Raynolds, Long, and Murray, "Regulating Corporate Responsibility in the American Market."

3. This distinction follows Kuklick's ("The Analytic-Synthetic and the Descriptive-Evaluative Distinctions," 92) definitions of descriptive meaning as "the meaning of words ... which describe or state facts" and evaluative meaning as "the meaning of words ... which are closely connected with choice, decision, and action" and includes "emotive," "laudatory," "commendatory," "prescriptive," and "normative" meaning.

4. Krippendorff, *Content Analysis*.

5. Beattie, McInnes, and Fearnley, "A Methodology for Analysing and Evaluating Narratives in Annual Reports"; Buhr, "Environmental Performance, Legislation and Annual Report Disclosure"; Gray, Kouhy, and Lavers, "Corporate Social and Environmental Reporting"; Milne, Walton, and Tregidga, "Words Not Actions! The Ideological Role of Sustainable Development Reporting."

6. Beck, Campbell, and Shrives, "Content Analysis in Environmental Reporting Research."

7. Bansal and Roth, "Why Companies Go Green."

8. Beck, Campbell, and Shrives, "Content Analysis in Environmental Reporting Research."

9. Ibid.

10. Landis and Koch, "The Measurement of Observer Agreement for Categorical Data."

## Appendix II

1. Mintel International Group Ltd, "Mintel Oxygen—Marketing Intelligence Reports."

2. Bednar and Westphal, "Surveying the Corporate Elite"; White and Luo, "Business Survey Response Rates."

3. Mintel International Group Ltd, "Mintel Oxygen—Marketing Intelligence Reports."

4. American Institute of Philanthropy, "Top-Rated Charities."

5. Public Broadcasting Service, "The Country's 20 Largest Environmental Organizations by Membership."

6. U.S. News and World Report, and GuideStar, "Largest Charities."

7. For more information about the RSVP Program, visit UC Berkeley Department of Psychology, "RSVP (Research Subject Volunteer Pool)."

# Bibliography

Achenbach, Joel. "The New Scientific Revolution: Reproducibility at Last." *Washington Post*, January 27, 2015. Accessed August 17, 2015. https://www.washingtonpost.com/national/health-science/the-new-scientific-revolution-reproducibility-at-last/2015/01/27/ed5f2076-9546-11e4-927a-4fa2638cd1b0_story.html?utm_term=.c6dc94081f45.

Aertsens, Joris, Wim Verbeke, Koen Mondelaers, and Guido Van Huylenbroeck. "Personal Determinants of Organic Food Consumption: A Review." *British Food Journal* 111 (10) (September 26, 2009): 1140–1167. doi:10.1108/00070700910992961.

Agar, Jon. *The Government Machine.* Cambridge, MA: MIT Press, 2003.

Ajzen, I. "The Theory of Planned Behavior." *Organizational Behavior and Human Decision Processes* 50 (2) (1991): 179–211.

Akerlof, George A. "The Market for 'Lemons': Quality Uncertainty and the Market Mechanism." *Quarterly Journal of Economics* 84 (3) (August 1970): 488–500.

Alagona, Peter. "Credibility." *Conservation Biology* 22 (6) (December 2008): 1365–1367.

Alexa. "Top Sites in United States." N.d. Accessed March 3, 2016. http://www.alexa.com/topsites/countries/US.

Algan, Yann, and Pierre Cahuc. "Trust and Growth." *Annual Review of Economics* 5 (1) (2013): 521–549. doi:10.1146/annurev-economics-081412-102108.

Allison-Hope, Dunstan. "How to Navigate the Maze of Materiality Definitions." *GreenBiz*, August 20, 2013. Accessed June 6, 2015. https://www.greenbiz.com/blog/2013/08/20/navigating-materiality.

Alter, Lloyd. "In Defence of LEED: Stop Bashing the Bike Racks!" *TreeHugger*, October 18, 2010. Accessed July 23, 2015. http://www.treehugger.com/bikes/in-defence-of-leed-stop-bashing-the-bike-racks.html.

American Institute of Philanthropy. "Top-Rated Charities." N.d. Accessed March 13, 2011. https://www.charitywatch.org/top-rated-charities#enviro.

Anderson, Terry. "Getting the Mix Right Again: An Updated and Theoretical Rationale for Interaction." *International Review of Research in Open and Distributed Learning* 4 (2) (October 1, 2003): 1–14. Accessed March 1, 2016. www.irrodl.org/index.php/irrodl/article/view/149.

Anheier, Helmut. *Civil Society: Measurement, Evaluation, Policy.* London: Earthscan, 2004.

Aubrey, Allison. "Sustainable Seafood Swims to a Big-Box Store Near You." *The Salt: What's on Your Plate*, NPR, January 20, 2012. Accessed March 27, 2016. http://www.npr.org/sections/thesalt/2012/01/19/145474067/sustainable-seafood-swims-to-a-big-box-store-near-you.

Auld, Graeme, and Lars H. Gulbrandsen. "Transparency in Nonstate Certification: Consequences for Accountability and Legitimacy." *Global Environmental Politics* 10 (August 2010): 97–119. doi:10.1162/GLEP_a_00016.

Autio, Minna, Rebecca Collins, Stefan Wahlen, and Marika Anttila. "Consuming Nostalgia? The Appreciation of Authenticity in Local Food Production." *International Journal of Consumer Studies* 37 (5) (September 1, 2013): 564–568. doi:10.1111/ijcs.12029.

Averill, Andrew. "Up the Green Mountain: The Battle between LEED and Green Globes." *Hardwood Floors Magazine*, August 12, 2014. Accessed July 23, 2015. http://www.woodfloorbusiness.com/green/up-the-green-mountain-talking-about-leed-and-green-globes.html.

Baby Milk Action. "The Nestlé Boycott." N.d. Accessed March 13, 2014. http://www.babymilkaction.org/nestlefree.

Bainfield, Kaid. "As Good and Important as It Is, LEED Can Be So Embarrassing." January 18, 2013. Accessed July 22, 2015. http://switchboard.nrdc.org/blogs/kbenfield/as_good_and_important_as_it_is.html.

Baker, Monya. "First Results from Psychology's Largest Reproducibility Test." *Nature News*, April 30, 2015. doi:10.1038/nature.2015.17433.

Bakos, Yannis, and Erik Brynjolfsson. "Bundling and Competition on the Internet." *Marketing Science* 19 (1) (2000): 63–82.

Balluck, Kyle. "Gore Warns of 'Climate Cliff,' Pushes Carbon Tax in 'Fiscal Cliff' Talks." *The Hill*, November 13, 2012. Accessed March 28, 2016. http://thehill.com/policy/energy-environment/267585-gore-climate-cliff-should-be-part-of-fiscal-cliff-negotiations.

Banerjee, Abhijit, and Barry D. Solomon. "Eco-Labeling for Energy Efficiency and Sustainability: A Meta-Evaluation of US Programs." *Energy Policy* 31 (2) (January 2003): 109–123. doi:10.1016/S0301-4215(02)00012-5.

Bansal, Pratima, and Kendall Roth. "Why Companies Go Green: A Model of Ecological Responsiveness." *Academy of Management Journal* 43 (4) (August 1, 2000): 717–736. doi:10.2307/1556363.

Barnhart, Phillip A. *The Guide to National Professional Certification Programs.* 2nd ed. Amherst, MA: HRD Press, 1997.

Bartley, T. "Certifying Forests and Factories: States, Social Movements, and the Rise of Private Regulation in the Apparel and Forest Products Fields." *Politics & Society* 31 (3) (2003): 433–464.

Bass, Bernard M., and Ronald E. Riggio. *Transformational Leadership.* New York: Psychology Press, 2006.

Bateman, Scott, Regan L. Mandryk, Carl Gutwin, Aaron Genest, David McDine, and Christopher Brooks. "Useful Junk? The Effects of Visual Embellishment on Comprehension and Memorability of Charts." In *Proceedings of the SIGCHI Conference on Human Factors in Computing Systems*, 2573–2582. New York: ACM, 2010. http://dl.acm.org/citation.cfm?id=1753716.

Bauer, Hans H., Daniel Heinrich, and Daniela B. Schäfer. ""The Effects of Organic Labels on Global, Local, and Private Brands: More Hype than Substance?" *Journal of Business Research.* Recent Advances in Globalization, Culture and Marketing Strategy 66 (8) (August 2013): 1035–1043. doi:10.1016/j.jbusres.2011.12.028.

Baur, Dorothea, and Guido Palazzo. "The Moral Legitimacy of NGOs as Partners of Corporations." *Business Ethics Quarterly* 21 (4) (2011): 579–604. doi:10.5840/beq201121437.

B Corporation. "About B Lab." N.d. Accessed March 13, 2016. http://www.bcorporation.net/what-are-b-corps/about-b-lab.

Beattie, Vivien, Bill McInnes, and Stella Fearnley. "A Methodology for Analysing and Evaluating Narratives in Annual Reports: A Comprehensive Descriptive Profile and Metrics for Disclosure Quality Attributes." *Accounting Forum* 28 (3) (September 2004): 205–236. doi:10.1016/j.accfor.2004.07.001.

Beck, A. Cornelia, David Campbell, and Philip J. Shrives. "Content Analysis in Environmental Reporting Research: Enrichment and Rehearsal of the Method in a British–German Context." *British Accounting Review* 42 (3) (September 2010): 207–222. doi:10.1016/j.bar.2010.05.002.

Bednar, Michael, and James Westphal. "Surveying the Corporate Elite: Theoretical and Practical Guidance on Improving Response Rates and Response Quality in

Top Management Survey Questionnaires." In *Research Methodology in Strategy and Management*, 1st ed., ed. David J. Ketchen and Donald D. Bergh, 3:37–56. Bingley, UK: JAI Press, 2006.

Beierlein, Constanze, Eldad Davidov, Peter Schmidt, Shalom H. Schwartz, and Beatrice Rammstedt. "Testing the Discriminant Validity of Schwartz' Portrait Value Questionnaire Items—A Replication and Extension of Knoppen and Saris (2009)." *Survey Research Methods* 6 (1) (April 25, 2012): 25–36.

Belson, Ken. "Technology Lets High-End Hotels Anticipate Guests' Whims." *New York Times*, November 16, 2005. Accessed March 28, 2016. http://www.nytimes.com/2005/11/16/technology/technology-lets-highend-hotels-anticipate-guests-whims.html?_r=0.

Bendell, Jem, and Anthony Kleanthous. "Deeper Luxury: Quality and Style When the World Matters." World Wildlife Fund (UK), November 2007. Accessed December 27, 2016. http://www.wwf.org.uk/deeperluxury/.

Bennear, Lori S. "What Do We Really Know? The Effect of Reporting Thresholds on Inferences Using Environmental Right-to-Know Data." *Regulation & Governance* 2 (3) (2008): 293–315. doi:10.1111/j.1748-5991.2008.00042.x.

Ben-Shahar, Omri, and Carl E. Schneider. *More Than You Wanted to Know: The Failure of Mandated Disclosure*. Princeton, NJ: Princeton University Press, 2014.

Berglund, Christer, and Simon Matti. "Citizen and Consumer: The Dual Role of Individuals in Environmental Policy." *Environmental Politics* 15 (4) (August 2006): 550–571. doi:10.1080/09644010600785176.

Berlo, David K., James B. Lemert, and Robert J. Mertz. "Dimensions for Evaluating the Acceptability of Message Sources." *Public Opinion Quarterly* 33 (4) (1969): 563–576.

Best, Jason. "The Surprising Truth about Who's Really Buying Organic." *TakePart*, April 21, 2015. Accessed July 13, 2016. http://www.takepart.com/article/2015/04/21/who-is-buying-organic.

Best Workplaces for Commuters. "Business Benefits Calculator." N.d. Accessed December 27, 2016. http://www.bestworkplaces.org/resource/calc.htm.

Bevir, Mark. *Key Concepts in Governance*. Los Angeles: SAGE Publications, 2009.

Bicchieri, Cristina, Erte Xiao, and Ryan Muldoon. "Trustworthiness Is a Social Norm, but Trusting Is Not." *Politics, Philosophy & Economics* 10 (2) (May 1, 2011): 170–187. doi:10.1177/1470594X10387260.

Big Room. "Ecolabel Index." *Ecolabel Index*. N.d. Accessed February 1, 2014. http://www.ecolabelindex.com/.

Bimber, Bruce. *Information and American Democracy: Technology in the Evolution of Political Power*. Cambridge: Cambridge University Press, 2003.

Biodiversity Indicators Project. "Area of Certified Forest." August 2013. Accessed March 26, 2016. http://www.bipindicators.net/forestcertification.

Bird, Monroe M., and James W. Robinson. "The Effectiveness of the Union Label and 'Buy Union' Campaigns." *Industrial & Labor Relations Review* 25 (4) (July 1972): 512–523.

Black, Julia. "Constructing and Contesting Legitimacy and Accountability in Polycentric Regulatory Regimes." *Regulation & Governance* 2 (2) (June 2008): 137–164. doi:10.1111/j.1748-5991.2008.00034.x.

Blackman, Allen, and Maria A. Naranjo. "Does Eco-Certification Have Environmental Benefits? Organic Coffee in Costa Rica." *Ecological Economics* 83 (November 2012): 58–66. doi:10.1016/j.ecolecon.2012.08.001.

Blamey, Russell. "The Activation of Environmental Norms: Extending Schwartz's Model." *Environment and Behavior* 30 (5) (September 1, 1998): 676–708. doi:10.1177/001391659803000505.

Bless, Herbert, Klaus Fiedler, and Fritz Strack. *Social Cognition: How Individuals Construct Social Reality*. New York: Psychology Press, 2004.

Bosley, Catherine. "Thomson Reuters Buys Swiss Data Provider ASSET4." *Reuters*, November 30, 2009. Accessed March 27, 2016. http://www.reuters.com/article/us-asset-idUSTRE5AT0OW20091130.

Boström, Magnus. "Establishing Credibility: Practising Standard-Setting Ideals in a Swedish Seafood-Labelling Case." *Journal of Environmental Policy and Planning* 8 (2) (2006): 135–158. doi:10.1080/15239080600772126.

Boström, Magnus, and Mikael Klintman. *Eco-Standards, Product Labelling and Green Consumerism*. New York: Palgrave Macmillan, 2011.

Bowen, Frances. *After Greenwashing: Symbolic Corporate Environmentalism and Society*. Cambridge: Cambridge University Press, 2014.

Broockman, David, and Joshua Kalla. "We Discovered One of Social Science's Biggest Frauds. Here's What We Learned." *Vox*, July 22, 2015. Accessed August 17, 2015. http://www.vox.com/2015/7/22/9009927/lacour-gay-homophobia-study.

Brown, John Seely, and Paul Duguid. *The Social Life of Information*. Boston: Harvard Business School Press, 2002.

Brownstone, Sydney. "Americans Will Pay More for Organic, but They Also Have No Idea What 'Organic' Means." *Co.Exist*, November 18, 2014. Accessed July 13, 2016.

http://www.fastcoexist.com/3038415/americans-will-pay-more-for-organic-but
-they-also-have-no-idea-what-organic-means.

Buhr, Nola. "Environmental Performance, Legislation and Annual Report Disclosure: The Case of Acid Rain and Falconbridge." *Accounting, Auditing & Accountability Journal* 11 (2) (1998): 163–190.

Bullock, Graham. "Green Grades: The Popularity and Perceived Effectiveness of Information-Based Environmental Governance Strategies." Ph.D. dissertation, University of California, Berkeley, 2011. Accessed January 13, 2013. https://escholarship.org/uc/item/65v7d6q6.

Bullock, Graham. "Independent Labels? The Power behind Environmental Information about Products and Companies." *Political Research Quarterly* 68 (1) (March 2015): 46–62. doi:10.1177/1065912914564685.

Bullock, Graham. "Information-Based Governance Theory." In *Handbook on Theories of Governance*, ed. Christopher Ansell and Jacob Torfing, 281–292. Cheltenham, UK: Edward Elgar, 2016.

Bullock, Graham. "Signaling the Credibility of Private Actors as Public Agents: Transparency, Independence, and Expertise in Environmental Evaluations of Products and Companies." *Business and Politics* 17 (2) (August 2015): 177–219.

Bullock, Graham, and Nicholas Wilder. "The Comprehensiveness of Competing Higher Education Sustainability Assessments." *International Journal of Sustainability in Higher Education* 17 (3) (May 3, 2016): 282–304. doi:10.1108/IJSHE-05-2014-0078.

Campbell-Kelly, Martin. "Information Technology and Organizational Change in the British Census, 1801–1911." In *Information Technology and Organizational Transformation: History, Rhetoric, and Practice*, ed. JoAnne Yates and John Van Maanen, 35–58. Thousand Oaks, CA: Sage Publications, Inc., 2001.

Carlson, Andrea, and Edward Jaenicke. "Changes in Retail Organic Price Premiums from 2004 to 2010." U.S. Department of Agriculture Economic Research Service, May 2016. Accessed July 13, 2016. http://www.ers.usda.gov/media/2091544/err209.pdf.

Case, Scot. "President Bush Requires Federal Agencies to Buy EPEAT Registered Green Electronic Products." January 24, 2007. Accessed March 8, 2011. http://www.epeat.net/Docs/Bush%20Requires%20EPEAT%20%281-24-07%29.pdf.

Casey, Tina. "Bloomberg's Sustainability Edge." *Triple Pundit: People, Planet, Profit*, May 9, 2011. Accessed March 27, 2016. www.triplepundit.com/2011/05/sustainability-bloomberg-csr-report/.

Cash, David, William Clark, Frank Alcock, Nancy Dickson, Noelle Eckley, and Jill Jäger. "Salience, Credibility, Legitimacy and Boundaries: Linking Research, Assessment and Decision Making." John F. Kennedy School of Government, Harvard

University, Faculty Research Working Paper Series RWP02-046. Cambridge, MA: Harvard Kennedy School of Government, 2002.

Cashore, Ben. "Towards a Better World? A Proposal to Enhance Market Support for Global Certification Systems." October 2008. Accessed March 27, 2016. https:// environment.yale.edu/files/biblio/YaleFES-00000015.pdf.

Cashore, Benjamin, Graeme Auld, and Deanna Newsom. *Governing through Markets: Forest Certification and the Emergence of Non-State Authority.* New Haven: Yale University Press, 2004.

Casteel, Chris. "Boren Targets Clinton Btu Tax with Own Plan." *NewsOK*, May 21, 1993. Accessed November 21, 2013. http://newsok.com/boren-targets-clinton-btu -tax-with-own-plan/article/2431196.

Charity Navigator. "How Do We Rate Charities?" N.d. Accessed July 29, 2016. http:// www.charitynavigator.org/index.cfm?bay=content.view&cpid=1284.

Charity Navigator. "Overview." N.d. Accessed December 27, 2016. http://www .charitynavigator.org/index.cfm?bay=content.view&cpid=628.

Chartier, Roger. *The Order of Books: Readers, Authors, and Libraries in Europe between the Fourteenth and Eighteenth Centuries.* Palo Alto, CA: Stanford University Press, 1994.

Chartier, Tim, Amy Langville, and Peter Simov. "March Madness to Movies." *Math Horizons*, April 2010, 16–19. doi:10.4169/194762110X495452.

Chartier, Timothy P., Erich Kreutzer, Amy N. Langville, and Kathryn E. Pedings. "Sensitivity and Stability of Ranking Vectors." *SIAM Journal on Scientific Computing* 33 (3) (January 2011): 1077–1102. doi:10.1137/090772745.

Chasek, Pamela S, David Leonard Downie, and Janet Welsh Brown. *Global Environmental Politics.* Boulder, CO: Westview Press, 2014.

Chatterji, Aaron K., David Levine, and Michael W. Toffel. "How Well Do Social Ratings Actually Measure Corporate Social Responsibility?" *Journal of Economics & Management Strategy* 18 (1) (2009): 125–169.

Chen, Mei-Fang. "Consumer Attitudes and Purchase Intentions in Relation to Organic Foods in Taiwan: Moderating Effects of Food-Related Personality Traits." *Food Quality and Preference* 18 (7) (October 2007): 1008–1021. doi:10.1016/ j.foodqual.2007.04.004.

Chi, M. T. H., R. Glaser, and M. J. Farr, eds. *The Nature of Expertise.* New York: Psychology Press, 1988.

Chinnici, Gaetano, Mario D'Amico, and Biagio Pecorino. "A Multivariate Statistical Analysis on the Consumers of Organic Products." *British Food Journal* 104 (3–5) (2002): 187–199.

Chong, Dennis, and James N. Druckman. "Framing Public Opinion in Competitive Democracies." *American Political Science Review* 101 (4) (November 2007): 637–655.

Chong, Dennis, and James N. Druckman. "Framing Theory." *Annual Review of Political Science* 10 (1) (2007): 103–126. doi:10.1146/annurev.polisci.10.072805.103054.

Choo, Chun Wei. *The Knowing Organization: How Organizations Use Information to Construct Meaning, Create Knowledge, and Make Decisions.* New York: Oxford University Press, 1998.

Clark, Lawrence. "What Owners Need to Know about ... Green-Building Rating Systems." *HPAC Engineering*, January 14, 2015, 20–23.

CMOsurvey.org. "CMO Survey Report: Highlights and Insights." February 2014. Accessed March 13, 2014. https://cmosurvey.org/wp-content/uploads/sites/11/2014/10/The_CMO_Survey-Highlights_and_Insights-Feb-2014.pdf.

Collins English Dictionary. "Public Sector." Definition. *Collins English Dictionary— Complete and Unabridged.* 12th ed. N.d. Accessed October 14, 2013. http://www.thefreedictionary.com/public+sector.

"Common Cause Communication: A Toolkit." 2015. Accessed June 29, 2015. http://valuesandframes.org/the-common-cause-communications-toolkit/.

Conroy, Michael E. *Branded! How the Certification Revolution Is Transforming Global Corporations.* Gabriola Island, Canada: New Society Publishers, 2007.

Consumer Reports. "About Us." N.d. Accessed December 27, 2016. http://www.consumerreports.org/cro/about-us/index.htm.

Consumer Reports. "Food Labels." N.d. Accessed December 21, 2016. http://greenerchoices.org/labels/.

Consumer Reports. "From Crop to Table Report." March 2015. Accessed July 29, 2015. http://www.consumerreports.org/content/dam/cro/magazine-articles/2015/May/Consumer%20Reports_From%20Crop%20to%20Table%20Report_March%202015.pdf.

Consumer Reports. "Greener Choices." N.d. Accessed December 30, 2016. http://greenerchoices.org/.

Consumer Reports. "When to Buy Organic Food: How to Shop Smarter and Healthier." March 17, 2014. Accessed August 5, 2016. http://www.consumerreports.org/cro/news/2014/03/when-to-buy-organic-food/index.htm.

Consumer Reports National Research Center. "Organic Food Labels Survey 2014 Nationally-Representative Phone Survey." March 2014. Accessed March 4, 2015. http://greenerchoices.org/wp-content/uploads/2016/08/CR2014OrganicFoodLabelsSurvey.pdf.

Consumers Union. "Mission." N.d. Accessed July 29, 2016. http://consumersunion
.org/about/mission/.

Costa, Sandine, Lisette Ibanez, Maria L. Loureiro, and Stéphan Marette. "Quality
Promotion through Eco-Labeling: Introduction to the Special Issue." *Journal of
Agricultural & Food Industrial Organization* 7 (2) (2009): 1–6.

Coster, Helen. "Ranking the World's Most Sustainable Companies." *Forbes*,
January 27, 2010. Accessed June 2, 2015. http://www.forbes.com/2010/01/26/most
-sustainable-companies-leadership-citizenship-100.html.

Cradle to Cradle Products Innovation Institute. "Get Cradle to Cradle Certified—
Platinum." N.d. Accessed March 14, 2016. www.c2ccertified.org/get-certified/levels/
platinum/v3_0.

Cramer, Theresa. "Should News Providers Focus on Mobile Browsers, Not Apps?"
*EContent*, January 25, 2016. Accessed March 6, 2016. www.econtentmag.com/
Articles/Editorial/Commentary/Should-News-Providers-Focus-on-Mobile-Browsers
-Not-Apps-108569.htm.

Crawford, Elizabeth. "Who Buys Organic and Where They Buy It Is Evolving."
*FoodNavigator-USA.com*, April 3, 2015. Accessed July 13, 2016. http://www
.foodnavigator-usa.com/Markets/Who-buys-organic-and-where-is-evolving.

Croxton, Rebecca A. "The Role of Interactivity in Student Satisfaction and Persistence in Online Learning." *Journal of Online Learning and Teaching/MERLOT* 10 (2)
(2014): 314–324.

CrunchBase. "TruValue Labs." N.d. Accessed March 27, 2016. https://www
.crunchbase.com/organization/truvalue#/entity.

CSRHub. "CSRHub Releases New Sustainable Ratings Widget WordPress Plugin."
N.d. *CSRHub*. Accessed March 13, 2016. https://blog.csrhub.com/2012/09/csrhub
-releases-new-sustainable-ratings-widget-wordpress-plugin.html.

Cullen, Michael J. *The Statistical Movement in Early Victorian Britain: The Foundations
of Empirical Social Research. Hassocks*. New York: Harvester Press, 1975.

Dando, Nicole, and Tracey Swift. "Transparency and Assurance: Minding the
Credibility Gap." *Journal of Business Ethics* 44 (2/3) (May 1, 2003): 195–200.
doi:10.2307/25075028.

Darby, Michael R., and Edi Karni. "Free Competition and the Optimal Amount of
Fraud." *Journal of Law & Economics* 16 (1) (1973): 67–88.

Darnall, Nicole, and Stephen Sides. "Assessing the Performance of Voluntary
Environmental Programs: Does Certification Matter?" *Policy Studies Journal: The Journal of the Policy Studies Organization* 36 (1) (February 1, 2008): 95–117. doi:10.1111/
j.1541-0072.2007.00255.x.

Darnton, Robert. "An Early Information Society: News and the Media in Eighteenth-Century Paris." *American Historical Review* 105 (1) (February 2000): 1–35. doi:10.2307/2652433.

Datler, Georg, Wolfgang Jagodzinski, and Peter Schmidt. "Two Theories on the Test Bench: Internal and External Validity of the Theories of Ronald Inglehart and Shalom Schwartz." *Social Science Research* 42 (3) (May 2013): 906–925. doi:10.1016/j.ssresearch.2012.12.009.

Dauvergne, Peter, and Jane Lister. *Eco-Business: A Big-Brand Takeover of Sustainability.* Cambridge, MA: MIT Press, 2013.

Davenport, Coral. "A Climate Deal, 6 Fateful Years in the Making." *New York Times,* December 13, 2015. Accessed July 25, 2016. http://www.nytimes.com/2015/12/14/world/europe/a-climate-deal-6-fateful-years-in-the-making.html?_r=0.

DC Department of Energy & Environment. "Green Building Act of 2006." October 11, 2011. Accessed July 28, 2016. doee.dc.gov/publication/green-building-act-2006.

de Groot, J. I. M., and L. Steg. "Value Orientations to Explain Beliefs Related to Environmental Significant Behavior: How to Measure Egoistic, Altruistic, and Biospheric Value Orientations." *Environment and Behavior* 40 (3) (August 2, 2007): 330–354. doi:10.1177/0013916506297831.

de Groot, Judith I. M., and Linda Steg. "Mean or Green: Which Values Can Promote Stable Pro-Environmental Behavior?" *Conservation Letters* 2 (2) (April 2009): 61–66. doi:10.1111/j.1755-263X.2009.00048.x.

de Groot, Judith I. M., and Linda Steg. "Value Orientations and Environmental Beliefs in Five Countries: Validity of an Instrument to Measure Egoistic, Altruistic and Biospheric Value Orientations." *Journal of Cross-Cultural Psychology* 38 (3) (May 1, 2007): 318–332. doi:10.1177/0022022107300278.

De Munck, Bert. "The Agency of Branding and the Location of Value. Hallmarks and Monograms in Early Modern Tableware Industries." *Business History* 54 (7) (2012): 1055–1076. doi:10.1080/00076791.2012.683422.

Dean, Moira, Monique M. Raats, and Richard Shepherd. "Moral Concerns and Consumer Choice of Fresh and Processed Organic Foods." *Journal of Applied Social Psychology* 38 (8) (2008): 2088–2107. doi:10.1111/j.1559-1816.2008.00382.x.

De Vos, Rosemary. "Certification and Eco-Labeling: What New Players Can Learn from Energy Star." *Jacquie Ottman's Green Marketing Blog,* July 10, 2009. Accessed September 1, 2009. http://www.greenmarketing.com/index.php/blog/comments/certification-and-eco-labeling-what-new-players-can-learn-from-energy-star.

Deegan, Craig. "The Legitimising Effect of Social and Environmental Disclosures—A Theoretical Foundation." *Accounting, Auditing & Accountability Journal* 15 (3) (2002): 282–311.

Delen, Erhan, Jeffrey Liew, and Victor Willson. "Effects of Interactivity and Instructional Scaffolding on Learning: Self-Regulation in Online Video-Based Environments." *Computers & Education* 78 (September 2014): 312–320. doi:10.1016/j.compedu.2014.06.018.

DeLeon, Peter, and Jorge E. Rivera. *Voluntary Environmental Programs: A Policy Perspective.* Lanham, MD: Rowman & Littlefield, 2009.

Delingpole, James. *The Little Green Book of Eco-Fascism: The Left's Plan to Frighten Your Kids, Drive Up Energy Costs, and Hike Your Taxes!* Washington, DC: Regnery Publishing, 2013.

Delmas, Magali A., and Vanessa Cuerel Burbano. "The Drivers of Greenwashing." *California Management Review* 54 (1) (Fall 2011): 64–87.

DesAutels, Philip, and Pierre Berthon. "The PC (Polluting Computer): Forever a Tragedy of the Commons?" *Journal of Strategic Information Systems* 20 (1) (March 2011): 113–122. doi:10.1016/j.jsis.2010.09.003.

Dettmann, Rachael L., and Carolyn Dimitri. "Who's Buying Organic Vegetables? Demographic Characteristics of U.S. Consumers." *Journal of Food Products Marketing* 16 (2010): 79–91. doi:10.1080/10454440903415709.

DeVinney, Timothy Michael, Pat Auger, and Giana M. Eckhardt. *The Myth of the Ethical Consumer.* Cambridge: Cambridge University Press, 2010.

Diaz, Robert J., and Rutger Rosenberg. "Spreading Dead Zones and Consequences for Marine Ecosystems." *Science* 321 (5891) (August 15, 2008): 926–929. doi:10.1126/science.1156401.

Dietz, Thomas, Amy Fitzgerald, and Rachael Shwom. "Environmental Values." *Annual Review of Environment and Resources* 30 (1) (2005): 335–372. doi:10.1146/annurev.energy.30.050504.144444.

Drost, Ellen A. "Validity and Reliability in Social Science Research." *Education Research and Perspectives* 38 (1) (June 1, 2011): 105–123.

"The Dozy Watchdogs." *The Economist*, December 13, 2014. Accessed July 13, 2015. http://www.economist.com/news/briefing/21635978-some-13-years-after-enron -auditors-still-cant-stop-managers-cooking-books-time-some.

Duguid, Paul. "Trade Marks, Innovation, and the Union Label." Unpublished manuscript, 2010.

E3. "U.S. Green Building Council." N.d. Accessed February 29, 2016. http://e3es.com/aboutus/usgbc/.

Earth Day Network. "Ecological Footprint Quiz." *Earth Day Network.* N.d. Accessed December 27, 2016. http://www.earthday.org/take-action/footprint-calculator/.

EconPort. "EconPort—Handbook—Public Goods—Classification Table." N.d. Accessed January 19, 2011. http://www.econport.org/econport/request?page=man _pg_table.

Edelman. "Trust in Institutions Drops to Level of Great Recession." *Edelman,* January 19, 2015. Accessed June 25, 2015. http://www.edelman.com/news/trust -institutions-drops-level-great-recession/.

Edelman. "2013 Edelman Trust Barometer Finds a Crisis in Leadership." *Edelman,* January 9, 2013. Accessed January 18, 2015. http://www.edelman.com/trust -downloads/press-release/.

Edwards, Jim. "These Maps Show That Android Is for People with Less Money." April 3, 2014. Accessed March 3, 2016. http://www.businessinsider.com/android-is -for-poor-people-maps-2014-4.

ENERGY STAR. "International Partners." N.d. Accessed December 22, 2016. https:// www.energystar.gov/products/office_equipment/displays.

ENERGY STAR. "Displays for Consumers." N.d.. Accessed January 16, 2016. https:// www.energystar.gov/products/office_equipment/displays.

ENERGY STAR. "Eligibility Criteria for the 1–100 ENERGY STAR Score." N.d. Accessed July 30, 2016. https://www.energystar.gov/buildings/facility-owners-and -managers/existing-buildings/use-portfolio-manager/understand-metrics/eligibility.

ENERGY STAR. "ENERGY STAR Certification for Your Building." N.d. Accessed July 30, 2016. https://www.energystar.gov/buildings/facility-owners-and-managers/ existing-buildings/earn-recognition/energy-star-certification.

ENERGY STAR. "ENERGY STAR Certified Buildings & Plants." N.d. Accessed July 30, 2016. https://www.energystar.gov/index.cfm?fuseaction=labeled_buildings .locator.

ENERGY STAR. "ENERGY STAR Certified Displays." N.d. Accessed January 25, 2016. https://www.energystar.gov/productfinder/product/certified-displays/results/.

ENERGY STAR. "ENERGY STAR Unit Shipment and Market Penetration Report Calendar Year 2012 Summary." N.d. Accessed December 22, 2016. https:// www.energystar.gov/ia/partners/downloads/unit_shipment_data/2012_USD _Summary_Report.pdf.

ENERGY STAR. "Facts and Stats." N.d. Accessed March 26, 2016. https://www .energystar.gov/buildings/about-us/facts-and-stats.

ENERGY STAR. "Learn about Benchmarking." N.d. Accessed July 30, 2016. https:// www.energystar.gov/buildings/about-us/how-can-we-help-you/benchmark-energy -use/benchmarking.

ENERGY STAR. "Product Retrospective: Computers and Monitors." 2012. Accessed January 16, 2016. https://www.energystar.gov/ia/products/downloads/CompMonitors_Highlights.pdf.

ENERGY STAR. "Third-Party Certification." N.d. Accessed January 25, 2016. https://www.energystar.gov/index.cfm?c=third_party_certification.tpc_index.

ENERGY STAR. "Version 3 Overview." N.d. Accessed July 30, 2016. https://www.energystar.gov/index.cfm?c=bldrs_lenders_raters.nh_benefits_utilities_1a.

Environmental Defense Fund. "Partnerships: The Key to Scalable Solutions." N.d. Accessed December 27, 2016. https://www.edf.org/approach/partnerships.

EPEAT. "Global Reach." N.d. Accessed December 22, 2016. http://www.epeat.net/about-epeat/global-reach/.

EPEAT. "Criteria." N.d. Accessed January 25, 2016. http://www.epeat.net/resources/criteria/#tabs-1=pcanddisplays.

EPEAT. "Environmental Benefits." 2013. Accessed January 26, 2016. http://www.epeat.net/about-epeat/environmental-benefits/.

EPEAT. "EPEAT Announces Findings in Ultrathin Notebook Investigations." *EPEAT*, October 12, 2012. http://www.epeat.net/ultrathin-investigation-findings/.

EPEAT. "EPEAT Registry Quick Search." N.d. Accessed January 25, 2016. http://ww2.epeat.net/searchoptions.aspx.

EPEAT. "Verification." N.d. Accessed March 25, 2016. http://www.epeat.net/resources/verification/.

EPEAT. "Environmental Benefits of 2012 EPEAT Purchasing." 2014. Accessed December 22, 2016. http://www.epeat.net/wp-content/uploads/2014/06/EPEAT_2012-Env-Benefits-Report.pdf.

Ericsson, K. A., and J. Smith, eds. *Toward a General Theory of Expertise: Prospects and Limits.* Cambridge: Cambridge University Press, 1991.

Erlandson, Dawn. "The Btu Tax Experience: What Happened and Why It Happened." *Pace Environmental Law Review* 12 (1994): 173–184.

Espach, Ralph. "Private Regulation amid Public Disarray: An Analysis of Two Private Environmental Regulatory Programs in Argentina." *Business and Politics* 7 (2) (2005): 1–36.

Evans, Laurel, Gregory R. Maio, Adam Corner, Carl J. Hodgetts, Sameera Ahmed, and Ulrike Hahn. "Self-Interest and Pro-Environmental Behaviour." *Nature Climate Change* 3 (2) (2013): 122–125. doi:10.1038/nclimate1662.

Fair Isaac Corporation. "What Is a Credit Score?" N.d. Accessed December 30, 2016. http://www.myfico.com/crediteducation/creditscores.aspx.

Fair Labor Association. "Charter Document." February 12, 2014. Accessed July 30, 2016. http://www.fairlabor.org/sites/default/files/fla_charter_2-12-14.pdf.

Fiorino, Daniel J. *The New Environmental Regulation*. Cambridge, MA: MIT Press, 2006.

Fishbein, Martin, and Icek Ajzen. *Predicting and Changing Behavior: The Reasoned Action Approach*. New York: Taylor & Francis, 2010.

Fleming, Neil. "The VARK Modalities." N.d. Accessed December 29, 2015. http://vark-learn.com/introduction-to-vark/the-vark-modalities/.

Floor Trends. "GreenGuard Part of UL's AQS Acquisition." February 3, 2011. Accessed March 27, 2016. http://www.floortrendsmag.com/articles/91307-greenguard-part-of-ul-s-aqs-acquisition.

Florini, Ann. *The Right to Know: Transparency for an Open World*. New York: Columbia University Press, 2007.

Fogg, B. J., and H. Tseng. "The Elements of Computer Credibility." In *Proceedings of the SIGCHI Conference on Human Factors in Computing Systems: The CHI Is the Limit* 87, 1999.

Forest Stewardship Council. "Board of Directors." *FSC Forest Stewardship Council*. N.d. Accessed July 13, 2015. https://ic.fsc.org/en/about-fsc/governance-01/board-of-directors.

Forest Stewardship Council. "Debrief of FSC General Assembly 2014." *FSC Forest Stewardship Council*. September 26, 2014. Accessed July 12, 2015. https://us.fsc.org/en-us/newsroom/newsletter/id/844.

Forest Stewardship Council. "Financial Reporting." N.d. Accessed June 21, 2015. https://us.fsc.org/en-us/who-we-are/financial-reporting.

Forest Stewardship Council. "Governance." N.d. Accessed June 19, 2015. https://us.fsc.org/en-us/who-we-are/governance.

Forest Stewardship Council. "Our History." N.d. Accessed June 22, 2015. https://us.fsc.org/en-us/who-we-are/our-history.

ForestEthics. "Follow the Greenwash Money: Details on the Logging Industry's 'Green' Label (SFI) and Its 'Environmental' Support." September 9, 2011. Accessed June 19, 2015. http://sfigreenwash.org/sustainable-forestry-initiative-follow-the-greenwash-money.

ForestEthics. "NGO Opposition Letter to SFI." September 7, 2011. Accessed June 21, 2015. *Scribd*. https://www.scribd.com/document/94048726/Forest-Ethics-NGO-Opposition-to-SFI-Signon-Letter.

ForestEthics. "SFI: Certified Greenwash." ForestEthics, November 2010. Accessed June 19, 2015. https://www.scribd.com/document/44016182/SFI-Certified -Greenwash-Report-Forest-Ethics.

Fortune. "10 Green Giants." N.d. Accessed December 27, 2016. http://archive .fortune.com/galleries/2007/fortune/0703/gallery.green_giants.fortune/index.html.

France, Kevin. "Drought Hurts Agriculture, Costs West Over $40 Billion." *Accu-Weather*, May 30, 2015. Accessed June 1, 2015. http://www.accuweather.com/en/ weather-news/drought-taking-a-big-hit-on-ag-1/47857167.

Frank, Robert H. *Microeconomics and Behavior*. Boston: McGraw-Hill, 2003.

Frankel, Oz. *States of Inquiry*. Baltimore: Johns Hopkins University Press, 2006.

Freedom House. "About Freedom in the World." N.d. Accessed July 29, 2016. https:// freedomhouse.org/report-types/freedom-world.

Freeman, R. Edward. *Strategic Management: A Stakeholder Approach*. Cambridge: Cambridge University Press, 2010.

Frey, Bruno S., and Felix Oberholzer-Gee. "The Cost of Price Incentives: An Empirical Analysis of Motivation Crowding-Out." *American Economic Review* 87 (4) (September 1, 1997): 746–755.

Friedman, M. "Consumer Boycotts: A Conceptual Framework and Research Agenda." *Journal of Social Issues* 47 (1) (1991): 149–168.

Friedman, Monroe. *Consumer Boycotts: Effecting Change through the Marketplace and the Media*. New York: Routledge, 1999.

FSC-Watch. "About." N.d. Accessed June 22, 2015. http://fsc-watch.com/about/.

FSC-Watch. "The 10 Worst Things about the Forest Stewardship Council." June 1, 2014. Accessed June 21, 2015. https://fsc-watch.com/2014/06/01/the-10-worst-things -about-the-forest-stewardship-council/.

Fukuyama, Francis. "Social Capital and the Global Economy: A Redrawn Map of the World." *Foreign Affairs* 74 (5) (September–October 1995): 89–103.

Fukuyama, Francis. *Trust: The Social Virtues and the Creation of Prosperity*. New York: Free Press, 1996.

Fung, Archon. "Infotopia: Unleashing the Democratic Power of Transparency." *Politics & Society* 41 (2) (June 1, 2013): 183–212. doi:10.1177/0032329213483107.

Fung, Archon, Mary Graham, and David Weil. *Full Disclosure: The Perils and Promise of Transparency*. Cambridge: Cambridge University Press, 2007.

Fung, Archon, and Dara O'Rourke. "Reinventing Environmental Regulation from the Grassroots Up: Explaining and Expanding the Success of the Toxics Release

Inventory." *Environmental Management* 25 (2) (March 2000): 115–127. doi:10.1007/s002679910009.

Gallup. "Environment." *Gallup*. N.d. Accessed July 29, 2016. http://www.gallup.com/poll/1615/Environment.aspx.

Gallup. "Forty-Five Percent of Americans Seek out Organic Foods." *Gallup*, August 7, 2014. Accessed July 13, 2016. http://www.gallup.com/poll/174524/forty-five-percent-americans-seek-organic-foods.aspx.

Gamlund, Espen. "Who Has Moral Status in the Environment? A Spinozistic Answer." *Trumpeter* 23, no. 1 (2007). Accessed September 21, 2013. http://trumpeter.athabascau.ca/index.php/trumpet/article/viewArticle/939.

GBI. "Green Building Initiative: About GBI." N.d. Accessed July 23, 2015. http://www.thegbi.org/about-gbi/.

Georgia-Pacific. "Forest Certification around the World." 2014. Accessed March 26, 2016. https://www.gp.com/~/media/Corporate/GPCOM/Files/Sustainability/Forestry_Certification_Around_world.ashx.

Gereffi, Gary. "The Global Economy: Organization, Governance, and Development." In *The Handbook of Economic Sociology*, 2nd ed., ed. Neil J. Smelser and Richard Swedberg, 160–182. Princeton, NJ: Princeton University Press, 2005.

Gereffi, Gary, Ronie Garcia-Johnson, and Erika Sasser. "The NGO-Industrial Complex." *Foreign Policy*, no. 125 (July 1, 2001): 56–65. doi:10.2307/3183327.

Gifford, E., and M. James. "Effective Shareholder Engagement: The Factors That Contribute to Shareholder Salience." *Journal of Business Ethics* 92 (Supplement 1) (2010): 79–97.

Gifford, Henry. "The Four Sins of LEEDwashing: LEED Green Buildings That Perhaps Aren't Really Green." *TreeHugger*. March 17, 2009. Accessed July 23, 2015. http://www.treehugger.com/sustainable-product-design/the-four-sins-of-leedwashing-leed-green-buildings-that-perhaps-arent-really-green.html.

Gilder, George F. *Knowledge and Power: The Information Theory of Capitalism and How It Is Revolutionizing Our World*. Washington, DC: Regnery Publishing, 2013.

Gillis, Justin, and Coral Davenport. "Leaders Roll Up Sleeves on Climate, but Experts Say Plans Don't Pack a Wallop." *New York Times*, April 21, 2016. Accessed July 25, 2016. http://www.nytimes.com/2016/04/22/science/united-nations-paris-climate-change-document.html.

Gleick, James. *The Information: A History, a Theory, a Flood*. New York: Vintage Books, 2012.

Global Ecolabelling Network. "GEN: Global Ecolabelling Network." N.d. Accessed December 30, 2016. http://www.globalecolabelling.net/about/gen-the-global -ecolabelling-network/.

Global Ecolabelling Network. "What Is Ecolabelling?" 2016. Accessed December 15, 2016. http://www.globalecolabelling.net/what-is-eco-labelling/.

Global Footprint Network. "Footprint Basics." April 11, 2016. Accessed December 27, 2016. http://www.footprintnetwork.org/en/index.php/gfn/page/footprint_basics _overview/.

Global Initiative for Sustainability Ratings. "Accreditation." N.d. Accessed March 28, 2016. http://ratesustainability.org/core/accreditation/.

Global Reporting Initiative. "G4 Sustainability Reporting Guidelines: Reporting Principles and Standard Disclosures." August 5, 2015. Accessed January 4, 2015. https://www.globalreporting.org/resourcelibrary/GRIG4-Part1-Reporting-Principles -and-Standard-Disclosures.pdf.

Goldman, Lawrence. *Science, Reform, and Politics in Victorian Britain*. Cambridge: Cambridge University Press, 2002.

Google. "Ngram Trends for 'Information Society' and 'Information Age.'" *Google Ngram Viewer*. N.d. Accessed March 15, 2014. https://books.google.com/ngrams/ graph?content=information+society%2C+information+age&year_start=1800&year _end=2000&corpus=15&smoothing=3&share=&direct_url=t1%3B%2Cinformation %20society%3B%2Cc0%3B.t1%3B%2Cinformation%20age%3B%2Cc0.

Gormley, William T., and David Leo Weimer. *Organizational Report Cards*. Cambridge, MA: Harvard University Press, 1999.

Graham, Mary. *Democracy by Disclosure: The Rise of Technopopulism*. Washington, DC: Governance Institute/Brookings Institution Press, 2002.

Grankvist, Gunne, Ulf Dahlstrand, and Anders Biels. "The Impact of Environmental Labelling on Consumer Preference: Negative vs. Positive Labels." *Journal of Consumer Policy* 27 (2) (June 2004): 213–230. doi:10.1023/B:COPO.0000028167.54739.94.

Gray, Rob, Reza Kouhy, and Simon Lavers. "Corporate Social and Environmental Reporting: A Review of the Literature and a Longitudinal Study of UK Disclosure." *Accounting, Auditing & Accountability Journal* 8 (2) (1995): 47–77.

Green, Penelope. "Fearing the Phase-Out of Incandescent Bulbs." *New York Times*, May 25, 2011. Accessed March 27, 2016. http://www.nytimes.com/2011/05/26/ garden/fearing-the-phase-out-of-incandescent-bulbs.html.

Green Building Initiative. "Certified Buildings Directory." N.d. Accessed July 30, 2016. http://www.thegbi.org/project-portfolio/certified-building-directory/.

Green Building Initiative. "Green Globes for New Construction: Technical Reference Manual (Version 1.3)." Green Building Initiative, 2014. Accessed September 6, 2015. http://www.thegbi.org/files/training_resources/Green_Globes_NC_Technical_Reference_Manual.pdf.

Green Building Initiative. "Green Globes Program Features: Weighted Criteria and Flexibility Features." December 18, 2014. Accessed July 30, 2016. http://www.thegbi.org/files/training_resources/Green-Globes-Weighted-Criteria--Flexibility-Features.pdf.

Green Building Initiative. "Overview." N.d. Accessed July 30, 2016. https://www.thegbi.org/guiding-principles-compliance-certification/overview/.

GreenBiz. "P&G Launches the Latest Supplier Sustainability Scorecard." May 12, 2010. Accessed April 2, 2011. https://www.greenbiz.com/news/2010/05/12/procter-gamble-launches-the-latest-suplier-sustainability-scorecard.

Greene, Alan, Bob Scowcroft, and Sylvia Tawse. "Top 10 Reasons to Support Organic in the 21st Century." N.d. Accessed August 11, 2014. http://www.organic.org/articles/showarticle/article-206.

Greenpeace. "Guide to Greener Electronics." November 2012. Accessed March 3, 2013. http://www.greenpeace.org/international/en/campaigns/climate-change/cool-it/Campaign-analysis/Guide-to-Greener-Electronics/.

Greenpeace. "Hazardous Substances Reduced but Not Eliminated from Laptops." October 23, 2007. Accessed November 15, 2015. http://www.greenpeace.org/international/en/news/features/hazardous-substances-laptops/.

Greenpeace. "Holding the Line with FSC." October 2008. Accessed June 21, 2015. http://www.greenpeace.org/international/Global/international/planet-2/report/2008/11/holding-the-line-with-fsc.pdf.

Greenpeace. "Leaders and Laggards." N.d. Accessed April 11, 2016. http://www.greenpeace.org/international/en/campaigns/climate-change/our_work/expeditions/leaders-and-laggards/.

Grigsby, Ellen. *Analyzing Politics: An Introduction to Political Science.* Belmont, CA: Cengage Learning, 2008.

Grimm, Jacob, and Wilhelm Grimm. "Little Snow-White." November 15, 2005. http://www.pitt.edu/~dash/grimm053.html.

Grover, Samir. "Friends of the Earth Backs Away from Forest Stewardship Council." *TreeHugger*, September 28, 2008. Accessed June 21, 2015. http://www.treehugger.com/corporate-responsibility/friends-of-the-earth-backs-away-from-forest-stewardship-council.html.

Grunert, Klaus G., Sophie Hieke, and Josephine Wills. "Sustainability Labels on Food Products: Consumer Motivation, Understanding and Use." *Food Policy* 44 (February 2014): 177–189. doi:10.1016/j.foodpol.2013.12.001.

Grzymala-Busse, Jerzy W. *Managing Uncertainty in Expert Systems*. New York: Springer Science & Business Media, 1991.

Gulbrandsen, Lars H. "The Effectiveness of Non-State Governance Schemes: A Comparative Study of Forest Certification in Norway and Sweden." *International Environmental Agreement: Politics, Law and Economics* 5 (2) (2005): 125–149.

Gunasekaran, A., C. Patel, and E. Ronald McGaughey. "A Framework for Supply Chain Performance Measurement." *International Journal of Production Economics* 87 (3) (February 18, 2004): 333–347. doi:10.1016/j.ijpe.2003.08.003.

Gunningham, Neil, Robert A. Kagan, and Dorothy Thornton. *Shades of Green: Business, Regulation, and Environment*. Palo Alto, CA: Stanford University Press, 2003.

Gutting, Gary. "Michel Foucault." *Stanford Encyclopedia of Philosophy*, September 17, 2008. http://plato.stanford.edu/entries/foucault/.

Hahnel, Ulf J. J., Céline Ortmann, Liridon Korcaj, and Hans Spada. "What Is Green Worth to You? Activating Environmental Values Lowers Price Sensitivity towards Electric Vehicles." *Journal of Environmental Psychology* 40 (December 2014): 306–319. doi:10.1016/j.jenvp.2014.08.002.

Hakim, Danny. "VW Admits Cheating in the U.S., but Not in Europe." *New York Times*, January 21, 2016. Accessed January 25, 2016. http://www.nytimes.com/2016/01/22/business/international/vw-admits-cheating-in-the-us-but-not-in-europe.html?_r=0.

Hamilton, James T. *Regulation through Revelation: The Origin, Politics, and Impacts of the Toxics Release Inventory Program*. Cambridge: Cambridge University Press, 2005.

Hance, Jeremy. "The FSC Is the 'Enron of Forestry' Says Rainforest Activist." *Mongabay*, April 17, 2008. https://news.mongabay.com/2008/04/the-fsc-is-the-enron-of-forestry-says-rainforest-activist/.

Hand, Stuart. "LEED vs. Green Globes: How to Choose." *Daily Journal of Commerce*, February 27, 2014. Accessed June 21, 2015. http://www.djc.com/news/en/12062746.html.

Haraway, Donna. "Situated Knowledges: The Science Question in Feminism and the Privilege of Partial Perspective." *Feminist Studies* 14 (3) (October 1, 1988): 575–599. doi:10.2307/3178066.

Harper, Douglas. "Value (N.)." *Online Etymology Dictionary*. N.d. Accessed June 7, 2015. http://www.etymonline.com/index.php?term=value&allowed_in_frame=0.

Harris Teeter. "Harris Teeter." N.d. Accessed December 27, 2016. https://www
.harristeeter.com/#/app/home.

Haufler, Virginia. "Self-Regulation and Business Norms: Political Risk, Political Activism." In *Private Authority and International Affairs*, ed. A. Claire Cutler, Virginia
Haufler, and Tony Porter, 199–222. New York: SUNY Press, 1999.

Hayek, Friedrich August. "The Use of Knowledge in Society." *American Economic
Review* 35 (4) (September 1945): 519–530.

Henry, Alan. "What Does Organic Really Mean, and Is It Worth My Money?"
*Lifehacker*, September 10, 2012. Accessed August 10, 2014. http://lifehacker.com/
5941881/what-does-organic-really-mean-and-should-i-buy-it.

Hensley, Scott. "Costs Temper Demand for Organic Foods." *NPR*, July 20, 2011.
Accessed July 5, 2015. www.npr.org/sections/health-shots/2011/07/20/138534183/
organic-foods-have-broad-appeal-but-costs-temper-demand.

Herrera, Tilde. "EPA Stuns Industry with Plans to Kill Climate Leaders Program."
*GreenBiz.com*. September 16, 2010. Accessed March 10, 2014. https://www.greenbiz
.com/news/2010/09/16/epa-stuns-industry-plans-kill-climate-leaders-program.

Herrera, Tilde. "Walmart Sustainability Index Means Big Business." *Greenbiz.com*.
September 24, 2009. Accessed September 25, 2009. https://www.greenbiz.com/
blog/2009/09/24/walmart-sustainability-index-means-big-business.

Hess, David. "Social Reporting and New Governance Regulation: The Prospects
of Achieving Corporate Accountability through Transparency." *Business Ethics
Quarterly* 17 (3) (July 2007): 453–476.

Hickman, Leo. "Does the Consumer Really Know Best?" *The Guardian*, October 25,
2007, sec. Environment. Accessed March 27, 2016. https://www.theguardian.com/
environment/2007/oct/25/ethicalliving.lifeandhealth1.

Higgins, E. Tory, and Arie W. Kruglanski. *Social Psychology: Handbook of Basic Principles*. New York: Guilford Press, 1996.

Higgs, Edward. *The Information State in England: The Central Collection of Information
on Citizens, 1500–2000*. Basingstoke, UK: Palgrave Macmillan, 2003.

Hixon, Todd. "What Kind of Person Prefers an iPhone?" *Forbes*. April 10,
2014. Accessed March 3, 2016. http://www.forbes.com/sites/toddhixon/2014/04/10/
what-kind-of-person-prefers-an-iphone/#1841fd823e5a.

Hjelmar, Ulf. "Consumers' Purchase of Organic Food Products: A Matter of Convenience and Reflexive Practices." *Appetite* 56 (2) (April 2011): 336–344. doi:10.1016/
j.appet.2010.12.019.

Honkanen, Pirjo, and Bas Verplanken. "Understanding Attitudes towards Genetically Modified Food: The Role of Values and Attitude Strength." *Journal of Consumer Policy* 27 (4) (2004): 401–420.

Houldsworth, Elizabeth, and Dilum Jirasinghe. *Managing and Measuring Employee Performance.* London: Kogan Page Limited, 2006.

Hovland, Carl I., Irving L. Janis, and Harold H. Kelley. *Communication and Persuasion.* New Haven, CT: Yale University Press, 1953.

Hruska, Joel. "Apple and the Environment Are Best Friends Again—But Mother Nature Blinked First." *ExtremeTech*, July 17, 2012. Accessed January 24, 2016. http://www.extremetech.com/computing/132902-apple-and-the-environment-are-best-friends-again-but-mother-nature-blinked-first.

Hruska, Joel. "Apple Bullies EPEAT into Greenwashing Its Impossible-to-Repair MacBook Pro." *ExtremeTech*, October 17, 2012. Accessed January 24, 2016. http://www.extremetech.com/computing/138015-apple-bullies-epeat-into-greenwashing-its-impossible-to-repair-macbook-pro.

Hsu, A., et al. "Environmental Performance Index." 2016. Accessed December 27, 2016. http://epi.yale.edu/.

Huang, Ivy B., Jeffrey Keisler, and Igor Linkov. "Multi-Criteria Decision Analysis in Environmental Sciences: Ten Years of Applications and Trends." *Science of the Total Environment* 409 (19) (September 2011): 3578–3594. doi:10.1016/j.scitotenv.2011.06.022.

Hudson, William. "Should You Buy Organic? Study Complicates Decision." *CNN.* September 4, 2012. Accessed August 5, 2016. http://www.cnn.com/2012/09/03/health/organics-versus-conventional/index.html.

Huffington Post. "Target to Sell 100% Sustainable Fish by 2015." *Huffington Post,* December 13, 2011. Accessed March 27, 2016. http://www.huffingtonpost.com/2011/10/13/target-sustainable-fish_n_1009192.html.

Hysing, Erik. "From Government to Governance? A Comparison of Environmental Governing in Swedish Forestry and Transport." *Governance: An International Journal of Policy, Administration and Institutions* 22 (4) (2009): 647–672. doi:10.1111/j.1468-0491.2009.01457.x.

Ilchman, Warren Frederick, and Norman Thomas Uphoff. *The Political Economy of Change.* Berkeley: University of California Press, 1969.

Integrity Research Associates. "Investment Banks Top Independents in ESG Research Survey." *Integrity Research*, March 15, 2012. Accessed March 27, 2016. http://www.integrity-research.com/integrity-researchfocus%c2%ae-esg-research-press-release/.

International Council of Chemical Associations. "Responsible Care." N.d. Accessed August 1, 2016. https://www.icca-chem.org/responsible-care/.

International Living Future Institute. "A Living Education: Living Building Challenge Projects as Ongoing Education Tools." N.d. Accessed July 30, 2016. http://living-future.org/living-education-living-building-challenge-projects -ongoing-education-tools.

International Living Future Institute. "Living Building Challenge 3.0." 2014. Accessed March 7, 2017. https://living-future.org/wp-content/uploads/2016/12/ Living-Building-Challenge-3.0-Standard.pdf.

International Living Future Institute. "The Certification Process." N.d. Accessed July 30, 2016. http://living-future.org/lbc/certification.

International Living Future Institute. "Registered Living Building Challenge Projects." N.d. Accessed July 30, 2016. http://living-future.org/projectmap.

International Standards Organization. "ISO 9241-11 Ergonomic requirements for office work with visual display terminals (VDTs)—Part 11: Guidance on Usability." 1998. Accessed December 23, 2016. https://www.iso.org/obp/ui/#iso:std:iso:9241 :-11:ed-1:v1:en.

International Trade Center. "Standards Map." N.d. Accessed March 28, 2016. http:// www.intracen.org/itc/market-info-tools/voluntary-standards/standardsmap/.

Investopedia. "Private Sector." Definition. *Investopedia*. N.d. Accessed October 14, 2013. http://www.investopedia.com/terms/p/private-sector.asp.

Investopedia. "Sunshine Laws." Definition. *Investopedia*. N.d. Accessed July 27, 2015. http://www.investopedia.com/terms/s/sunshinelaws.asp.

ISEAL Alliance. "About Us." N.d. Accessed March 27, 2016. http://www .isealalliance.org/about-us.

Jackson, Lisa. "Next Steps for the National Environmental Performance Track Program and the Future of Environmental Leadership Programs." March 16, 2009. Accessed March 10, 2014. https://archive.epa.gov/performancetrack/web/pdf/ performancetracknextstepsmemoexternal-text.pdf.

Jacobson, Maryann Tomovich. "Is Organic Milk Worth the Price?" *WebMD*, January 17, 2014. Accessed August 10, 2014. http://blogs.webmd.com/food-and-nutrition/ 2014/01/is-organic-milk-worth-the-price.html.

Jacoby, William G. "The AJPS Replication Policy: Innovations and Revisions." *American Journal of Political Science*, March 26, 2015. Accessed August 17, 2015. https:// ajps.org/2015/03/26/the-ajps-replication-policy-innovations-and-revisions/.

Jasanoff, Sheila. "(No?) Accounting for Expertise." *Science & Public Policy* 30 (3) (2003): 157–162. doi:10.3152/147154303781780542.

Jordan, Andrew, Rüdiger K. W. Wurzel, and Anthony Zito. "The Rise of 'New' Policy Instruments in Comparative Perspective: Has Governance Eclipsed Government?" *Political Studies* 53 (3) (2005): 477–496. doi:10.1111/j.1467-9248.2005.00540.x.

Jose, Anita, and Shang-Mei Lee. "Environmental Reporting of Global Corporations: A Content Analysis Based on Website Disclosures." *Journal of Business Ethics* 72 (4) (June 1, 2007): 307–321. doi:10.2307/25075385.

Juicy Studio. "Readability Test." N.d. Accessed December 20, 2016. http://juicystudio.com/services/readability.php.

Kaczynski, Andrew T., Sonja A. Wilhelm Stanis, and J. Aaron Hipp. "Point-of-Decision Prompts for Increasing Park-Based Physical Activity: A Crowdsource Analysis." *Preventive Medicine* 69 (December 2014): 87–89. doi:10.1016/j.ypmed.2014.08.029.

Kahneman, Daniel. *Thinking, Fast and Slow.* 1st ed. New York: Farrar, Straus and Giroux, 2011.

Kaplan, Robert S., and David P. Norton. *The Balanced Scorecard: Translating Strategy into Action.* Boston: Harvard Business School Press, 1996.

Karp, David Gutierrez. "Values and Their Effect on Pro-Environmental Behavior." *Environment and Behavior* 28 (1) (January 1, 1996): 111–133. doi:10.1177/0013916596281006.

Khanna, Madhu, and Lisa A. Damon. "EPA's Voluntary 33/50 Program: Impact on Toxic Releases and Economic Performance of Firms." *Journal of Environmental Economics and Management* 37 (1) (1999): 1–25.

Kiker, Gregory A., Todd S. Bridges, Arun Varghese, Thomas P. Seager, and Igor Linkov. "Application of Multicriteria Decision Analysis in Environmental Decision Making." *Integrated Environmental Assessment and Management* 1 (2) (2005): 95–108. doi:10.1897/IEAM_2004a-015.1.

King, Andrew A., and Michael J. Lenox. "Industry Self-Regulation without Sanctions: The Chemical Industry's Responsible Care Program." *Academy of Management Journal* 43 (4) (August 1, 2000): 698–716. doi:10.2307/1556362.

King, Bart. "Materiality Assessments: The Missing Link for Sustainability Strategy." *GreenBiz*, September 10, 2013. Accessed June 5, 2015. https://www.greenbiz.com/blog/2013/09/10/materiality-assessments-missing-link-sustainability-strategy.

Kingdon, John W., and James A. Thurber. *Agendas, Alternatives, and Public Policies.* Vol. 45. Boston: Little, Brown, 1984.

Kjær, Anne Mette. *Governance.* Malden, MA: Polity Press, 2004.

Klein, Naomi. *No Logo: Taking Aim at the Brand Bullies.* New York: Picador, 1999.

KnowMore.org. "KnowMore Extension." September 29, 2008. Accessed March 13, 2016. https://addons.mozilla.org/en-US/firefox/addon/knowmore-extension/.

Koetsier, John. "Wait, What? Mobile Browser Traffic Is 2X Bigger than App Traffic, and Growing Faster." *VentureBeat*, September 25, 2015. Accessed March 6, 2016. http://venturebeat.com/2015/09/25/wait-what-mobile-browser-traffic-is-2x-bigger -than-app-traffic-and-growing-faster/.

Koppell, Jonathan GS. "Pathologies of Accountability: ICANN and the Challenge of 'Multiple Accountabilities Disorder.'" *Public Administration Review* 65 (1) (January 1, 2005): 94–108. doi:10.1111/j.1540-6210.2005.00434.x.

Korsgaard, Christine M. "Two Distinctions in Goodness." *Philosophical Review* 92 (2) (1983): 169–195.

Kotchen, Matthew J. "Green Markets and Private Provision of Public Goods." *Journal of Political Economy* 114 (4) (2006): 816–834.

KPMG. "The Essentials of Materiality Assessment." *KPMG*, November 3, 2014. Accessed June 5, 2015. https://home.kpmg.com/xx/en/home/insights/2014/10/ materiality-assessment.html.

Kraft, Michael E., Mark Stephan, and Troy D. Abel. *Coming Clean: Information Disclosure and Environmental Performance*. Cambridge, MA: MIT Press, 2011.

Kramer, Roderick M. "Trust and Distrust in Organizations: Emerging Perspectives, Enduring Questions." *Annual Review of Psychology* 50 (1) (1999): 569–598. doi:10.1146/annurev.psych.50.1.569.

Krippendorff, Klaus. *Content Analysis: An Introduction to Its Methodology*. Thousand Oaks, CA: SAGE, 2012.

Kriwy, Peter, and Rebecca-Ariane Mecking. "Health and Environmental Consciousness, Costs of Behaviour and the Purchase of Organic Food." *International Journal of Consumer Studies* 36 (1) (January 1, 2012): 30–37. doi:10.1111/j.1470-6431. 2011.01004.x.

Kropp, Robert. "Jantzi–Sustainalytics Merger the Most Recent as the Consolidation of ESG Research Sector Continues." *SocialFunds*, September 23, 2009. Accessed March 27, 2016. http://dev.socialfunds.com/news/article.cgi/article2786.html.

Kropp, Robert. "SRI Field Continues to Shift with RiskMetrics' Acquisition of KLD." *GreenBiz*, November 6, 2009. Accessed March 27, 2016. https://www.greenbiz.com/ news/2009/11/06/riskmetrics-acquires-kld-0.

Krug, Steve. *Don't Make Me Think, Revisited: A Common Sense Approach to Web Usability*. Berkeley, CA: New Riders, 2014.

Krystallis, Athanasios, Marco Vassallo, and George Chryssohoidis. "The Usefulness of Schwartz's 'Values Theory' in Understanding Consumer Behaviour towards

Differentiated Products." *Journal of Marketing Management* 28 (11/12) (October 2012): 1438–1463.

Krystallis, Athanassios, Marco Vassallo, George Chryssohoidis, and Toula Perrea. "Societal and Individualistic Drivers as Predictors of Organic Purchasing Revealed through a Portrait Value Questionnaire (PVQ)-Based Inventory." *Journal of Consumer Behaviour* 7 (2) (April 1, 2008): 164–187. doi:10.1002/cb.244.

Kuklick, Bruce. "The Analytic-Synthetic and the Descriptive-Evaluative Distinctions." *Journal of Value Inquiry* 3 (2) (1969): 91–99. doi:10.1007/BF00137405.

Lamoureux, David. "Advertising: How Many Marketing Messages Do We See in a Day?" February 23, 2012. http://www.fluiddrivemedia.com/advertising/marketing -messages/.

Landis, J. Richard, and Gary G. Koch. "The Measurement of Observer Agreement for Categorical Data." *Biometrics* 33 (1) (1977): 159–174.

Langville, Amy N., and Carl D. Meyer. *Who's #1? The Science of Rating and Ranking.* Princeton, NJ: Princeton University Press, 2012.

Larson, Charles. *Persuasion: Reception and Responsibility.* 12th ed. Boston: Cengage Learning, 2010.

Lasswell, Harold D. *Politics: Who Gets What, When and How.* Gloucester, MA: Peter Smith Publisher, 1990.

Law, Steve. "LEED vs. Green Globes." *Daily Journal of Commerce,* February 27, 2014. http://portlandtribune.com/sl/201025-leed-vs-green-globes-.

Layzer, Judith A. *Open for Business: Conservatives' Opposition to Environmental Regulation.* Cambridge, MA: MIT Press, 2012.

Layzer, Judith A. *The Environmental Case: Translating Values into Policy.* 3rd ed. Washington, DC: CQ Press, 2012.

Layzer, Judith A. *The Environmental Case; Translating Values into Policy.* 4th ed. Thousand Oaks, CA: CQ Press, 2015.

Lee, Younghwa, and Kenneth A. Kozar. "Understanding of Website Usability: Specifying and Measuring Constructs and Their Relationships." *Decision Support Systems* 52 (2) (January 2012): 450–463. doi:10.1016/j.dss.2011.10.004.

Leong, C. K., and B. S. Randhawa. *Understanding Literacy and Cognition: Theory, Research, and Application.* New York: Plenum Press, 1989.

Levi, Margaret, and Laura Stoker. "Political Trust and Trustworthiness." *Annual Review of Political Science* 3 (1) (2000): 475–507. doi:10.1146/annurev.polisci.3.1.475.

Levitan, Kathrin. *A Cultural History of the British Census: Envisioning the Multitude in the Nineteenth Century.* New York: Palgrave Macmillan, 2011.

Lipschutz, Ronnie D., and Cathleen Fogel. "'Regulation for the Rest of Us?' Global Civil Society and the Privatization of Transnational Regulation." In *The Emergence of Private Authority in Global Governance*, ed. Rodney Bruce Hall and Thomas J. Biersteker. 115–161. Cambridge: Cambridge University Press, 2002.

Lockie, Stewart, Kristen Lyons, Geoffrey Lawrence, and Kerry Mummery. "Eating 'Green': Motivations behind Organic Food Consumption in Australia." *Sociologia Ruralis* 42 (1) (2002): 23–40. doi:10.1111/1467-9523.00200.

Loureiro, M. L., J. J. McCluskey, and R. C. Mittelhammer. "Assessing Consumer Preferences for Organic, Eco-Labeled, and Regular Apples." *Journal of Agricultural and Resource Economics* 26 (2001): 404–416.

Lyon, Thomas P. "Green Firms Bearing Gifts." *Regulation* 26 (3) (Fall 2003): 36–41.

Lyon, Thomas P., and John W. Maxwell. "Environmental Public Voluntary Programs Reconsidered." *Policy Studies Journal: The Journal of the Policy Studies Organization* 35 (4) (2007): 723–750.

Ma, Yoon Jin, and Hyun-Hwa Lee. "Understanding Consumption Behaviours for Fair Trade Non-Food Products: Focusing on Self-Transcendence and Openness to Change Values." *International Journal of Consumer Studies* 36 (6) (November 1, 2012): 622–634. doi:10.1111/j.1470-6431.2011.01037.x.

MacFarquhar, Larissa. "The Gilder Effect." *New Yorker* 29 (2000): 102–111.

Maniates, Michael. "Editing Out Unsustainable Behavior." In *State of the World: Transforming Cultures From Consumerism to Sustainability*, ed. The Worldwatch Institute, 119–126. New York: W. W. Norton & Company, 2010.

Mann, Stefan, and Henry Wüstemann. "Public Governance of Information Asymmetries—The Gap between Reality and Economic Theory." *Journal of Socio-Economics* 39 (2) (April 2010): 278–285. doi:10.1016/j.socec.2009.10.009.

Marsh, Stephen, and Mark R. Dibben. "The Role of Trust in Information Science and Technology." *Annual Review of Information Science & Technology* 37 (2003): 465–498.

Marsh, Stephen, and Mark R. Dibben. "Trust, Untrust, Distrust and Mistrust—An Exploration of the Dark(er) Side." In *Trust Management*, ed. P. Herrmann, V. Issarny, and S. Shiu, 17–33. Berlin: Springer, 2005. Accessed February 18, 2010. http://link.springer.com/chapter/10.1007/11429760_2.

Massie, Caroline. "The Top 10 Countries for LEED Projects." *Architect*, July 22, 2015. Accessed March 26, 2016. www.architectmagazine.com/Design/the-top-10-countries-for-leed-projects_o.

Mattli, Walter, and Tim Büthe. "Accountability in Accounting? The Politics of Private Rule-Making in the Public Interest." *Governance: An International Journal* of *Policy, Administration and Institutions* 18 (3) (July 2005): 399–429. doi:10.1111/j.1468-0491.2005.00282.x.

Maury, Mary D. "A Circle of Influence: Are All the Stakeholders Included?" *Journal of Business Ethics* 23 (1) (January 1, 2000): 117–121. doi:10.2307/25074228.

Maynor, John. "Civic Republicanism." *Encyclopedia Britannica*. March 4, 2013. Accessed March 28, 2016. https://www.britannica.com/topic/civic-republicanism.

McDermott, Constance L. "Trust, Legitimacy and Power in Forest Certification: A Case Study of the FSC in British Columbia." *Geoforum* 43 (3) (May 2012): 634–644. doi:10.1016/j.geoforum.2011.11.002.

McKnight, D. Harrison, Vivek Choudhury, and Charles Kacmar. "The Impact of Initial Consumer Trust on Intentions to Transact with a Web Site: A Trust Building Model." *Journal of Strategic Information Systems* 11 (3–4) (December 2002): 297–323. doi:10.1016/S0963-6687(02)00020-3.

Medcalf, Bradley D. "Marketing Green to Grab Green: FTC More Aggressive in Pursuit of Unsubstantiated Environmental Marketing Claims." *Journal of Environmental and Sustainability Law* 21 (2015): 435–454.

Mena, Sébastien, and Guido Palazzo. "Input and Output Legitimacy of Multi-Stakeholder Initiatives." *Business Ethics Quarterly* 22 (3) (July 2012): 527–556. doi:10.5840/beq201222333.

Menger, Carl. *Principles of Economics*. Auburn, AL: Ludwig von Mises Institute, 2004. Accessed December 23, 2016. https://mises.org/library/principles-economics.

Merriam-Webster. "Credibility." Definition. *Merriam-Webster Online Dictionary*. N.d. Accessed February 1, 2012. http://www.merriam-webster.com/dictionary/credibility.

Merriam-Webster. "Information." Definition. *Merriam-Webster Online Dictionary*. N.d. Accessed April 21, 2010. http://www.merriam-webster.com/dictionary/information.

Meyers, Robert Cornelius V. *Theodore Roosevelt, Patriot and Statesman: The True Story of an Ideal American*. Philadelphia: P. W. Ziegler & Co., 1902.

Milne, Markus J., Sara Walton, and Helen Tregidga. "Words Not Actions! The Ideological Role of Sustainable Development Reporting." *Accounting, Auditing & Accountability Journal* 22 (8) (2009): 1211–1257.

Mintel International Group Ltd. "Mintel Oxygen—Marketing Intelligence Reports." 2009. Accessed August 6, 2009. http://reports.mintel.com/homepages/guest/.

Mol, Arthur P. J. "Environmental Governance in the Information Age: The Emergence of Informational Governance." *Environment and Planning. C, Government & Policy* 24 (4) (2006): 497–514. doi:10.1068/c0508j.

Mol, Arthur P. J. *Environmental Reform in the Information Age.* Cambridge: Cambridge University Press, 2008.

Morgenstern, Richard D., and William A. Pizer. *Reality Check: The Nature and Performance of Voluntary Environmental Programs in the United States, Europe, and Japan.* New York: Routledge, 2012.

Morningstar. "Equity Research Methodology." March 6, 2015. Accessed December 27, 2016. http://corporate1.morningstar.com/WorkArea/DownloadAsset.aspx?id =11725.

Morse, Robert. "The Birth of the College Rankings." *U.S. News & World Report,* May 16, 2008. Accessed March 13, 2014. http://www.usnews.com/news/national/ articles/2008/05/16/the-birth-of-college-rankings.

Mouawad, Jad. "Beyond VW Scandal: Home Appliance Industry No Stranger to Tricks." *New York Times,* October 9, 2015. Accessed January 25, 2016. http:// www.nytimes.com/2015/10/10/business/makers-of-consumer-products-have-long -history-of-cheating.html?_r=0.

Mountford, Sam. "Why Businesses Suffer from a Trust Gap." *GreenBiz,* April 5, 2012. Accessed June 29, 2015. http://www.greenbiz.com/blog/2012/04/05/why-businesses -are-suffering-trust-gap.

"MSCI Buys RiskMetrics for $1.55 Billion." *New York Times,* March 1, 2010. Accessed March 27, 2016. http://dealbook.nytimes.com/2010/03/01/msci-buys-riskmetrics -for-1-55-billion/.

Mueller, Martin, Virginia Gomes Dos Santos, and Stefan Seuring. "The Contribution of Environmental and Social Standards towards Ensuring Legitimacy in Supply Chain Governance." *Journal of Business Ethics* 89 (4) (November 2009): 509–523.

Myers, James H., and Mark I. Alpert. "Semantic Confusion in Attitude Research: Salience vs. Importance vs. Determinance." *Advances in Consumer Research* 4 (1) (1977): 106–110.

Nadaï, Alain. "Conditions for the Development of a Product Ecolabel." *European Environment: The Journal of European Environmental Policy (Wiley)* 9 (5) (October 1999): 202–211.

Nannestad, Peter. "What Have We Learned about Generalized Trust, If Anything?" *Annual Review of Political Science* 11 (1) (2008): 413–436. doi:10.1146/annurev. polisci.11.060606.135412.

National Research Council. "A Data-Based Assessment of Research-Doctorate Programs in the United States." 2011. Accessed March 13, 2014. https://www.nap.edu/ rdp/.

National Research Council. "A Data-Based Assessment of Research-Doctorate Programs in the United States: Excel Data Table." April 29, 2011. Accessed February 29, 2016. https://www.nap.edu/rdp/#download.

Natural Resources Defense Council. "NRDC: A Shopper's Guide to Home Tissue Products." August 5, 2009. Accessed June 22, 2015. https://www.nrdc.org/resources/shoppers-guide-home-tissue-products.

Natural Resources Defense Council. "Policy Library." N.d. Accessed December 27, 2016. https://www.nrdc.org/policy-library.

Nature. "Availability of Data, Material and Methods." April 15, 2015. http://www.nature.com/authors/policies/availability.html.

Nelson, Phillip. "Information and Consumer Behavior." *Journal of Political Economy* 78 (2) (1970): 311–329.

Newsweek. "2016 Green Rankings." *Newsweek*, 2016. Accessed December 27, 2016. http://www.newsweek.com/green-2016/top-green-companies-us-2016.

Newton, Philip M. "The Learning Styles Myth Is Thriving in Higher Education." *Frontiers in Psychology* 6 (December 15, 2015): 1–5. doi:10.3389/fpsyg.2015.01908.

Nielsen. "Green Generation: Millennials Say Sustainability Is a Shopping Priority." November 5, 2015. Accessed July 22, 2016. http://www.nielsen.com/us/en/insights/news/2015/green-generation-millennials-say-sustainability-is-a-shopping-priority.html.

Nielsen. "Health and Wellness through the Eyes of the Diverse Consumer." June 23, 2014. Accessed December 23, 2016. http://www.nielsen.com/content/dam/corporate/us/en/docs/consumer-360-extra-docs/2014/health-wellness-6-23-14-1pm-FINAL.pptx.

Nielsen. "TV Ratings." N.d. Accessed December 27, 2016. http://www.nielsen.com/us/en/solutions/measurement/television.html.

Nielsen, Jakob. "PDF: Unfit for Human Consumption." July 14, 2003. Accessed December 23, 2016. https://www.nngroup.com/articles/pdf-unfit-for-human-consumption/.

Nielsen, Jakob. "10 Heuristics for User Interface Design." January 1, 1995. Accessed December 11, 2011. https://www.nngroup.com/articles/ten-usability-heuristics/.

Nilsson, Helen, Burcu Tunçer, and Åke Thidell. "The Use of Eco-Labeling like Initiatives on Food Products to Promote Quality Assurance—Is There Enough Credibility?" *Journal of Cleaner Production* 12 (5) (June 2004): 517–526. doi:10.1016/S0959-6526(03)00114-8.

Nimmo, Kurt. "New Eco-Fascist Light Bulbs to Cost $50 Each." InfoWars. May 17, 2011. http://www.infowars.com/new-eco-fascist-light-bulbs-to-cost-50-each/.

Nunberg, Geoffrey. "Farewell to the Information Age." In *The Future of the Book*, ed. Geoffrey Nunberg, 103–138. Berkeley: University of California Press, 1996.

Obach, Brian K. *Organic Struggle: The Movement for Sustainable Agriculture in the United States*. Cambridge, MA: MIT Press, 2015.

Obama, Barack. "Executive Order—Planning for Federal Sustainability in the Next Decade." The White House, March 19, 2015. Accessed March 24, 2016. https://www.whitehouse.gov/the-press-office/2015/03/19/executive-order-planning-federal-sustainability-next-decade.

O'Brien, Sara Ashley. "New App Lets Barbers Remember Your Haircut." *New York Post*, November 3, 2013. March 28, 2016. http://nypost.com/2013/11/02/new-app-helps-hairstylists-remember-your-haircut/.

O'Connor, Thomas, and David Whitall. "Linking Hypoxia to Shrimp Catch in the Northern Gulf of Mexico." *Marine Pollution Bulletin* 54 (4) (April 2007): 460–463. doi:10.1016/j.marpolbul.2007.01.017.

Odom, Che. "Investors Want Sustainability Disclosures in SEC Overhaul." *Bloomberg BNA*, July 20, 2016. Accessed July 28, 2016. http://www.bna.com/investors-sustainability-disclosures-n73014445099/.

Office of the EPA Inspector General. "Performance Track Could Improve Program Design and Management to Ensure Value." Report No. 2007-P-00013, March 29, 2007. Accessed March 10, 2014. https://www.epa.gov/office-inspector-general/report-performance-track-could-improve-program-design-and-management-ensure.

Olsen, Janeen, Liz Thach, and Liz Hemphill. "The Impact of Environmental Protection and Hedonistic Values on Organic Wine Purchases in the US." *International Journal of Wine Business Research* 24 (1) (March 16, 2012): 47–67. doi:10.1108/17511061211213783.

Oreg, Shaul, and Tally Katz-Gerro. "Predicting Proenvironmental Behavior Cross-Nationally: Values, the Theory of Planned Behavior, and Value-Belief-Norm Theory." *Environment and Behavior* 38 (4) (July 1, 2006): 462–483. doi:10.1177/0013916505286012.

Organic Trade Association. "U.S. Organic Sales Post New Record of $43.3 Billion in 2015." May 19, 2016. Accessed July 13, 2016. https://www.ota.com/news/press-releases/19031.

O'Rourke, Dara. *Shopping for Good*. Cambridge, MA: MIT Press, 2012.

Ottman, Jacquelyn A. "Focus on Consumer Self-Interest to Win Today's Green Customer." *The Guardian*, September 23, 2011, sec. Guardian Sustainable Business. Accessed September 14, 2014. https://www.theguardian.com/sustainable-business/blog/green-marketing-consumer-behaviour-change.

Ottman, Jacquelyn A., Edwin R. Stafford, and Cathy Hartman. "Avoiding Green Marketing Myopia: Ways to Improve Consumer Appeal for Environmentally Preferable Products." *Environment* 48 (5) (2006): 22–36.

Owens, Brendan, Chrissy Macken, Adam Rohloff, and Heather Rosenberg. "LEED v4 Impact Category and Point Allocation Process." U.S. Green Building Council. N.d. Accessed September 6, 2015. http://www.usgbc.org/sites/default/files/LEED%20 v4%20Impact%20Category%20and%20Point%20Allocation%20Process _Overview_0.pdf.

Oxfam. "Battle of the Brands: The Annual Scorecard Update." 2015. Accessed June 2, 2015. https://blogs.oxfam.org/en/blogs/15-03-31-battle-brands-annual-scorecard -update.

Oxford English Dictionary. "Feasibility, N." Definition. *OED Online*. Oxford: Oxford University Press. N.d. Accessed December 21, 2015. http://www.oed.com.

Oxford English Dictionary. "Ideology, N." Definition. *OED Online*. Oxford: Oxford University Press. N.d. Accessed August 3, 2016. http://www.oed.com.

Oxford English Dictionary. "Intelligibility, N." Definition. *OED Online*. Oxford: Oxford University Press. N.d. Accessed December 20, 2016. http://www.oed.com.

Oxford English Dictionary. "Politics, N." Definition. *OED Online*. Oxford: Oxford University Press. N.d. Accessed August 3, 2016. http://www.oed.com.

Oxford English Dictionary. "Prominence, N." Definition. *OED Online*. Oxford: Oxford University Press. N.d. Accessed August 3, 2016. http://www.oed.com.

Oxford English Dictionary. "Reliability, N." Definition. *OED Online*. Oxford: Oxford University Press, N.d. Accessed August 3, 2016. http://www.oed.com.

Peattie, Ken. "Green Consumption: Behavior and Norms." *Annual Review of Environment and Resources* 35 (1) (2010): 195–228. doi:10.1146/annurev-environ -032609-094328.

Peattie, Ken, and Andrew Crane. "Green Marketing: Legend, Myth, Farce or Prophesy?" *Qualitative Market Research* 8 (4) (2005): 357–370.

Petty, Richard E., and John T. Cacioppo. "The Elaboration Likelihood Model of Persuasion." In *Advances in Experimental Social Psychology*, ed. Leonard Berkowitz, vol. 19 (1986): 123–205. Accessed May 5, 2015. http://www.sciencedirect.com/ science/article/pii/S0065260108602142.

Pew Research Center. "Beyond Distrust: How Americans View Their Government." November 23, 2015. Accessed July 15, 2016. http://www.people-press.org/2015/ 11/23/1-trust-in-government-1958-2015/.

Pew Research Center. "The Demographics of Social Media Users." August 19, 2015. Accessed March 3, 2016. http://www.pewinternet.org/2015/08/19/the-demographics -of-social-media-users/.

Pew Research Center. "Offline Adults." September 30, 2013. Accessed March 3, 2016. http://www.pewinternet.org/data-trend/internet-use/offline-adults/.

Pew Research Center. "U.S. Smartphone Use in 2015." April 1, 2015. Accessed March 3, 2016. http://www.pewinternet.org/2015/04/01/us-smartphone-use-in-2015/.

Phillips, H., and R. Bradshaw. "How Customers Actually Shop: Customer Interaction with the Point of Sale." *Journal of the Market Research Society* 35 (1) (1993): 51–62.

Pinchot Institute for Conservation. "National Forest Certification Study: An Evaluation of the Applicability of Forest Stewardship Council (FSC) and Sustainable Forest Initiative (SFI) Standards on Five National Forests." October 22, 2007. Accessed July 28, 2016. http://www.fs.fed.us/projects/forestcertification/executive-summary .pdf.

Pinker, Steven. *Learnability and Cognition: The Acquisition of Argument Structure.* Cambridge, MA: MIT Press, 2013.

Pino, Giovanni, M. Alessandro Peluso, and Gianluigi Guido. "Determinants of Regular and Occasional Consumers' Intentions to Buy Organic Food." *Journal of Consumer Affairs* 46 (1) (March 1, 2012): 157–169. doi:10.1111/j.1745-6606.2012.01223.x.

Pizer, William A., Richard Morgenstern, and Jhih-Shyang Shih. "The Performance of Industrial Sector Voluntary Climate Programs: Climate Wise and 1605(b)." *Energy Policy* 39 (12) (December 2011): 7907–7916. doi:10.1016/j.enpol.2011.09.040.

Plato. *The Republic.* New York: W. W. Norton & Company, 1985.

Popkin, Samuel L. *The Reasoning Voter.* Chicago: University of Chicago Press, 1991.

Porter, Theodore M. *Trust in Numbers: The Pursuit of Objectivity in Science and Public Life.* Princeton: Princeton University Press, 1996.

Portney, Kent E. *Taking Sustainable Cities Seriously: Economic Development, the Environment, and Quality of Life in American Cities.* 2nd ed. Cambridge, MA: MIT Press, 2013.

Prakash, Aseem, and Matthew Potoski. *The Voluntary Environmentalists: Green Clubs, ISO 14001, and Voluntary Regulations.* Cambridge: Cambridge University Press, 2006.

Prescott, Jaques. "The Rio Summit—Success or Failure?" *Canadian Biodiversity* 2 (3) (Fall 1992): 35–38.

Public Broadcasting Service. "The Country's 20 Largest Environmental Organizations by Membership." *Earth on Edge.* N.d. Accessed March 13, 2011. http:// www.pbs.org/earthonedge/resources2.html.

Rabotyagov, S. S., C. L. Kling, P. W. Gassman, N. N. Rabalais, and R. E. Turner. "The Economics of Dead Zones: Causes, Impacts, Policy Challenges, and a Model of the Gulf of Mexico Hypoxic Zone." *Review of Environmental Economics and Policy* 8 (1) (January 1, 2014): 58–79. doi:10.1093/reep/ret024.

Rahman, Noushi, and Corinne Post. "Measurement Issues in Environmental Corporate Social Responsibility (ECSR): Toward a Transparent, Reliable, and Construct Valid Instrument." *Journal of Business Ethics* 105 (3) (February 2012): 307–319.

Rainforest Alliance. "Certification and Assurance Services." N.d. Accessed December 27, 2016. http://www.rainforest-alliance.org/business/certification-verification/.

Randazzo, Tracy. "Sustainable Options: LEED™ and Green Globes™—a Comparison—Part One of a Three Part Series." June 9, 2015. http://bermanwright.com/sustainable-options-leed-and-green-globes-a-comparison-part-one-of-a-three-part-series/.

Raymond, Joad. *News, Newspapers, and Society in Early Modern Britain*. Portland, OR: Frank Cass Publishers, 1999.

Raynolds, Laura T., Michael A. Long, and Douglas L. Murray. "Regulating Corporate Responsibility in the American Market: A Comparative Analysis of Voluntary Certifications." *Competition & Change* 18 (2) (April 2014): 91–110. doi:10.1179/1024529414Z.00000000050.

Remler, Dahlia K., and Gregg G. Van Ryzin. *Research Methods in Practice: Strategies for Description and Causation*. Thousand Oaks, CA: SAGE Publications, 2015.

Ricardo, David. *On the Principles of Political Economy, and Taxation*. London: J. Murray, 1817.

Richardson, Gary. "Brand Names before the Industrial Revolution." Working Paper, National Bureau of Economic Research, April 2008. Accessed March 12, 2014. http://www.nber.org/papers/w13930.

Rindova, V., I. O. Williamson, A. P. Petkova, and J. M. Sever. "Being Good or Being Known: An Empirical Examination of the Dimensions, Antecedents, and Consequences of Organizational Reputation." *Academy of Management Journal* 48 (6) (2005): 1033–1049.

Rivera, Jorge E. "Institutional Pressures and Voluntary Environmental Behavior in Developing Countries: Evidence from the Costa Rican Hotel Industry." *Society & Natural Resources* 17 (9) (October 2004): 779–797. doi:10.1080/08941920490493783.

Roberts, Robin. "Determinants of Corporate Social Responsibility Disclosure: An Application of Stakeholder Theory." *Accounting, Organizations and Society* 17 (6) (1992): 595–612.

Roberts, Tristan, and Paula Melton. "LEED vs. Green Globes: The Definitive Analysis." BuildingGreen.com, May 29, 2014. https://www.buildinggreen.com/continuing-education/leed-vs-green-globes.

Rodriguez Garcia, J. M. "Scientia Potestas Est–Knowledge Is Power: Francis Bacon to Michel Foucault." *Neohelicon* 28 (1) (2001): 109–121.

Roe, Brian, and Ian Sheldon. "Credence Good Labeling: The Efficiency and Distributional Implications of Several Policy Approaches." *American Journal of Agricultural Economics* 89 (4) (2007): 1020–1033.

Rohrer, Doug, and Harold Pashler. "Learning Styles: Where's the Evidence?" *Medical Education* 46 (7) (2012): 634–635.

Romaniuk, Jenni, and Byron Sharp. "Conceptualizing and Measuring Brand Salience." *Marketing Theory* 4 (4) (December 1, 2004): 327–342. doi:10.1177/1470593104047643.

Romzek, Barbara S., and Melvin J. Dubnick. "Accountability." In *International Encyclopedia of Public Policy and Administration*, vol. 1, ed. Jay M. Shafritz, 6–11. Boulder: Westview Press, 1998.

Romzek, Barbara S, and Patricia Wallace Ingraham. "Cross Pressures of Accountability: Initiative, Command, and Failure in the Ron Brown Plane Crash." *Public Administration Review* 60 (3) (May 1, 2000): 240–253. doi:10.1111/0033-3352.00084.

Rondinelli, Dennis, and Gyula Vastag. "Panacea, Common Sense, or Just a Label? The Value of ISO 14001 Environmental Management Systems." *European Management Journal* 18 (5) (October 2000): 499–510. doi:10.1016/S0263-2373(00)00039-6.

Rosenau, James N. "Governance in a New Global Order." In *Governing Globalization: Power, Authority and Global Governance*, ed. David Held and Anthony McGrew, 70–86. Cambridge: Polity Press, 2002.

Rosenbaum, Walter A. *Environmental Politics and Policy*. 9th ed. Thousand Oaks, CA: CQ Press, 2014.

Roundtable on Sustainable Palm Oil. "Certification." N.d. Accessed March 14, 2016. http://www.rspo.org/certification.

Ruiter, Rac, C. Abraham, and G. Kok. "Scary Warnings and Rational Precautions: A Review of the Psychology of Fear Appeals." *Psychology & Health* 16 (6) (November 2001): 613–630.

Sadowski, Michael, Kyle Whitaker, and Alicia Ayars. "Rate the Raters Phase 3: Uncovering Best Practices." SustainAbility, February 2011. Accessed May 1, 2011. http://sustainability.com/our-work/reports/rate-the-raters-phase-three/.

Sadowski, Michael, Kyle Whitaker, and Frances Buckingham. "Rate the Raters Phase Two: Taking Inventory of the Ratings Universe." SustainAbility, October

2010. Accessed December 5, 2010. http://sustainability.com/our-work/reports/rate-the-raters-phase-two/.

Sandlund, Chris. "Trust Is a Must." *Entrepreneur*. October 1, 2002. Accessed June 24, 2015. https://www.entrepreneur.com/article/55354.

Scharpf, Fritz W. "Economic Integration, Democracy and the Welfare State." *Journal of European Public Policy* 4 (1) (January 1997): 18–36. doi:10.1080/135017697344217.

Scherer, Ron. "Investors Turn Up Heat over Carbon Emissions." *Christian Science Monitor*, February 14, 2007. http://www.csmonitor.com/2007/0214/p02s02-usec.html.

Schmidt, Eric, and Jared Cohen. *The New Digital Age: Reshaping the Future of People, Nations, and Business*. New York: Alfred A. Knopf, 2013.

Schnaars, Christopher, and Hannah Morgan. "In U.S. Building Industry, Is It Too Easy to Be Green?" *USA Today*, June 13, 2013. Accessed July 23, 2015. http://www.usatoday.com/story/news/nation/2012/10/24/green-building-leed-certification/1650517/.

Schneider, Nancy. "11 Steps to Mainstream Your Green Products." *GreenBiz*, January 31, 2012. Accessed May 27, 2015. https://www.greenbiz.com/blog/2012/01/31/11-steps-mainstream-your-green-products.

Schroeder, Mark. "Value Theory." In *The Stanford Encyclopedia of Philosophy*, edited by Edward N. Zalta. Stanford: Metaphysics Research Lab, Stanford University, 2016. Accessed December 27, 2016. https://plato.stanford.edu/archives/fall2016/entries/value-theory/.

Schudson, Michael. "The Objectivity Norm in American Journalism." *Journalism* 2 (2) (August 1, 2001): 149–170. doi:10.1177/146488490100200201.

Schuldt, Jonathon P., and Sungjong Roh. "Of Accessibility and Applicability: How Heat-Related Cues Affect Belief in 'Global Warming' versus 'Climate Change.'" *Social Cognition* 32 (3) (2014): 217–238.

Schwartz, Ariel. "GoodGuide's Transparency Toolbar Tells You When You're about to Buy Something Toxic." *Fast Company*, August 25, 2011. Accessed March 13, 2016. https://www.fastcompany.com/1775890/goodguides-transparency-toolbar-tells-you-when-youre-about-buy-something-toxic.

Schwartz, Heidi. "Drexel University Releases Green Globes vs. LEED Cost Comparison Study." *Facility Executive*, June 5, 2014. Accessed September 6, 2015. http://facilityexecutive.com/2014/06/green-globes-vs-leed-drexel-study/.

Schwartz, John. "Another Inconvenient Truth: It's Hard to Agree How to Fight Climate Change." *New York Times*, July 11, 2016. Accessed July 25, 2016. http://www.nytimes.com/2016/07/12/science/climate-change-movement.html.

Schwartz, Shalom. "An Overview of the Schwartz Theory of Basic Values." *Online Readings in Psychology and Culture* 2 (1) (December 1, 2012): 1–20. doi:10.9707/2307-0919.1116.

Schwartz, Shalom H. "Normative Influences on Altruism." *Advances in Experimental Social Psychology* 10 (1977): 221–279.

Schwartz, Shalom H. "Universals in the Content and Structure of Values: Theoretical Advances and Empirical Tests in 20 Countries." *Advances in Experimental Social Psychology* 25 (1) (1992): 1–65.

Schwartz, Shalom H., and Wolfgang Bilsky. "Toward a Universal Psychological Structure of Human Values." *Journal of Personality and Social Psychology* 53 (3) (1987): 550–562. doi:10.1037/0022-3514.53.3.550.

Schwartz, Shalom H., and Klaus Boehnke. "Evaluating the Structure of Human Values with Confirmatory Factor Analysis." *Journal of Research in Personality* 38 (3) (June 2004): 230–255. doi:10.1016/S0092-6566(03)00069-2.

Schwartz, Shalom H., Gila Melech, Arielle Lehmann, Steven Burgess, Mari Harris, and Vicki Owens. "Extending the Cross-Cultural Validity of the Theory of Basic Human Values with a Different Method of Measurement." *Journal of Cross-Cultural Psychology* 32 (5) (September 1, 2001): 519–542. doi:10.1177/0022022101032005001.

Scofield, John H. "Efficacy of LEED-Certification in Reducing Energy Consumption and Greenhouse Gas Emission for Large New York City Office Buildings." *Energy and Building* 67 (December 2013): 517–524. doi:10.1016/j.enbuild.2013.08.032.

SCS Global Services. "Certification Manual: Fair Trade USA V2.0." June 2014. Accessed April 3, 2016. https://fairtradeusa.org/sites/all/files/wysiwyg/filemanager/standards/FTUSA_MAN_CertificationManual_V2-0__061314.pdf.

Sen, Sankar, Zeynep Gürhan-Canli, and Vicki Morwitz. "Withholding Consumption: A Social Dilemma Perspective on Consumer Boycotts." *Journal of Consumer Research* 28 (3) (December 1, 2001): 399–417. doi:10.1086/323729.

Shapin, Steven. "Cordelia's Love: Credibility and the Social Studies of Science." *Perspectives on Science* 3 (3) (1995): 255–275.

Shapiro, Lila. "Richer, More Equal Countries Are More Trusting, Study Finds." *Huffington Post*, April 21, 2011. Accessed July 4, 2015. www.huffingtonpost.com/2011/04/20/trust-wealth_n_851519.html.

Shaw, Deirdre, Emma Grehan, Edward Shiu, Louise Hassan, and Jennifer Thomson. "An Exploration of Values in Ethical Consumer Decision Making." *Journal of Consumer Behaviour* 4 (3) (March 1, 2005): 185–200. doi:10.1002/cb.3.

Sheppard, Kate. "The GOP: Defending Your Right to Life, Liberty, and Inefficient Lighting." *Mother Jones*, September 17, 2010. Accessed March 28, 2016. http://www.motherjones.com/blue-marble/2010/09/gop-defending-your-right-life-liberty-and-inefficient-lighting.

Shill, Harold B. "Information 'Publics' and Equitable Access to Electronic Government Information: The Case of Agriculture." *Government Information Quarterly* 9 (3) (1992): 305–322.

Sierra Club. "Politics and Elections." N.d. Accessed December 27, 2016. http://content.sierraclub.org/politics-elections.

Simon, Herbert A. *Economic Analysis and Public Policy, Volume 1: Models of Bounded Rationality*. Cambridge, MA: MIT Press, 1982.

Sinnott-Armstrong, Walter. "Consequentialism." *Stanford Encyclopedia of Philosophy*, February 9, 2006. Accessed January 20, 2011. http://plato.stanford.edu/entries/consequentialism/.

Smith, Alison, Paul Watkiss, Geoff Tweddle, Alan McKinnon, Mike Browne, Alistair Hunt, Colin Treleven, Chris Nash, and Sam Cross. "The Validity of Food Miles as an Indicator of Sustainable Development." UK Department of Environment, Food, and Rural Affairs, July 2005. http://trid.trb.org/view.aspx?id=770092.

Smith, N. Craig, Guido Palazzo, and C. B. Bhattacharya. "Marketing's Consequences: Stakeholder Marketing and Supply Chain Corporate Social Responsibility Issues." *Business Ethics Quarterly* 20 (4) (October 2010): 617–641.

Smith, T. V. "Power: Its Ubiquity and Legitimacy." *American Political Science Review* 45 (3) (September 1, 1951): 693–702. doi:10.2307/1951158.

Soler, R. E., K. D. Leeks, L. R. Buchanan, R. C. Brownson, G. W. Heath, D. H. Hopkins, and the Task Force on Community Preventive Services. "Point-of-Decision Prompts to Increase Stair Use. A Systematic Review Update." *American Journal of Preventive Medicine* 38 (2) (2010): 292–300.

Sønderskov, Kim, and Carsten Daugbjerg. "The State and Consumer Confidence in Eco-Labeling: Organic Labeling in Denmark, Sweden, the United Kingdom and the United States." *Agriculture and Human Values* 28 (4) (December 2011): 507–517. doi:10.1007/s10460-010-9295-5.

Yzer, Marco C., Brian G. Southwell, and Michael T. Stephenson. "Inducing Fear as a Public Communication Campaign Strategy." In *Public Communications Campaigns*, ed. Ronald E. Rice and Charles K. Atkin, 163–176. Thousand Oaks, CA: SAGE, 2013.

Spedden, Ernest Radcliffe. *The Trade Union Label*. Baltimore: The Johns Hopkins Press, 1910.

Spence, Ewan. "The Mobile Browser Is Dead, Long Live the App." *Forbes*, April 2, 2014. Accessed March 6, 2016. http://www.forbes.com/sites/ewanspence/2014/04/02/the-mobile-browser-is-dead-long-live-the-app/#44be1d6f6e7d.

Spence, Michael. "Job Market Signaling." *Quarterly Journal of Economics* 87 (3) (August 1973): 355–374.

Stahlberg, Markus, and Ville Maila, eds. *Shopper Marketing: How to Increase Purchase Decisions at the Point of Sale.* 2nd ed. London; Philadelphia: Kogan Page, 2012.

Stanwick, Sarah D., and Peter A. Stanwick. "Exploring Voluntary Environmental Partnerships." *Journal of Corporate Accounting & Finance* 10 (3) (Spring 1999): 111–125.

Starobin, Shana, and Erika Weinthal. "The Search for Credible Information in Social and Environmental Global Governance: The Kosher Label." *Business and Politics* 12 (3) (2010): 1–35. doi:10.2202/1469-3569.1322.

Stasavage, David. "Transparency, Democratic Accountability, and the Economic Consequences of Monetary Institutions." *American Journal of Political Science* 47 (3) (July 1, 2003): 389–402. doi:10.1111/ajps.2003.47.issue-3, 10.1111/1540-5907 .00028.

Steelman, Toddi A., and Jorge Rivera. "Voluntary Environmental Programs in the United States: Whose Interests Are Served?" *Organization & Environment* 19 (4) (December 1, 2006): 505–526. doi:10.1177/1086026606296393.

Steinmetz, Katy. "Why Many Consumers Think 'Organic' Labels Are Hogwash." *Time*, May 14, 2015. Accessed July 5, 2015. http://time.com/3857799/organic-label -standards-poll/.

Stenmark, Dick. "The Relationship between Information and Knowledge." *Proceedings of IRIS* 24 (2001): 11–14.

Sterling, Greg. "No, Apps Aren't Winning. The Mobile Browser Is." September 27, 2015. Accessed March 6, 2016. http://marketingland.com/morgan-stanley-no -apps-arent-winning-the-mobile-browser-is-144303.

Stern, Paul C., Thomas Dietz, Troy Abel, Gregory A. Guagnano, and Linda Kalof. "A Value-Belief-Norm Theory of Support for Social Movements: The Case of Environmentalism." *Human Ecology Review* 6 (2) (1999): 81–98.

Stern, Paul C., Thomas Dietz, and Gregory A. Guagnano. "A Brief Inventory of Values." *Educational and Psychological Measurement* 58 (6) (December 1, 1998): 984–1001. doi:10.1177/0013164498058006008.

Sternberg, Robert J., and Li-fang Zhang. *Perspectives on Thinking, Learning, and Cognitive Styles.* New York: Routledge, 2011.

Stiglitz, Joseph E. "The Theory of 'Screening,' Education, and the Distribution of Income." *American Economic Review* 65 (3) (June 1975): 283–300.

Stirling, Andrew. "Analysis, Participation and Power: Justification and Closure in Participatory Multi-Criteria Analysis." *Land Use Policy*, Resolving Environmental Conflicts: Combining Participation and Multi-Criteria Analysis 23 (1) (January 2006): 95–107. doi:10.1016/j.landusepol.2004.08.010.

Suchman, Mark C. "Managing Legitimacy: Strategic and Institutional Approaches." *Academy of Management Review* 20 (3) (July 1, 1995): 571–610. doi:10.2307/258788.

Suh, Sangwon, Shivira Tomar, Matthew Leighton, and Joshua Kneifel. "Environmental Performance of Green Building Code and Certification Systems." *Environmental Science & Technology* 48 (5) (March 4, 2014): 2551–2560. doi:10.1021/es4040792.

Sun, Jui-ni, and Yu-chen Hsu. "Effect of Interactivity on Learner Perceptions in Web-Based Instruction." *Computers in Human Behavior* 29 (1) (January 2013): 171–184. doi:10.1016/j.chb.2012.08.002.

Sunstein, Cass R. *Infotopia: How Many Minds Produce Knowledge*. New York; Oxford: Oxford University Press, 2008.

Sustainable Brands. "Poll Finds US Shoppers Willing to Spend 31% More Per Week on Responsibly Produced Food." October 16, 2014. Accessed October 16, 2014. http://www.sustainablebrands.com/news_and_views/stakeholder_trends_insights/sustainable_brands/poll_finds_us_shoppers_willing_spend_3.

Sustainable Consumption Roundtable. "Looking Back, Looking Forward: Lessons in Choice Editing for Sustainability: 19 Case Studies into Drivers and Barriers to Mainstreaming More Sustainable Products." 2006. Accessed March 27, 2016. https://research-repository.st-andrews.ac.uk/handle/10023/2314.

Sustainable Forestry Initiative. "SFI Board Members." N.d. Accessed July 13, 2015. http://www.sfiprogram.org/about-us/sfi-governance/sfi-board-members/.

Swearingen, Anastasia. "LEED-Certified Buildings Are Often Less Energy-Efficient Than Uncertified Ones." *Forbes*, April 30, 2014. Accessed July 23, 2015. http://www.forbes.com/sites/realspin/2014/04/30/leed-certified-buildings-are-often-less-energy-efficient-than-uncertified-ones/.

Szasz, Andrew. *Shopping Our Way to Safety: How We Changed from Protecting the Environment to Protecting Ourselves*. Minneapolis: University of Minnesota Press, 2009.

Sztompka, Piotr. *Trust: A Sociological Theory*. Cambridge: Cambridge University Press, 2000.

Szwajkowski, Eugene, and Raymond E. Figlewicz. "Evaluating Corporate Performance: A Comparison of the Fortune Reputation Survey and the Socrates Social Rating Database." *Journal of Managerial Issues* 11 (2) (July 1, 1999): 137–154.

Tangherlini, Dan. "GSA Green Building Certification Systems Review Letter to Secretary of Energy." October 25, 2013. Accessed September 6, 2015. http://www.gsa.gov/portal/mediaId/180467/fileName/GSA_Green_Building_Certification_Systems_Review_Letter_to_Sec_Energy.action.

TCO. "Criteria in TCO Certified." TCO Development. N.d. Accessed March 23, 2016. http://tcodevelopment.com/tco-certified/criteria-in-tco-certified/.

Teisl, Mario F., Jonathan Rubin, and Caroline L. Noblet. "Non-Dirty Dancing? Interactions between Eco-Labels and Consumers." *Journal of Economic Psychology* 29 (2) (April 2008): 140–159. doi:10.1016/j.joep.2007.04.002.

TerraChoice. "The Sins of Greenwashing: Home and Family Edition 2010." TerraChoice, 2010. Accessed October 22, 2013. http://sinsofgreenwashing.com/.

Thaler, Richard H., and Cass R. Sunstein. *Nudge: Improving Decisions about Health, Wealth, and Happiness.* New Haven: Yale University Press, 2008.

Thøgersen, John, Pernille Haugaard, and Anja Olesen. "Consumer Responses to Ecolabels." *European Journal of Marketing* 44 (11/12) (December 15, 2010): 1787–1810. doi:10.1108/03090561011079882.

Todorov, Alexander. "The Accessibility and Applicability of Knowledge: Predicting Context Effects in National Surveys." *Public Opinion Quarterly* 64 (4) (2000): 429–451.

Tognazzini, Bruce. "First Principles of Interaction Design." www.asktog.com. March 5, 2014. Accessed December 23, 2016. http://asktog.com/basics/firstPrinciples.html.

Tolkien, J. R. R. *The Fellowship of the Ring.* Boston: Houghton Mifflin, 1954.

Tong, Scott. "Pope Francis Makes Climate Change a Moral Issue." September 24, 2015. Accessed March 28, 2016. http://www.marketplace.org/2015/09/24/sustainability/vatican-omics/pope-francis-makes-climate-change-moral-issue.

Transparency International. "What We Do—Research." N.d. Accessed July 29, 2016. http://www.transparency.org/research.

Treib, Oliver, Holger Bähr, and Gerda Falkner. "Modes of Governance: Towards a Conceptual Clarification." *Journal of European Public Policy* 14 (1) (2007): 1–20. doi:10.1080/13501760601071406.

TruValue Labs. "Insight360: The Transparent Approach to Sustainability Data." *Insight360.* N.d. Accessed March 27, 2016. https://www.insight360.io/.

Tyler, Gus. *Look for the Union Label: A History of the International Ladies' Garment Workers' Union*. Armonk, NY: M. E. Sharpe, 1995.

UC Berkeley Department of Psychology. "RSVP (Research Subject Volunteer Pool)." N.d. Accessed December 27, 2016. http://psychology.berkeley.edu/rsvp/index.html.

UL. "UL Company Fact Sheet." 2012. Accessed March 27, 2016. http://www.ul.com/global/documents/corporate/newsroom/ulfactsheet2012.pdf.

UL. "UL Environment Certified/Validated Products." N.d. Accessed March 27, 2016. http://productguide.ulenvironment.com/quickSearch.aspx.

Uphoff, Norman. "Distinguishing Power, Authority & Legitimacy: Taking Max Weber at His Word by Using Resources-Exchange Analysis." *Polity* 22 (2) (December 1, 1989): 295–322. doi:10.2307/3234836.

U.S. Census Bureau. "NAICS—North American Industry Classification System." 2007. Accessed December 21, 2010. http://www.census.gov/cgi-bin/sssd/naics/naicsrch?chart=2007.

U.S. Consumer Product Safety Commission. "About CPSC." N.d. Accessed December 23, 2016. https://www.cpsc.gov/About-CPSC/.

USDA Economic Research Service. "Table 2—U.S. Certified Organic Farmland Acreage, Livestock Numbers, and Farm Operations, 1992–2011." September 27, 2013. Accessed December 23, 2016. https://www.ers.usda.gov/webdocs/DataFiles/Organic_Production__18002/Farmlandlivestockandfarm.xls?v=41544.

USDA Economic Research Service. "Table 3—Certified Organic and Total U.S. Acreage, Selected Crops and Livestock, 1995–2011." August 26, 2015. Accessed December 23, 2016. https://www.ers.usda.gov/webdocs/DataFiles/Organic_Production__18002/CertifiedandtotalUSacreageselectedcropslivestock.xls?v=41571.

U.S. Department of Energy. "Vehicle Cost Calculator." May 3, 2016. Accessed December 27, 2016. http://www.afdc.energy.gov/calc/.

U.S. Department of Agriculture. "What Is BioPreferred?" N.d. Accessed March 14, 2016. https://www.biopreferred.gov/BioPreferred/faces/pages/AboutBioPreferred.xhtml.

U.S. Department of Energy. "History and Impacts of Appliance and Equipment Standards." N.d. Accessed January 25, 2016. http://energy.gov/eere/buildings/history-and-impacts.

U.S. Energy Information Agency. "A Look at the U.S. Commercial Building Stock: Results from EIA's 2012 Commercial Buildings Energy Consumption Survey (CBECS)." 2015. Accessed March 26, 2016. https://www.eia.gov/consumption/commercial/reports/2012/buildstock/.

U.S. Environmental Protection Agency. "EPA Environmentally Preferable Purchasing Program Pilot to Assess Standards and Ecolabels for EPA's Recommendations to Federal Agencies: Final PILOT Assessment Guidelines." December 2016. Accessed December 26, 2016. https://www.epa.gov/greenerproducts/final-pilot-assessment-guidelines-epas-recommendations-standards-and-ecolabels.

U.S. Environmental Protection Agency. "ENERGY STAR." N.d. Accessed February 19, 2010. https://www.energystar.gov/.

U.S. Environmental Protection Agency. "National Air Pollutant Emission Estimates 1970–1979." March 1981.

U.S. Environmental Protection Agency. "National Air Quality and Emissions Trends Report, 1987." March 1989.

U.S. Environmental Protection Agency. "Executive Orders: ENERGY STAR." N.d. Accessed March 8, 2011. https://www.energystar.gov/index.cfm?c=pt_reps_purch_procu.pt_reps_exec_orders.

U.S. Environmental Protection Agency. "Life Cycle Assessment: Principles and Practice." May 2006. Accessed March 7, 2017. https://nepis.epa.gov/Exe/ZyPURL.cgi?Dockey=P1000L86.txt.

U.S. Environmental Protection Agency. "Green Power Partnership National Top 100." Overviews and Factsheets, July 25, 2016. Accessed July 29, 2016. https://www.epa.gov/greenpower/green-power-partnership-national-top-100.

U.S. Environmental Protection Agency. "List of Voluntary Partnership Programs." N.d. Accessed March 29, 2011. https://www.epa.gov/partners/programs/.

U.S. Environmental Protection Agency. "National Environmental Performance Track." January 12, 2011. Accessed July 29, 2016. https://archive.epa.gov/performancetrack/web/html/index.html.

U.S. Environmental Protection Agency. "National Water Quality Inventory Report 1988 Report to Congress." April 1990.

U.S. Federal Trade Commission. "Enforcement." N.d. Accessed March 28, 2016. https://www.ftc.gov/enforcement.

U.S. Federal Trade Commission. "Proposed Revisions to the Green Guides." 2010. Accessed December 5, 2010. https://www.gpo.gov/fdsys/pkg/FR-2010-10-15/html/2010-25000.htm.

U.S. General Services Administration. "Environmentally Preferred Products." N.d. Accessed March 27, 2016. http://www.gsa.gov/portal/category/27119.

U.S. General Services Administration. "Green Building Certification System Review." N.d. Accessed July 23, 2015. http://www.gsa.gov/portal/content/131983.

U.S. General Services Administration. "Green Building Certification System Review Co-Chairs Memorandum." October 29, 2012. Accessed September 6, 2015. http://www.gsa.gov/portal/mediaId/162655/fileName/GBCS_CoChair_memo__GBCSv_10-29-12_508_.action.

U.S. Government Accountability Office. "ENERGY STAR PROGRAM: Covert Testing Shows the Energy Star Program Certification Process Is Vulnerable to Fraud and Abuse." Washington, DC, March 5, 2010. Accessed April 1, 2010. http://www.gao.gov/products/GAO-10-470.

U.S. Green Building Council. "Building Life-Cycle Impact Reduction." N.d. Accessed July 31, 2016. http://www.usgbc.org/node/2614363?return=/credits/new-construction/v4/material-%26-resources.

U.S. Green Building Council. "Distinguishing Your Expertise." N.d. Accessed April 27, 2014. http://www.usgbc.org/credentials.

U.S. Green Building Council. "Fallsway Housing and Service Center." N.d. Accessed September 15, 2015. http://www.usgbc.org/projects/fallsway-housing-and-service-center?view=scorecard.

U.S. Green Building Council. "Guide to LEED Certification: Commercial." N.d. Accessed July 30, 2016. http://www.usgbc.org/cert-guide/commercial.

U.S. Green Building Council. "LEED." N.d. Accessed July 23, 2015. http://www.usgbc.org/leed.

U.S. Green Building Council. "LEED Dynamic Plaque." N.d. Accessed July 31, 2016. https://www.leedon.io/.

U.S. Green Building Council. "LEED: Leadership in Energy and Environmental Design." N.d. Accessed July 22, 2015. http://leed.usgbc.org/leed.html.

U.S. Green Building Council. "LEED Technical Advisory Groups Charge." N.d. Accessed September 6, 2015. http://www.usgbc.org/Docs/Archive/General/Docs6198.pdf.

U.S. Green Building Council. "LEED v4, the Newest Version of LEED Green Building Program Launches at USGBC's Annual Greenbuild Conference." November 20, 2013. Accessed July 30, 2016. http://www.usgbc.org/articles/leed-v4-newest-version-leed-green-building-program-launches-usgbc%E2%80%99s-annual-greenbuild-confe.

U.S. Green Building Council. "Terms and Conditions." N.d. Accessed July 31, 2016. http://www.usgbc.org/terms.

U.S. News and World Report. "Best Colleges." N.d. Accessed December 27, 2016. http://www.usnews.com/best-colleges.

U.S. News and World Report, and GuideStar. "Largest Charities: Conservation and Environmental Education." N.d. Accessed March 13, 2011. http://www.usnews.com/usnews/biztech/charities/lists/conservation.htm.

US SIF Foundation. "Report on US Sustainable, Responsible and Impact Investing Trends 2014." 2014. Accessed August 4, 2016. http://www.ussif.org/Files/Publications/SIF_Trends_14.F.ES.pdf.

Utting, Peter. "Regulating Business via Multistakeholder Initiatives: A Preliminary Assessment." In *Voluntary Approaches to Corporate Responsibility: Readings and a Resource Guide*, ed. Rhys Jenkins, Peter Utting, and Renato Alva Pino, 61–130. Geneva: United Nations Research Institute for Social Development and United Nations Non-Governmental Liaison Service, 2002. http://www.unrisd.org/80256B3C005BCCF9/(httpAuxPages)/35F2BD0379CB6647C1256CE6002B70AA/$file/uttngls.pdf.

Valiant, Leslie G. "A Theory of the Learnable." *Communications of the ACM* 27 (11) (1984): 1134–1142.

Van Amstel, Mariëtte, Claar de Brauw, Peter Driessen, and Pieter Glasbergen. "The Reliability of Product-Specific Eco-Labels as an Agrobiodiversity Management Instrument." *Biodiversity and Conservation* 16 (14) (December 15, 2007): 4109–4129.

Van Dam, Ynte K., and Janneke De Jonge. "The Positive Side of Negative Labelling." *Journal of Consumer Policy* 38 (1) (March 2015): 19–38. doi:10.1007/s10603-014-9274-0.

van Dam, Ynte K., and Hans C. M. van Trijp. "Relevant or Determinant: Importance in Certified Sustainable Food Consumption." *Food Quality and Preference* 30 (2) (December 2013): 93–101. doi:10.1016/j.foodqual.2013.05.001.

Van Duyne, Douglas K., James A. Landay, and Jason I. Hong. *The Design of Sites: Patterns, Principles, and Processes for Crafting a Customer-Centered Web Experience.* Boston: Addison-Wesley, 2003.

van Noort, Guda, Hilde A. M. Voorveld, and Eva A. van Reijmersdal. "Interactivity in Brand Web Sites: Cognitive, Affective, and Behavioral Responses Explained by Consumers' Online Flow Experience." *Journal of Interactive Marketing* 26 (4) (November 2012): 223–234. doi:10.1016/j.intmar.2011.11.002.

Vaughn, Jacqueline. *Environmental Politics: Domestic and Global Dimensions.* 6th ed. Boston: Cengage Learning, 2011.

Verplanken, Bas, and Rob W. Holland. "Motivated Decision Making: Effects of Activation and Self-Centrality of Values on Choices and Behavior." *Journal of Personality and Social Psychology* 82 (3) (March 2002): 434–447. doi:10.1037/0022-3514.82.3.434.

Vig, Norman J. "Presidential Powers and Environmental Policy." In *Environmental Policy: New Directions for the Twenty-First Century*, ed. Norman J. Vig and Michael E. Craft, 84–108. Thousand Oaks, CA: CQ Press, 2013.

Vogel, David. *The Market for Virtue: The Potentials and Limits of Corporate Social Responsibility*. Washington, DC: Brookings Institution Press, 2005.

Vogel, David, and Robert A. Kagan. *Dynamics of Regulatory Change*. Berkeley: University of California Press, 2004.

Volcovici,Valerie, and Emmanuel Jarry. "Microsoft's Gates to Start Multi-Billion-Dollar Clean Tech Initiative." *Reuters*, November 27, 2015. Accessed December 23, 2016. http://www.reuters.com/article/climatechange-summit-technology -idUSL1N13M0XG20151127.

Wald, Matthew L., and Leslie Kaufman. "U.S. Tightens Standards for EnergyStar Label." *New York Times*, April 14, 2010, sec. Business/Energy & Environment. Accessed April 14, 2010. http://www.nytimes.com/2010/04/15/business/energy -environment/15star.html?src=un&feedurl=http%3A%2F%2Fjson8.nytimes .com%2Fpages%2Fbusiness%2Fenergy-environment%2Findex.jsonp.

Walker, Leon. "Consumers Don't Trust Green Product Claims, Survey Says." *Environmental Leader*, March 28, 2012. Accessed August 25, 2013. http://www .environmentalleader.com/2012/03/28/consumers-dont-trust-green-product-claims -survey-says/?graph=full&id=1.

Walker, Leon. "UL Environment Buys GoodGuide." *Environmental Leader*, Environmental & Energy Management News, August 6, 2012. Accessed March 27, 2016. http://www.environmentalleader.com/2012/08/06/ul-environment-buys -goodguide/.

Walls, Judith L., Phillip H. Phan, and Pascual Berrone. "Measuring Environmental Strategy: Construct Development, Reliability, and Validity." *Business & Society* 50 (1) (March 1, 2011): 71–115. doi:10.1177/0007650310394427.

Walsh, Richard G., John B. Loomis, and Richard A. Gillman. "Valuing Option, Existence, and Bequest Demands for Wilderness." *Land Economics* 60 (1) (February 1, 1984): 14–29. doi:10.2307/3146089.

Wang, N., K. M. Fowler, and R. S. Sullivan. "Green Building Certification System Review." U.S. Department of Energy, March 2012. Accessed September 6, 2015. http://www.gsa.gov/portal/mediaId/197819/fileName/GBCS2012_Cert_Sys _Review.action.

WebpageFX. "Readability Test Tool." N.d. Accessed December 20, 2016. http:// www.webpagefx.com/tools/read-able/.

Weber, Max. *From Max Weber: Essays in Sociology*. Oxford: Oxford University Press (Galaxy imprint), 1958.

Weber, Max. *The Theory of Social and Economic Organization*. New York: The Free Press, 1947.

Weeks, Katie. "Anti-Greenwashing Group Launches with Letter to GBI." *EcoBuilding Pulse*, May 15, 2014. Accessed July 23, 2015. www.ecobuildingpulse.com/news/anti-greenwashing-group-launches-with-letter-to-gbi_o.

Wei, Wenjuan, Olivier Ramalho, and Corinne Mandin. "Indoor Air Quality Requirements in Green Building Certifications." *Building and Environment* 92 (October 2015): 10–19. doi:10.1016/j.buildenv.2015.03.035.

Wheeland, Matthew. "UL Acquires TerraChoice in Green Standard Consolidation." *GreenBiz*, August 31, 2010. Accessed March 27, 2016. https://www.greenbiz.com/news/2010/08/31/ul-acquire-terrachoice-green-standard-consolidation.

Whelan, Jennifer. "Competition for LEED: GBI's Green Globes Shakes Up Building Certification." *ArchDaily*, May 26, 2014. Accessed July 23, 2015. http://www.archdaily.com/509690/competition-for-leed-gbi-s-green-globes-shakes-up-building-certification/.

Whitaker, Albert C. *History and Criticism of the Labor Theory of Value in English Political Economy*. Kitchener, Ont.: Batoche, 2001.

White, G. D., and A. Luo. "Business Survey Response Rates: Can They Be Improved." In *The American Statistical Association Proceedings of the Section on Survey Research Methods*, Alexandria, VA: American Statistical Association, 2005, 3666–3368. Accessed July 11, 2012. http://ww2.amstat.org/sections/srms/Proceedings/y2005/Files/JSM2005-000499.pdf.

Whole Foods. "Seafood." *Whole Foods Market*. N.d. Accessed March 27, 2016. http://www.wholefoodsmarket.com/department/seafood.

Whole Foods. "Whole Foods Market Empowers Shoppers to Make Sustainable Seafood Choices with Color-Coded Rating System." September 13, 2010. Accessed March 13, 2016. http://media.wholefoodsmarket.com/news/whole-foods-market-partners-with-blue-ocean-institute-and-monterey-bay-aqua.

Wiens, Kyle. "The Retina MacBook Pro Was 'Verified' EPEAT Gold, but It's Not Green." *iFixit*, October 16, 2012. Accessed January 24, 2016. http://ifixit.org/blog/3525/the-retina-macbook-pro-was-verified-epeat-gold-but-its-not-green/.

Wikidiff. "Misinformation vs Disinformation—What's the Difference?" August 18, 2013. Accessed June 3, 2016. http://wikidiff.com/disinformation/misinformation.

Williams, Cynthia A. "The Securities and Exchange Commission and Corporate Social Transparency." *Harvard Law Review* 112 (6) (April 1999): 1197–1311.

Willingham, Daniel T., Elizabeth M. Hughes, and David G. Dobolyi. "The Scientific Status of Learning Styles Theories." *Teaching of Psychology* 42 (3) (July 1, 2015): 266–271. doi:10.1177/0098628315589505.

Willis, Margaret M., and Juliet B. Schor. "Does Changing a Light Bulb Lead to Changing the World? Political Action and the Conscious Consumer." *Annals of the American Academy of Political and Social Science* 644 (1) (November 1, 2012): 160–190. doi:10.1177/0002716212454831.

Wilner, Tamar. "LEED Buildings 'Less Energy Efficient'—Says Who?" *Environmental Leader*, March 4, 2014. Accessed July 23, 2015. www.environmentalleader.com/2014/03/04/leed-buildings-less-energy-efficient-says-who/.

Wiser, Ryan H., and Steven Pickle. "Green Marketing, Renewables, and Free Riders: Increasing Customer Demand for a Public Good." Berkeley: Environmental Energy Technologies Division, Ernest Orlando Lawrence Berkeley National Laboratory, University of California, 1997. Accessed March 16, 2014. https://emp.lbl.gov/sites/all/files/REPORT%20%20lbnl%20-%2040632.pdf.

Wong, Kate. "Mother Nature's Medicine Cabinet." *Scientific American*, April 9, 2001. Accessed June 1, 2015. http://www.scientificamerican.com/article/mother-natures -medicine-c/.

Yale School of Forestry & Environmental Studies. "Forest Certification." *Global Forest Atlas*. Yale University. N.d. Accessed March 26, 2016. http://globalforestatlas.yale.edu/conservation/forest-certification.

Yeo, Richard. "Reading Encyclopedias: Science and the Organization of Knowledge in British Dictionaries of Arts and Sciences, 1730–1850." *Isis* 82 (1) (1991): 24–49.

Young, Oran R. *On Environmental Governance: Sustainability, Efficiency, and Equity.* Boulder, CO: Paradigm Publishers, 2013.

Zelman, Kathleen. "Organic Food—Is 'Natural' Worth the Extra Cost?" N.d. Accessed August 11, 2014. http://www.webmd.com/food-recipes/features/organic-food-is -natural-worth-the-extra-cost.

Zimmerman, Michael J. "Intrinsic vs. Extrinsic Value." In *The Stanford Encyclopedia of Philosophy*, ed. Edward N. Zalta, Spring 2015. Accessed February 13, 2016. http://plato.stanford.edu/archives/spr2015/entries/value-intrinsic-extrinsic/.

# Index

Page numbers followed by t refer to tables; page numbers followed by f refer to figures.

Printed in the United States
by Baker & Taylor Publisher Services